SECRETS OF
COMPANION PLANTING
FOR SUCCESSFUL GARDENING

CARROTS

love

TOMATOES

&

ROSES

love

GARLIC

LOUISE RIOTTE

Storey

Cover and interior designed by Anthony Serge
Line drawings by the author

Library of Congress Cataloging-in-Publication Data

Riotte, Louise.
    Carrots love tomatoes & roses love garlic : secrets of companion
planting for successful gardening / Louise Riotte.
        p.   cm.
    Combination of two books that have been reflowed and redesigned.
    Includes bibliographical references and index.
    ISBN 1–58017–829–4 hardcover
    1. Companion planting.   2. Companion crops.   3. Gardening.   I. Title:
Carrots love tomatoes and roses love garlic.   II. Title.
SB603.5 .R56 2004
631.5'8—dc22   2003026573

        8   10   9   hardcover

CONTENTS

PART 1: CARROTS LOVE TOMATOES

THE MYSTERY OF COMPANION PLANTING

The magic and mystery of companion planting have intrigued and fascinated humans for centuries, yet it is a part of the gardening world that has never been fully explored. Even today we are just on the threshold. In years to come, I hope that scientists, gardeners, and farmers everywhere will work together in making more discoveries that will prove of great value in augmenting the world's food supply.

Plants that assist each other to grow well, plants that repel insects, even plants that repel other plants—all are of great practical use. They always have been, but we are just beginning to find out why. Delving deeply into this fascinating aspect of gardening can provide both pleasure and very useful information for us. I hope that what I have written here will give you many of the tools to work with.

Vegetable growers find that companion planting provides many benefits, one of which is protection from pests. A major enemy of the carrot is the carrot fly, whereas the leek suffers from the onion fly and leek moth. Yet when leek and carrot live together in companionship, the strong and strangely different smell of the partner plant repels the insects so much that they do not even attempt to lay their eggs on the neighbor plant. They take off speedily to get away from the smell. This is why mixed plantings give better insect control than a monoculture, where many plants of the same type are planted together in row after

row. Even when plants are affected by diseases, a mixed plant culture can usually alleviate the situation.

It is important to remember that not all "protective" botanicals act quickly. For example, marigolds, to be effective in nematode control, should be grown over at least one full season, and more is better, for their effect is cumulative. One should also realize that certain companion plants will diminish each other's natural repelling ability as they grow together. All through this book, you will find "what to grow with" and "what *not* to grow with." Both are equally important to gardening success.

The effects of plants on one another are important outside the vegetable garden, among trees and shrubs as well as grains, grasses, and field crops. These have chapters to themselves, as do herbs, the group of plants most widely used as protective companions.

Wild plants also play a vital part in the plant community. Some are accumulator plants—those that have the ability to collect trace minerals from the soil. They actually can store in their tissues up to several hundred times the amount contained in an equal amount of soil. These plants, many of which are considered weeds, are useful as compost, green manure, or mulch. Some are "deep diggers," sending their roots deep into the ground to penetrate hardpan and help to condition the soil, and some have value as protectors of garden plants.

An entirely different type of community life is that of fruit and nut trees and the bush and bramble small fruits. For many of this group, the choice of good companions is not only helpful but also essential. Have you ever experienced the disappointment of having a beautiful fruit tree blossom, be visited by the bees, and yet fail to bear? There is a reason, of course, and it lies in pollination. Pollen is the dust from blossoms that is needed to make the plant fruitful. If the tree is self-unfruitful and there is no pollenizer of the correct type growing near, it is doubtful that the tree will ever bear well. In the chapters on fruit and nut growing, I'll attempt to unravel some of this mystery, which seems to plague new gardeners and orchardists particularly.

A note on the chapter devoted to poisonous plants: This information is not meant to frighten but to warn, for most of the nursery catalogs do not tell us which plants are poisonous or to what degree. Even some of the gardening encyclopedias do not.

Cases of death resulting from poisonous plants are rare, but they do happen. In this book I refer to poisonous plants that are useful in the garden for various reasons. It is only fair to tell you that some of those most commonly used may be harmful to children, livestock, or even you.

Many of our loveliest and most decorative plants are poisonous—oleanders, daffodils, scillas, lily of the valley, hyacinths, and larkspurs. Other equally poisonous plants are of value for medicines or as insect repellents. To know is to be forewarned, and because we know, we may use these plants safely. Poisonous plants, unlike poisonous insects or animals, are never aggressive. You are in control of them at all times.

All of the suggestions given in this book for companion planting are only a beginning. I have included practical information on soil improvement and garden techniques, as well as some sample garden plans, to help you put companion plants to work for you. Your own experiments will lead you to make many exciting discoveries.

VEGETABLES

ASPARAGUS (*Asparagus officinalis*)

Parsley planted with asparagus gives added vigor to both. Asparagus also does well with basil, which itself is a good companion for tomatoes. Tomatoes will protect asparagus against asparagus beetles because they contain a substance called solanine. But if asparagus beetles are present in great numbers, they will attract and be controlled by their natural predators, making spraying unnecessary. A chemical derived from asparagus juice also has been found effective on tomato plants as a killer of nematodes, including root-knot, sting, stubby root, and meadow nematodes.

In my garden I grow asparagus in a long row at one side. After the spears are harvested in early spring I plant tomatoes on either side, and find that both plants prosper from the association. Cultivating the tomatoes also keeps down the weeds from the asparagus. The asparagus fronds should never be cut, if at all, until very late in fall, as the roots need this top growth to enable them to make spears the following spring.

BEAN (*Phaseolus vulgaris* and *Vicia faba*)

Many different kinds of beans have been developed, each with its own lore of "good" and "bad" companions. Generally speaking, however, all will thrive when interplanted with carrots and cauliflower; the carrots especially help the beans grow. Beans grow well with beets, too, and are of aid to cucumbers and cabbages.

A moderate quantity of beans planted with leek and celeriac will help all, but planted too thickly they have an inhibiting effect, causing all three to make poor growth. Marigolds in bean rows help repel the Mexican bean beetle.

Summer savory with green beans improves their growth and flavor as well as deterring bean beetles. It is also very good to cook with beans.

Beans are inhibited by any member of the Onion family—garlic, shallots, or chives—and they also dislike being planted near gladiolus.

Broad beans are excellent companions with corn, climbing diligently up the cornstalks to reach the light. Not only does it anchor the corn more firmly, acting as a protection against the wind, but a heavy vine growth may also act as a deterrent to raccoons. Beans also increase the soil's nitrogen, which is needed by the corn.

BEAN, BUSH (*Phaseolus vulgaris*)

Included with bush beans are those known as green, snap, string, and wax beans. All will do well when planted with a moderate amount of celery, about one celery plant to every six or seven of beans.

Bush beans do well also when planted with cucumbers. They are mutually beneficial. Bush beans planted in strawberry rows are mutually helpful, both advancing more rapidly than if planted alone.

Bush beans will aid corn if planted in alternate rows. They grow well with summer savory but never should be planted near fennel. They also dislike onions, as do all beans.

BEAN, LIMA (*Phaseolus lunatus*)

Nearby locust trees have a good effect on the growth of lima and butter beans. Other plants give them little or no assistance in repelling insects. Never cultivate lima beans when they are wet, because this will cause anthracnose, if present, to spread. If the ground has sufficient lime and phosphorus, there will probably be little trouble from anthracnose and mildew.

BEAN, POLE (*Phaseolus vulgaris*)

Like others of the family, pole beans do well with corn and summer savory. They also have some pronounced dislikes, such as kohlrabi and sunflower. Beets do not grow well with them, but radishes and pole beans seem to derive mutual benefit.

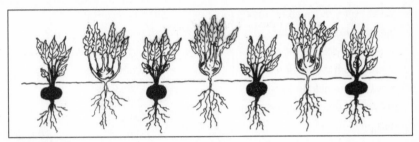

Beets and kohlrabi make good companions. Both take the same kind of culture, and they take soil nourishment at different levels.

BEET (*Beta vulgaris*)

Beets grow well near bush beans, onions, and kohlrabi but are turned off by pole beans. Field mustard and charlock inhibit their growth. Lettuce and most members of the Cabbage family are friendly to them.

BROCCOLI (*Brassica oleracea,* Botrytis Group)

Like all members of the Cabbage family, broccoli does well with such aromatic plants as dill, celery, camomile, sage, peppermint, and rosemary and with other vegetables such as potatoes, beets, and onions. Do not plant it with tomatoes, pole beans, or strawberries. Use pyrethrum on broccoli against aphids, before the flower buds open. (See the Pest Control chapter.)

CABBAGE (*Brassica* spp.)

The Cabbage family includes not only cabbage but also cauliflower, kale, kohlrabi, broccoli, collards, and Brussels sprouts—even rutabaga and turnip. While each plant of this group has been developed in a special way, they

If rabbits dig your cabbage patch, plant any member of the Onion family among them. Or you can dust with ashes, powdered aloes, or cayenne pepper. Rabbits also shun dried blood and blood meal.

are all pretty much subject to the same likes and dislikes, insects and diseases. Hyssop, thyme, wormwood, and southernwood are helpful in repelling the white cabbage butterfly.

All members of the family are greatly helped by aromatic plants or those that have many blossoms. Good companions are celery, dill, chamomile, sage, peppermint, rosemary, onions, and potatoes. Cabbages dislike strawberries, tomatoes, and pole beans.

Butterflies themselves do no harm and can help pollinate plants. It is their caterpillars that do much damage to the orchard and field crops. The white cabbage butterfly is perhaps the most destructive. Herbs will repel them: Try hyssop, peppermint, rosemary, sage, thyme, and southernwood.

All members of the family are heavy feeders and should have plenty of compost or well-decomposed cow manure worked into the ground before they are planted. Mulching will help if the soil has a tendency to dry out in hot weather, and water should be given if necessary.

Cabbage and cauliflower are subject to clubroot, and if this occurs, try new soil in a different part of the garden. Dig to a depth of 12 inches and incorporate plenty of well-rotted manure into the soil. Rotate cabbage crops every two years.

If cabbage or broccoli plants do not head up well, it is a sign that lime, phosphorus, or potash is needed. Boron deficiency may cause the heart of cabbage to die out.

CARROT (*Daucus carota* spp. *sativus*)

For sweet-tasting carrots your soil must have sufficient lime, humus, and potash. Too much nitrogen will cause poor flavor as will a long period of hot weather.

Onions, leeks, and herbs such as rosemary, worm-wood, and sage act as repellents to the carrot fly (*Psila rosae*), whose maggot or larva often attacks the rootlets of young plants. Black salsify (*Scorzonera hispanica*), sometimes called black oyster plant, also is effective in repelling the carrot fly. Use as a mixed crop.

Apples and carrots should be stored a distance from each other to prevent the carrots from taking on a bitter flavor.

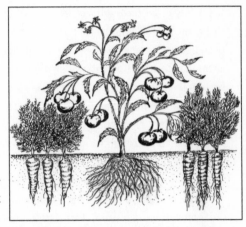

Carrots are good to grow with tomatoes—also with leaf lettuce, chives, onions, leeks, radishes, rosemary, and sage. They have a pronounced dislike for dill. Carrot roots themselves contain an exudate beneficial to the growth of peas.

CAULIFLOWER (*Brassica oleracea*, Botrytis Group)
The white cabbage butterfly (*Pieris rapae*) is repelled if celery plants are grown near the cauliflower, but cauliflower does not like tomatoes or strawberries. Extract from cauliflower seeds inactivates the bacteria causing black rot.

CELERIAC (*Apium graveolens* var. *rapaceum*)
A sowing of winter vetch before planting celeriac is helpful, for the plant needs a rich, loose soil with plenty of potassium. The leek, also a potassium lover, is a good companion in alternating rows, as are scarlet runner beans.

Celeriac does not need as much attention as celery since blanching is not necessary, but as the root starts to enlarge, the crown may be helped to better development and higher quality by removing the fine roots and the soil attached to them. Many lateral roots close to the top of the crown tend to make the fleshy part irregular and coarse.

CELERY (*Apium graveolens* var. *dulce*)

Celery grows well with leeks, tomatoes, cauliflower, and cabbage, while bush beans and celery seem to give mutual assistance. One gardener believes that celery is particularly benefited if grown in a circle so that the lacy, loosely interwoven roots may make a more desirable home for earthworms and soil microbes. Celery and leeks both grow well when trenched.

Both celery and celeriac are reported to have a hormone that has an effect similar to insulin, making them an excellent seasoning for diabetics or for anyone on a salt-reduced diet.

Celery dinant or French celery dinant is a unique type that sends out a multitude of narrow stalks. I have found it easy to grow here in southern Oklahoma. It has a much fuller flavor than common celery and less should be used in cooking.

This celery is completely insect-free and grows well with all garden vegetables. Plants will freeze in winter, but the root does not and will put out new leaves from the center with the advent of warm spring weather. In a cold climate the leaves may be dried for winter use.

CHAYOTE (*Sechium edule*)

This is a perennial tropical vine, an annual in colder climates, which bears a delicious, light green, pear-shaped fruit in the fall. Two vines must be grown or it will not bear well. Chayotes in a cream sauce are a dish "fit for the gods." I grow them on my garden fence along with cucumbers, where both do exceptionally well. They apparently have no insect enemies and seem to be protective to the cucumbers.

CHILE PEPPERS (*Capsicum* spp.)

The Aztec "chilli" or the Spanish "chile" is known to have existed as early as 700 B.C.E. (B.C.), but its birthplace remains a mystery, says Anne Lindsay Greer, writing in *Cuisine of the American Southwest*. She also claims that all chiles contain an element of unpredictability but insists that not all Mexican food or all chiles are hot. More chiles are produced and consumed than any other seasoning in the world. The hotness of each depends on the climate in which it is grown.

Early South and Central American cultures used chiles for medicinal purposes, currency, and as a discipline for disobedient children. Chiles are thought to be an aid to digestion, to give protection from colds, and to cure everything from toothache to colic to an indifference toward romance. But after separating fact from folklore, some of the mystique is lost, yet certain facts about chiles remain undisputed:

- Chiles are high in vitamins A and C. Just 1 ounce of dried chile has two times the daily requirement of vitamin A.

- Chiles serve as a natural meat preservative by retarding the oxidation of fats.

- Chiles added personality and flavor to the otherwise bland diet of early southwesterners.

- Chiles can be used to reduce the amount of salt in many dishes or to eliminate salt altogether for those on a low-sodium diet.

Columbus is said to have discovered chiles in the West Indies (though he called them peppers). Therefore they were enthusiastically adopted by the rest of the world. Indian food, Szechuan sauces, Cajun cooking, and southern African cuisines all favor chiles for their assertive flavors. Southwestern and Asian recipes are becoming increasingly popular and many call for chile peppers.

Chiles can be direct seeded after the last spring frost in long, warm-season areas or sown indoors six to eight weeks before transplanting out. Chile peppers are easily transplanted, and most produce fruits with almost reckless abandon!

A CHILE LINEUP

'Habanero'. Hottest! Lantern-shaped, lime green to orange fruit. Aromatic and tasty in sauces or pickled. Wear gloves when preparing this chile.

'Piquin'. Very hot, round, or slightly pointed, perennial in Mexico and south Texas.

'Thai Pepper'. Tiny peppers (1 inch long), red when ripe, bred in Thailand, hot and the heat lingers. Use in Oriental dishes.

'Anaheim "M" Chili'. An 8-inch-long, tapering chile, which can be used green for rellenos and green chile sauce or dry in red enchilada sauce. A medium-hot, long-bearing pepper.

'Aji'. Hot, fruity Peruvian, a thin orange to red 3 to 5 inches long.

'Centennial'. Tiny ornamental purple to white to red simultaneously. Hot pods are edible at all stages. (Named for our flag's colors.)

'De Arbol'. Treelike plant; 3 to 4 feet tall; thin 3- to 4-inch chiles; hot and smoky. Good for drying.

'Serrano'. Hot, small, slender, rounded chile. Serve fresh, dry, or pickled. Good in salsas. May be frozen if blanched first.

'Cayenne'. Hot, short, slender. Best dried and ground for red-hot sauces. Used in Asian as well as Mexican and southwestern food.

'Sandia Hot'. Very hot, thin-walled, best dried and ground. A standard for hot chile lovers. May be used green or red, fresh or dried.

'Chimayo'. Pick green for your favorite stew or salsa. Dried and ground for red chile powder, it makes an excellent enchilada sauce.

'Española' Improved. A special short-season chile that will turn red and get hot even in northern climates. Narrow, pointed, and thin-walled. Use green or red.

'Jalapeño'. Everybody's favorite. Easily grown, prolific producer. A 3-inch, dark green, almost black chile with thick, meaty walls. Good for pickling or fresh. Finely chopped, adds a new dimension to corn bread.

'Santa Fe Grande'. Short, thick, hot, yellow pepper, good in condiments or pickled. Very attractive as an ornamental.

'Cascabel'. Medium hot, tough, round, and dark red, the 1-inch pod makes a dark, smoky chile pepper.

'NuMex Eclipse'. Dark brown when ripe, medium hot. Adds color to vegetable dishes.

'NuMex Sunrise'. Bright yellow when ripe, medium hot. May be used as a green chile.

'Big Jim'. Five 'Big Jims' may weigh as much as 1 pound. This pepper is a perfect candidate for chile rellenos.

'Vallero'. Dark red, medium hot. Use dried for red chile sauces and soups.

'Mulatto'. Dark brown when dried, excellent for rellenos, very smoky flavor. The 5-inch-long, 2-inch-wide pods are mild.

'Guajillo'. Used dry in soups and salsas.

'Andho' ('Poblano'). Large, heart-shaped, mild chile of excellent flavor. Use peeled 'Poblano' for rellenos, or cut into strips for soups and stews. Good dried.

'Pasilla'. Almost black when dried, use for mole. Pods are mild.

'Pimento'. Hungarians use for paprika. Heart-shaped, 4 inches long, red, mild, good stuffed or as a stuffing. Mild to hot.

CHINESE CELERY CABBAGE
(*Brassica rapa*, Pekinensis Group)

This easy-to-grow vegetable, deserving to be better known in America, is one of the oldest vegetable crops in China. I grow both types: the tall, slender heads and the huge, deeply savoyed hybrid type, which makes a round head. I find that celery cabbage grows well, making enormous

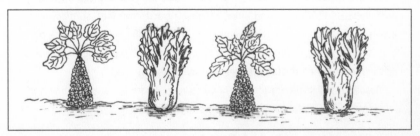

Chinese celery cabbage does well in the fall garden, interplanted with Brussels sprouts.

heads, if planted in my fall garden about 2 feet apart. I alternate the plants in the row with Brussels sprouts or cauliflower.

Celery cabbage has few insect enemies but should not be grown near corn, as the corn worms will infest it.

COLLARD (*Brassica oleracea,* Acephala Group)

Cornell University's College of Agriculture has shown that collards benefit from interplanting with tomatoes, since the flea beetle, the prime pest of collards, is significantly decreased.

Collards, widely grown in the southern states as a source of greens, are more nutritious than heading cabbage, and their taste is improved by freezing.

CORN (*Zea mays*)

Sweet corn does well with potatoes, peas, beans, cucumbers, pumpkin, and squash. Peas and beans help corn by restoring to the soil the nitrogen used up by the corn. Is there anyone who hasn't heard the story of Indians putting a fish in every corn hill?

Melons, squash, pumpkins, and cucumbers like the shade provided by corn. In turn they benefit the corn, protecting it from the depredations of raccoons, which do not like to travel through the thick vines. Similarly, pole beans may be planted with corn to climb on the stalks. But don't plant tomatoes near corn because the tomato fruitworm and corn earworm are identical.

An experiment with odorless marigold showed that when it was planted next to corn, the Japanese beetle did not chew off the corn silks.

An experiment reported in the British *New Scientist* in 1970 states that "Reduced incidence of fall armyworm on maize [corn] and a correspond-ingly increased yield were obtained by growing the crop with sunflower in alternating strips. . . . There were also large reductions in the numbers of *Carpophilus* beetles in the sunflower strips, compared with unbroken areas of the crop. Some infestations were cut by more than half."

Research has shown that to remove corn suckers is a waste of time and is, in fact, detrimental to the development of the ears.

CUCUMBER (*Cucumis sativus*)

Cucumbers apparently are offensive to raccoons, so they are a good plant with corn. And corn seemingly protects the cucumber against the virus that causes wilt. Thin strips of cucumber will repel ants.

Cucumbers also like beans, peas, radishes, and sunflower, and, preferring some shade, they will grow well in young orchards.

Sow two or three radish seeds in cucumber hills to protect against cucumber beetles. Do not pull the radishes, but let them grow as long as they will, even blossoming and going to seed. Cucumber beetles also may be trapped by filling shallow containers about three-fourths full of water into which some cooking oil has been poured.

If cucumbers are attacked by nematodes, try a sugar spray. I boil $1/2$ cup of sugar in 2 cups of water, stirring until completely dissolved. Let cool and dilute with 1 gallon of water. Strange as it seems, sugar kills nematodes by drying them out. This will also attract honeybees, ensuring pollination and resulting in a bumper crop of cucumbers, so the spray is worth trying even if you don't suspect the presence of nematodes.

Beneficial fungi are another enemy of nematodes. If you suspect their presence, build up the humus content of your soil.

A chive spray is helpful for downy mildew on cucumbers, as is a spray made of horsetail (*Equisetum arvense*). (See *Horsetail* in the chapter on Wild Plants.)

Cucumbers dislike potatoes, while potatoes grown near cucumbers are more likely to be affected by phytophthora blight, so keep the two apart. Cucumbers also have a dislike for aromatic herbs.

DISEASE-RESISTANT VEGETABLE VARIETIES

Many vegetable varieties have been specially bred to resist specific plant diseases, and more are constantly being developed.

It's important to know what diseases are most troublesome in your area. Your county's Cooperative Extension Service is a good source of information on this; it may have lists of varieties recommended for local conditions.

Seed catalogs normally indicate which of their offerings have inbred resistance to various problems, often using a code (for instance, "VFT" to show that a tomato cultivar is resistant to verticillium wilt, fusarium wilt, and tobacco mosaic). You can greatly increase your chances of success by heeding this information and matching your choices to the troubles you're most likely to run into.

EGGPLANT (*Solanum melongena*)

Redroot pigweed makes eggplants more resistant to insect attack. During dry weather, mulching and irrigation will help prevent wilt disease. Dry cayenne pepper sprinkled on plants while still wet with dew will repel caterpillars. Eggplants growing among green beans will be protected from the Colorado potato beetle. The beetles like eggplant even more than potatoes, but they find the beans repellent.

HIGH-VITAMIN VEGETABLES

A new tomato, 'Doublerich', containing as much vitamin C as citrus fruit, was introduced in 1956. Professor A. F. Yeager of the University of New Hampshire developed it using crosses of the tiny wild Peruvian tomato, which is 4 times richer in vitamin C than our ordinary garden types.

A few years later 'Caro-Red', containing about 10 times the amount of vitamin A found in standard varieties, was perfected at the Indiana Experiment Station. 'Caro-Red' owed its richness to its orange pigment, beta-carotene, and a single fruit could supply up to twice the minimum daily requirement of vitamin A for an adult. Perhaps best of all, this is a very delicious tomato to eat. Later came 'Caro-Rich', containing even more vitamin A.

These are just a few examples of the ongoing effort to improve the nutritional content of garden vegetables. Announcements of new vitamin-rich introductions appear in seed catalogs and gardening literature, and those who want the greatest possible food value from their growing space should take note of them.

HORSERADISH (*Amoracia rusticana*)

Horseradish and potatoes have a symbiotic effect on each other, causing the potatoes to be healthier and more resistant to disease.

Plants should be set only at the corners of the potato plot and should be dug after each season to prevent spreading. Horseradish does not seem to protect against the potato beetle, but it is effective against the blister beetle. A tea made from horseradish is beneficial against monilia on apple trees.

JERUSALEM ARTICHOKE (*Helianthus tuberosus*)

In Italy these are called *girasole,* meaning "turn with the sun." They really are a type of sunflower and should not be confused with the globe arti-choke, which is an entirely different plant.

Jerusalem artichokes, native American plants, were known to and used by American Indians. They are a good companion to corn. The tuber is the edible portion, for this sunflower has its surprise at the bottom, the flowers being attractive but not large.

The principal food content of the Jerusalem artichoke is inulin, a taste-less, white polysaccharide dissolved in the sap of the roots, which can be converted into levulose sugar. This is of special interest to diabetics, for lev-ulose is highly nutritious and the sweetest of all known natural sugars. Levulose also occurs in most fruits, in the company of dextrose, which dia-betics must avoid, but in the Jerusalem artichoke it is present alone. The artichokes are high in food value and rich in vitamins. They may be cooked or eaten raw in salads.

KALE (*Brassica oleracea,* Acephala Group)

This cool-weather crop is fine to grow in the fall garden, and it will stand most average winters if given a little protection.

Kale does well in the same rows as late cabbage or potatoes. If planted about the first of August following late beans or peas, it will continue to grow until a hard freeze. A light freeze does not hurt it and even improves its flavors.

Wild mustard and kale sometimes are a problem in oat fields. Rolling is the best method of control. It should be practiced early in the morning while the plants are still wet with dew. The springy oats will pop back up again, but the mustard or kale will be broken.

KOHLRABI (*Brassica oleracea*, Gongylodes Group)

Kohlrabi is mutually beneficial with onions or beets, with aromatic plants, and surprisingly with cucumbers, in part because they occupy different soil strata. It dislikes strawberries, tomatoes, and pole beans but helps protect the garden members of the mustard family.

It is a demanding plant, needing plenty of water but good drainage, as well as good supplies of compost. It grows best in filtered sunlight.

LEEK (*Allium porrum*)

Leek is one of the "heavy feeders" and should be planted in soil well fertilized with rotted manure. Leeks are usually sold in the grocery store (at least where I live) with the root still attached. I once bought several bunches and planted them; they grew well and propagated, and I've had leeks ever since.

Leeks are good plants to grow with celery and onions and also are benefited by carrots. Returning the favor, leeks repel carrot flies.

LETTUCE (*Lactuca sativa*)

In spring I keep a supply of small lettuce plants growing in cold frames. When I pull every other green onion for table use, I pop in lettuce plants. They will aid the onions, and the compost in the onion row will still be in good supply for the lettuce to feed on, while the onion will repel any rabbits.

Lettuce grows well with strawberries, cucumbers, and carrots and it has long been considered good to team with carrots and radishes. Radishes grown with lettuce in summer are particularly succulent.

Lettuce needs cool weather and ample moisture to make its best growth, and I find that the seed will not germinate in very hot weather. Already-started lettuce should have some summer shade.

MELON (*Cucurbitaceae*)

Crop rotation can be one of your best weapons against garden pests, but do not rotate melon, squash, and cucumber with each other, since all are cucurbits.

Timing is another weapon. Most cucurbits are not very susceptible to borers once they are past the seedling stage, so try either earlier or later plantings. I find that cucumbers and squash planted in midsummer for a fall crop are almost entirely insect-free.

Do not plant melons near potatoes, though they will grow well with corn and sunflowers. Morning glory is thought to stimulate the germination of melon seeds.

Heavy waxed paper placed under melons helps keep worms from entering, while sabadilla dust is effective, too. (See *Insecticides, Botanical,* in the Pest Control chapter.) Melon leaves, rich in calcium, are good to place on the compost pile.

OKRA (*Abelmoschus esculentus*)

This native of the Old World tropics is grown for its immature pods, which are called okra or gumbo. It's a warm-weather plant that will grow wherever melons or cucumbers thrive. I plant two rows, dig a trench between, and cover it with mulch. On the north side of my okra I plant a row of sweet bell peppers and on the south side a row of eggplant. All are well mulched as the season advances. When the weather becomes dry in midsummer, I lay the hose in the trench and flood it so that all three companions grow well.

ONION (*Allium* spp.)

Onions and all members of the Cabbage family get along well with each other. They also like beets, strawberries, tomatoes, lettuce, summer savory, and chamomile (sparsely), but do *not* like peas and beans. Ornamental relations of the onion are useful as protective companions for roses.

Since onion maggots travel from plant to plant when set in a row, scatter your onion plants throughout the garden.

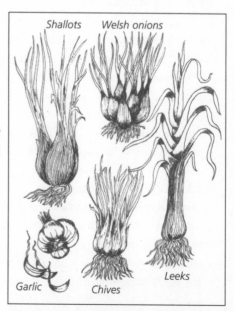

The Onion family is the gardener's best friend.

PARSLEY (*Petroselinum crispum*)

Parsley mixed with carrot seed helps repel carrot flies by its masking aroma. It protects roses against rose beetles. Planted with tomatoes or with asparagus, it will give added vigor to both.

Poultry are sometimes turned loose at intervals in parsley patches where there are many parsley worms, which are the larvae of the black swallow-tail butterfly.

A number of different strains of parsley, including the Hamburg (*Petroselinum crispum* var. *tuberosum*), are grown solely for the fleshy roots, which are cooked and eaten in the same way as parsnips.

PARSNIP (*Pastinaca sativa*)

The parsnip is of ancient culture but remains a vegetable for the special palate. The parsnips have few insect enemies and suffer from few diseases, but both the foliage and roots make a safe insect spray. They are not injured by freezing and are often left in the ground over winter. The seeds germinate slowly and unevenly and should not be used if over a year old.

PEA (*Pisum sativum*)

For large crops, treat pea and bean seed with inoculant, a natural bacterial agent available under various brand names from garden centers and seed catalogs. (See *Inoculants* in the Soil Improvement chapter.)

Peas grow well with carrots, turnips, radishes, cucumbers, corn, beans, and potatoes as well as many aromatic herbs. They do *not* grow well with onions, garlic, and gladiolus.

Always plow pea vines under or return them to the compost pile. Wood ashes used around the base of pea vines help control aphids.

PEANUT (*Arachis hypogaea*)

As members of the Legume family, peanuts are good soil builders. In many areas of the South and Southwest, they may be grown as a second crop after an earlier one, such as carrots or beets, has been harvested. They make a good groundcover in an orchard of young nut trees. (See *Legumes* in the Soil Improvement chapter.)

PEPPER, SWEET (*Capsicum annuum*, Grossum Group)

The general requirements of sweet peppers are surprisingly like those of basil, so plant them together. Sweet peppers also grow well with okra, and since they are very brittle plants, the okra, growing taller, serves as a windbreak.

PUMPKIN (*Cucurbita pepo*)

Pumpkins grow well when jimsonweed (*Datura*), sometimes called thorn apple, is in the vicinity. (See the Wild Plants chapter.) Pumpkins grow well with corn, a practice followed by Native Americans, but pumpkins and potatoes have an inhibiting effect on each other.

Middle Eastern peoples consider the seeds an inexhaustible source of vigor offered by a bountiful nature. While we know today that there are no mysterious potions for tired lovers, we also know that some of the old formulas did perform seeming miracles—not through magic but through good nutrition—and pumpkin seeds are really vitamin-rich.

Pumpkin varieties have been developed with seeds that lack the shell of normal pumpkin and squash. The hull-less seeds may be removed and simply washed and dried; they are a delicious snack when roasted and lightly salted.

RADISH (*Raphanus sativus*)

Radish is aided by redroot pigweed, which loosens soil; by nasturtiums; and by mustard's protective oils. Do not rotate radish with cabbage, cauliflower, Brussels sprouts, kohlrabi, broccoli, or turnip, since all are members of the Cabbage family.

Early radishes are good to sow with beets, spinach, carrots, and parsnips to mark the row. Sow radishes with cucumbers, squash, and melons to repel the striped cucumber beetle, and with tomatoes to rout the two-spotted spider mite. Radishes grow well with kohlrabi, bush beans, and pole beans. The presence of leaf lettuce in summer will make radishes more tender. Tobacco dust protects them from flea beetles, and garlic juice from many diseases.

Radish and hyssop should never be sown near each other.

RHUBARB
(*Rheum rhabarbarum*)

This very ornamental as well as useful plant is a good companion to columbines (*Aquilegia*), helping protect them against red spider.

Rhubarb leaves contain oxalic acid. They may be boiled in water and made into a spray that, watered into drillings before sowing plants of the Cabbage family, wallflowers, and other seeds, is helpful in preventing clubroot. It

Rhubarb stems make delicious pies, but the leaves are very toxic and sometimes also cause skin irritation. They are safe on the compost pile, however.

is also useful on roses against greenfly and black spot.

Rhubarb, often called the pie plant, is technically a vegetable but is mostly used for dessert. It also has been long recognized as a laxative. This is one of our oldest garden plants, which Marco Polo found growing in China centuries ago.

SALSIFY (*Tragopogon porrifolius*)

Sometimes this is called oyster plant. To achieve a delicate and different flavor, the milky, oyster-flavored roots need a moist, cool soil for at least four months before reaching maturity.

Salsify grows well with mustard greens, and try growing it with watermelons. Plant the warm-weather watermelon several weeks later than the cool-weather salsify. Let the melons fill the middles of the rows before hot, dry weather arrives. Acting as a living mulch and tending to rest on the ground, the melon vines will still leave the salsify foliage exposed to light and air.

Never use salsify seed over a year old.

SHALLOT (*Allium cepa,* Aggregatum Group)

Shallots, more delicate in flavor than onions, are propagated by planting the sections or cloves that make up the large bulb. They are good to grow with most garden vegetables but, like onions and garlic, should not be located near peas or beans.

SPINACH (*Spinacia oleracea*)

Because of its saponin content, spinach is a useful pre-crop and does well planted with strawberries. (See *Saponin* in the Soil Improvement chapter.) *Bacillus thuringiensis* (See Pest Control) may be used as an insect control.

SQUASH (*Curcubita* spp.)

Two or three icicle radishes planted in each hill will help prevent insects on squash as on cucumbers. Let the radishes grow and go to seed. Nasturtiums will repel squash bugs and so will cigarette ash and other tobacco residue if placed with the seed when it is planted. Squash planted either earlier or later than usual often will escape insect damage. Here in Oklahoma I find squash planted in midsummer is almost entirely insect-free.

Early in the day, before the sun is strong, squash stinkbugs are sluggish, and in the small garden may be picked off. There also are insect-resistant strains of squash.

SWEET POTATO (*Ipomoea batatas*)

Sweet potatoes generally have high energy value, only peas and beans yielding more. They have a common enemy, the fungus disease or wilt called stem rot, which can be controlled with disease-free seed and by rotating the crop. White hellebore controls a number of leaf-eating insects. Nematodes can be a problem, and some varieties have been bred with built-in resistance.

If rabbits bother your sweet potato patch, spray with a diluted fish emulsion.

TOMATO (*Lycopersicon esculentum*)

Tomatoes and all members of the Brassica (Cabbage) family repel each other and should be kept apart. Tomatoes also dislike potatoes and fennel.

Tomatoes will protect asparagus against the asparagus beetle. Since they are tender plants, put tomatoes in during late spring after the early

crop of asparagus spears has been harvested. Tomatoes protect gooseberries against insects.

Tomatoes are compatible with chives, onion, parsley, marigold, nasturtium, and carrot, and for several years I have planted garlic bulbs between
my tomato plants to protect them from red spider mites. Stinging nettle
growing nearby improves their keeping qualities, and redroot pigweed, in
small quantities, is also beneficial.

Though not containing fungicidal elements, tomatoes will protect roses
against black spot. The active principle of tomato leaves is solanine, a
volatile alkaloid that at one time was used as an agricultural insecticide. To
make a spray for roses: Make a solution of tomato leaves in your vegetable
juicer, adding 4 or 5 pints of water and 1 tablespoon of cornstarch. Strain
and spray on roses where it is not convenient to plant tomatoes as companions. Keep any unused spray refrigerated.

Root excretions of tomatoes have an inhibiting effect on young apricot
trees, and don't plant tomatoes near corn, since the tomato fruitworm is
identical to the corn earworm. Don't plant near potatoes, either, since
tomatoes render them more susceptible to potato blight.

Unlike most other vegetables, tomatoes prefer to grow in the same
place year after year. This is all right unless you have a disease problem, in
which case plant your tomatoes in a new area. Since they are heavy
feeders, give them ample quantities of compost or decomposed manure.
Mulch and water in dry weather to maintain soil moisture and stave off
wilt disease and blossom end rot. But never water tomatoes from the top.
Water from below and water deeply.

If you smoke, be sure to wash your hands thoroughly before you work
in your garden, for tomatoes are susceptible to diseases transmitted
through tobacco.

TURNIP, RUTABAGA
(*Brassica rapa,* Rapiferra Group;
and *Brassica napus,* Naprobrassica Group)

An accident revealed that hairy vetch and turnips are excellent companions. Turnip seeds became mixed with the vetch that a gardener planted,

and they came up as volunteer plants. He found that the turnip greens were completely free of the aphids that usually infest them, apparently because the vetch provided shelter for ladybugs, which feast on aphids. Elsewhere it has been found that wood ashes around the base of turnip plants will control scab.

I find that peas planted near turnips are mutually benefited. Turnip and radish seed mixed with clover will bolster the nitrogen content of the soil. In your crop rotation it is good to follow the heavy feeders with light feeders such as turnips and rutabagas.

Turnips dislike hedge mustard and knotweed, and do not rotate them with other members of the Cabbage family, such as broccoli and kohlrabi. A naturally occurring chemical compound in turnips when synthesized is deadly to aphids, spider mites, houseflies, German cockroaches, and bean beetles.

Rutabagas take much the same culture as turnips but require a longer growing season.

WATERMELON (*Citrullus lanatus*)

Watermelons are good to interplant with potatoes, particularly if the potatoes are mulched with straw. The hybrid seedless watermelons, which set no pollen, will produce better if planted with a good pollinator such as 'Sugar Baby'. Watermelons need plenty of sunshine, so do not plant them with or near tall-growing vegetables.

HERBS

ABSINTHIUM (*Artemisia absinthium*)

Also called wormwood, this plant is grown as a border to keep animals from the garden. Ornamental species such as *A. pontica* have leaves of great delicacy and are good to plant in flowerbeds and around choice evergreens. (See *Wormwood* in this chapter.)

ALOE VERA, MEDICINAL ALOE
(*Aloe barbadensis*, syn. *A. vera*)

Nature's own medicine plant, known and used for centuries, is a vegetable belonging to the Lily family. The name "aloe vera" means "true aloe," and it is so named because, among 200 species of aloe, it has the best medicinal properties.

The *Aloe* genus belongs to a larger class of plants known as the xeroids, which possess the ability to close their stomata completely to avoid the loss of water. The plants are easy to grow outside in warm, frost-free climates and equally easy to grow indoors in pots. Possibly because of the bitter taste of the gel, they appear to be completely disease- and insect-free. Almost all xeroids are laxative in nature and have a bitter flavor.

Aloin is the thick, mucilaginous yellow juice that occurs at the base of the aloe leaf, and is also present in the rind. The gel may be removed from the leaf as one would fillet a fish. It is a very slippery, clear, viscous

juice useful for sunburns and for healing cutaneous ulcers of radioactive origin, as well as burns and scalds of various types. Many fishermen carry aloe vera aboard their boats to stop the pain of a sting from a Portuguese man-of-war. It also will stop the sting often experienced when gathering okra.

Aloe vera is known as nature's medicine plant.

The gel may be used instead of tree wound dressing if it becomes necessary to cut a tree limb. The surface will heal over quickly and insects are repelled by the bitter taste. The juice may be mixed with water to make a spray for plants. Powdered aloes dusted on young plants will repel rabbits, but this must be repeated after rain. Aloe vera plants, thrown in the drinking water for chickens, are said to cure the birds of certain diseases.

ANISE (*Pimpinella anisum*)

The spicy seeds of this annual herb, related to caraway and dill, are used to flavor licorice as well as pastry, cookies, and certain kinds of cheese. The oil extracted from the seeds is used to make absinthe, and it is also used in medicine. The flower, powdered and infused with vermouth, is used for flavoring muscatel wine. Anise is antiseptic and is useful as an ointment (when mixed with lard) for lice and itching from insect bites.

When sown with coriander, aniseed will germinate better, grow more vigorously, and form better heads.

BASIL (*Ocimum basilicum*)

Basil helps tomatoes overcome both insects and disease, also improving growth and flavor. Since this is a small plant, 1 to 2 feet tall, grow it parallel to the tomatoes rather than among them. It repels mosquitoes and flies, and when laid over tomatoes in a serving bowl will deter fruit flies.

Sweet basil has inch-long, dark green leaves and a clove-pepperish odor and taste. Pinch out the plant tops and they will grow into little bushes, the dwarf varieties especially becoming beautifully compact. As a kitchen herb, basil is used in vinegar, soup, stew, salad, chopped meat, and sausage as well as in cottage cheese, egg, and tomato dishes; it may be sprinkled over vegetables. 'Dark Opal' makes a very handsome houseplant.

Though it is often said that herbs enhance everything except dessert, sweet basil is one that may be used to give a subtle, indefinable, but delicious flavor to pound cake. It is also one of the culinary herbs that may be used in certain dishes to replace black pepper. (See *Pepper Substitutes* in this chapter.)

It has been known since ancient times that basil and rue dislike each other intensely. Perhaps this is because basil is sweet and rue is very bitter.

BAY (*Laurus nobilis*)

Bay (or laurel) leaves put in stored grains such as wheat, rice, rye, beans, oats, and corn will eliminate weevils. The bay belongs to the same family as the cinnamon, camphor, avocado, and sassafras trees. I have tried sassafras leaves in grains and flours and find them also effective against insects and weevils.

BEE BALM (*Monarda* spp.)

This plant improves both the growth and flavor of tomatoes.

Borage

BORAGE (*Borago officinalis*)

This is an excellent provider of organic potassium, calcium, and other natural minerals of benefit to plants. Grow this herb in orchards and as a border for strawberry beds. Honeybees like to feast on the blossoms.

CARAWAY (*Carum carvî*)

Since it is difficult to sprout caraway seed, sow it with a companion crop of peas. After harvesting the peas, harrow the area and the caraway will come up. It's good to plant on wet, heavy land, the long roots making an excellent substitute for subsoiling. Do not grow fennel near it.

Europeans like and use caraway seed more than we do. It is put in rye bread for its aromatic flavor and to make it more digestible. It is also used in cakes, cheeses, and apple and cabbage recipes.

CATNIP (*Nepeta cataria*)

Catnip contains an insect-repellent oil, nepetalactone, and fresh catnip steeped in water and sprinkled on plants will send flea beetles scurrying.

The catnip compound is chemically allied to those found in certain insects. Two of these occur in ants and another in the walkingstick insect, which ejects a spray against such predators as ants, beetles, spiders, birds, and even humans. Freshly picked catnip placed on infected shelves will drive away black ants. But my cats love catnip, and I like to eat it in salads.

CHAMOMILE (*Matricaria recutita*)

The real plant, the German or wild chamomile (*Matricaria recutita*), recognizable by the hollow bottom of the blossom and its highly aromatic odor, is often confused with Roman chamomile (*Chamaemelum nobile*). This is an excellent companion to cabbages and onions, improving growth and flavor of both. But it should be grown sparingly, only one plant every 150 feet.

Chamomile

Wheat grown with chamomile in the proportion of 100 to 1 will grow heavier and with fuller ears, but too much will harm field crops.

Chamomile contains the substance chamazulene, which has anti-allergic and anti-inflammatory properties when used in the form of tea. It is also used for diarrhea or scour in calves. A tea of one-third each chamomile, lemon balm, and chervil applied as a warm compress will cure hoof rot in animals.

Chamomile flowers may be used in the dog's bed against fleas. When using herbs in pet pillows, simply add the dried form to the stuffing, occasionally adding more to freshen up.

The blossoms soaked in cold water for a day or two can be used as a spray for treating many plant diseases and to control damping-off in greenhouses and cold frames.

A chamomile rinse is excellent for blond hair. Use 3 or 4 tablespoons of dried flowers to 1 pint of water. Boil 20 to 30 minutes, straining when cool. Shampoo the hair before using, since it must be free of oil. Pour rinse over the hair several times and do not rinse with clear water after using. It will leave the hair smelling like sweet clover.

Chamomile contains a hormone that stimulates the growth of yeast. Grown with peppermint in very small quantities, it increases the essential oil.

CHERVIL (*Anthriscus cerefolium*)

This is one of the few herbs that will grow better in partial shade, which can be provided by taller plants growing near it. It does not take kindly to transplanting. Chervil is a good companion to radishes, improving their growth and flavor.

CHIVE (*Allium schoenoprasum*)

Chives are a good companion to carrots, improving both growth and flavor. Planted in apple orchards they are of benefit in preventing apple scab, or made into chive tea may be used as a spray for apple scab or against powdery mildew on gooseberries and cucumbers.

COFFEE SUBSTITUTES

The price of America's favorite hot brew fluctuates from one year to the next, depending on weather conditions in the countries where it is grown. It is interesting to know that there are a number of herbal substitutes that can be made into an acceptable hot drink and are almost universally available. Like any food or beverage, however, the product should be weighed on its own merits. To taste parched and perked seeds, roots, or nuts with the thought of "coffee" in mind is unfair to the substitute.

Chicory (*Cichorium intybus*), which grows on roadsides and in waste clay soils from Canada southward, is identifiable when young by its leaves,

which resemble those of the dandelion. But as the plants mature, a rigid, loosely branched, 2-foot stem develops, and blue flowers bloom in midsummer. The tubular roots, which grow horizontally, should be dug in September after flowering is over. To use chicory as a substitute or as a coffee additive, the roots should be washed, coarsely ground, dried in a very slow oven for two or three hours, then roasted in a clean skillet. This should be done very slowly and the granulated chicory repeatedly stirred until the proper color and flavor are reached.

Chicory

Chicory does not taste like coffee, but when used as an additive, some people believe, it improves the color and flavor of South American coffee as well as extending the number of cups. In France, and in our own South, subject to French influence, chicory is added to coffee, not as an adulterant but for its distinctive flavor.

According to Virginia Scully in *A Treasury of American Indian Herbs,* the Native Americans roasted the roots and used them as they did the dandelion roots. Nelson Coon, author of *Using Plants for Healing,* writes of chicory as "a plant of ancient usage," the name tracing back through Arabic medicinal language to Greek and Egyptian, and mention of the use of chicory is found in Roman writings.

Cleavers (*Galium aparine*), after chicory, seems to be the most popular coffee substitute. Cleavers grows in Alaska, southward across Canada, and on down into Texas, and is found on seashores and in rocky woods. Although cleavers sprouts may be eaten in spring, the tiny twin burr-seeds ripen later. Cleavers grows best in damp thickets. Cleavers, also known as goose grass, is eagerly sought by animals and poultry for its medicinal qualities. Cleavers also, according to all the old herbalists, has a reputation as a reducing diet par excellence, painlessly paring pounds from plump persons.

The usefulness of cleavers is not confined to its value as a reducing diet or as a palatable cooked vegetable. According to Euell Gibbons in *Stalking the Healthful Herbs,* the little two-lobed seeds make the best coffee substitute of almost any plant growing in our range. Perhaps this is because cleavers belongs to Rubiaceae, the natural order of plants to which the coffee tree also belongs.

The little hard fruits should be gathered during summer when they are full size and then roasted in an oven until they are very brown. They can then be made into a beverage that tastes much like coffee and also has a definite coffee aroma. Cleavers contains no caffeine.

Cleavers has long been considered a medicinal herb, its properties being listed as diuretic, tonic, alterative, and somewhat laxative. Gather the herb in May or June when in flower and dry in a warm room.

Nut grass (*Cyperus esculentus*) is an herb of almost universal distribution. If nut grass tubers are roasted until they are a very dark brown all through, then pulverized in a blender or coffee mill, they make a very palatable hot drink when brewed exactly as you do coffee. Nut grass coffee contains no harmful stimulants and can freely be given to children who insist on having "coffee" when the grown-ups do.

Sunflower (*Helianthus annuus*) has long been used by Native Americans, who parched the highly nutritious seeds and made them into a meal for gruel and cakes. Frequently, they added water to the meal as a drink and crushed roasted seeds to make a drink like coffee.

Sassafras (*Sassafras albidum*) is also called wild cinnamon or mitten plant. It likes to grow in rich woods in humusy soil and both leaves and bark may be used, the leaves being gathered in spring and the bark in fall. The name sassafras is a corruption of the Spanish word *saxifrage,* which in turn is derived from two Latin words, *saxum,* meaning "a rock," and *frango,* meaning "I break." It is interesting to note that herbalists recommend that sassafras be always included in a formula of kidney-stimulating herbs.

A hot drink of sassafras may be prepared by stirring $1/2$ teaspoonful of the ground bark in 1 cup of boiling water. Cover with a saucer for about 5 minutes. Stir, strain, and add a little honey for sweetness.

Dandelion roots (*Taraxacum officinale*), roasted, have long been used as a coffee substitute and taste much like chicory. The young leaves often appear in salads, and Native Americans chewed the young stems like gum.

Dried persimmon seeds (*Diospyros*), **wild senna seeds** (*Cassia occidentalis*), and **roasted acorn beans** have been ground and used as coffee. The black or red oak (*Quercus*) acorns were used.

Cereals, nuts, peas, soya beans, and even okra seeds have been roasted and used as coffee substitutes, dilutants (thinners), or additives. Again, when it comes to extending the family coffee supply, it is well to remember that some sacrifice of flavor must be made.

COMFREY (*Symphytum officinale*)

Comfrey, also called knitbone or healing herb, is high in calcium, potassium, and phosphorus, and rich in vitamins A and C. It was an ancient belief that comfrey preparations taken internally or as a poultice bound to injured parts hastened the healing of broken bones.

It is possible that the nutrients present in comfrey actually do assist in the healing process, since we now know that the herb also contains a drug called allantoin, which promotes the strengthening of the lining of hollow internal organs. However, this herb also contains alkaloids that cause liver damage and is no longer considered safe for internal use.

The leaves of Russian comfrey are ideal for the compost heap, having a carbon-nitrogen ratio similar to that of barnyard manure.

CORIANDER (*Coriandrum sativum*)

Coriander has a reputation for repelling aphids while being immune to them itself. It helps anise germinate but hinders the seed formation of fennel. In blossom, the herb is very attractive to bees.

Many people think the foliage and fresh seed of coriander have a disagreeable smell, but as the seeds ripen they gain a delicious fragrance that intensifies as they dry. The savory seeds, sometimes sugar-coated as a confection, are baked in breads or used to flavor meats.

Coriander has four times more carotene than parsley, three times as much calcium, more protein and minerals, more riboflavin, and more vitamin B_1 and niacin. Oil of coriander is used medicinally to correct nausea.

DILL (*Anethum graveolens*)

Dill is a good companion to cabbage, improving its growth and health. It does not do well by carrots and, if allowed to mature, will greatly reduce that crop, so pull it before it blooms.

Dill will do well if sowed in empty spaces where early beets have been harvested, and light sowings may be made with lettuce, onions, or cucumbers. Honeybees like to visit dill blossoms.

DRYING HERBS

Leaf herbs should be cut, washed, tied in loose bunches, and allowed to drip dry. Place upside down in large brown paper bags that have been labeled. Close the mouth of the bag about the stems; let the herbs hang

free inside the bag. Hang
the bags where they will
have good air circulation.
In this way none of the oil
is absorbed by contact
with the paper, as may be
the case if dried herbs are
stored in cardboard boxes.

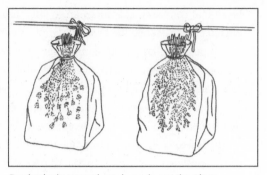

Dry herbs in paper bags hung in an airy place.

With seed herbs, let
the plants mature until
the ripe seeds part from
the dry umbels with a little pressure. This occurs after they lose their
greenish color but before they will drop of their own accord. Cut the
heads on a dry morning and spread them out on brown paper in the sun
for the rest of the day, stirring occasionally. Do this for several days, taking
them in at night, until they are thoroughly dry. Seeds may be stored in
opaque glass bottles or clear ones away from the light.

ELECAMPANE (*Inula helenium*)

My German heritage bids me have great respect for elecampane, sometimes called horseheal or horse alder. This herb was under the protection
of the goddess Hulda, who first taught mortals the art of spinning and
weaving flax. Candied elecampane, according to a 17th-century herbal,
was thought to "cause mirth."

Elecampane came to America as a healer, being introduced into gardens as a home remedy. Finding the climate congenial, it went native and
now grows wild in many places.

The substance most abundantly contained in elecampane root is
inulin, a sort of invert starch usable as a replacement for ordinary starch
in the diet of diabetics. It also contains a volatile oil and several identifiable crystalline substances. The thick, yellow taproot has the odor of
camphor.

Elecampane once was used in England for the heaves in horses. As far
back as 1885, a Dr. Korab demonstrated that the active bitter principle of

the plant, called helenin, was a powerful antiseptic and bactericide, particularly destructive to the tubercle bacillus.

Elecampane in the garden provides a 6-foot-tall accent of bright yellow in midsummer when the large-rayed flowers stand above the enormous felty leaves. The plants are useful in providing a bit of shade for lower-growing mints.

FENNEL (*Foeniculum vulgare*)

Most plants dislike fennel, and it is one herb that should be planted well away from the vegetable garden, since it has an inhibiting effect on bush beans, caraway, kohlrabi, and tomatoes. Fennel planted away from the garden is valuable for its masses of fringed foliage. At one time the fragrant seeds, which smell and taste like licorice, were made into a tea soothing to colicky babies. Mixed with peppermint leaves, it also makes a delicious tea for adults. Fennel is inhibited by the presence of coriander and will not form seed. It also dislikes wormwood.

GARLIC (*Allium sativum*)

The Babylonians and Hindus knew of garlic's medicinal power 3,000 years ago, and it was well known to the ancient Egyptians, who fed great quantities of garlic to the workers who built the pyramids. The Greek physicians, fathers of present-day medicine, used garlic in their prescriptions, and it was rationed to the soldiers of the mighty Roman armies.

The 800-year-old medical school of Salerno included garlic in its *materia medica,* and it always has been one of the standbys of folk medicine practitioners.

Garlic is an effective control for many insects. Try this recipe for garden use: Take 3 to 4 ounces of chopped garlic bulbs and soak in 2 tablespoons of mineral oil for a day. Add 1 pint of water in which 1 teaspoon of fish emulsion has been dissolved. Stir well. Strain the liquid and store in a glass or china container, as it reacts with metals. Dilute this, starting with 1 part to 20 parts of water, and use as a spray against your worst insect pests.

If sweet potatoes or other garden plants are attracting rabbits, try this spray. Rabbits dislike the smell of fish, too, so fish emulsion may help.

Garlic sprays are useful in controlling late blight on tomatoes and potatoes. Garlic is an effective destroyer of the diseases that damage stone
fruits, cucumbers, radishes, spinach, beans, nuts, and tomatoes.

A garlic-based oil sprayed on breeding ponds showed a 100 percent kill
of mosquito larvae in a University of California experiment. Garlic cloves
stored in the grain are good against grain weevils.

Garlic grown in a circle around fruit trees is good against borers. It is
beneficial to the growth of vetch and is protective planted with roses. All
alliums, however, inhibit the growth of peas and beans. Plant garlic with
tomatoes against red spider. I have done this for three successive years
with good results.

HERBS FOR TEA

Herb teas have both aided digestion and given pleasure to untold generations. Here is how to make tea:

Pick some of the leaves, using only those that are undamaged (or dried
leaves may be used). You will need four or five leaves for a cup of tea and
about a handful for a pot. Wash the leaves in cool water, put in a cup or
pot, and pour boiling water over them. Let steep, covered, for three to five
minutes. Add sugar or honey to sweeten.

Please note: Research on herbs is ongoing, and some that were formerly used internally are now thought to be potentially harmful. Make
teas only from plants suggested for this purpose in up-to-date guides to
the use of herbs.

HOREHOUND (*Marrubium vulgare*)

In centuries past, virtues attributed to horehound included the power
to cure snakebite and merit as a fly repellent, vermifuge, and an ointment for wounds and itches. The Hebrew name for the plant, *marrob,*
means "a bitter juice." It was one of the five bitter herbs required to
be eaten at the Passover feast. The Romans considered it a good and
sometimes magical herb. Horehound's real value relates to pulmonary
ailments, and it is widely used as an ingredient in lozenges for coughs
and colds.

For small stems and better quality, grow the plants intended for candy making close together. Either the fresh or dried herb can be used for this purpose.

Grasshoppers and other insects dislike the taste of horehound. The plant grows well with tomatoes, improving their quality, causing them to bear more abundantly and continue later in the season.

HORSETAIL
See the Wild Plants chapter.

HYSSOP (*Hyssopus officinalis*)
It's hard to find anything more delightful than a hyssop hedge in full sun. The blue, white, and pink flowered hyssop makes an intriguing design grown with gray Roman wormwood.

LAVENDER (*Lavandula augustifolia*)
In a 2 percent emulsion spray for cotton pests, lavender kills somewhere between 50 and 80 percent within a period of 24 hours.

Few ticks are found in lavender plantations, although neighboring woods and shrubs may harbor many. It has been used effectively as a mouse repellent, and lavender sachets are often put in woolen clothing to repel moths, while leaves scattered under woolen carpets are helpful for the same purpose.

The plant grows from seed very slowly. Both plants and seeds are obtainable from Nichols Garden Nursery. (See Sources on page 442.)

LEMON BALM (*Melissa officinalis*)
Lemon balm, often called the bee herb, has long been famous for its delightful, lemon-scented foliage and honeyed sweetness. *Melissa*, the generic name, is Greek for "honeybee," and there is a very old belief that bees will not leave the hive area if melissa grows near it. Pliny wrote, "When bees have stayed away they do find their way back home by it."

Melissa tea calms the nervous system and stimulates the heart, it is very relaxing, and may even dispel headache or migraine. In pastures this plant increases the flow of cows' milk, and it is very good to give cows after calving in a tea with marjoram.

LOVAGE (*Levisticum officinale*)

Lovage planted here and there will improve the health and flavor of other plants. It is one of the herbs that may be used to reduce the amount of

---

Freckle Remedies

I do not dislike freckles—my mother always told me they were the scars left by the angels' kisses—but some lucky people seem to receive more kisses than they want. If freckles trouble you, elder flowers added to facial steam baths will clear and soften the skin and are also good against freckles and faulty pigmentation, especially when they are used in conjunction with whey and yogurt. Such face packs are not only soothing but also tonic and stimulating. Parsley water used externally also is said to remove freckles or moles.

Freshly crushed leaves or freshly pressed juice of lady's-mantle (*Alchemilla vulgaris*) is helpful against inflammation of the skin and acne, as well as freckles. Externally used, lime flowers (*Tilia*) are a fine cosmetic against freckles, wrinkles, and impurities of the skin, and they also are stimulating to hair growth.

Sow the hyssop seeds in late fall so they will germinate first thing in spring. Planted near grapevines they will increase their yield; near cabbages, they will lure away the cabbage butterfly. Bees are very fond of visiting hyssop blossoms, yet many other insects find the plant repellent. Radishes will not do well if hyssop is nearby.

Hyssop leaves have a peculiar fragrance reminiscent of civet, yet some people use them the same way as savory. A compress of hyssop leaves is good for removing black and blue spots from bruises.

This is another of the bitter herbs used in Jewish ceremonies but it is not the true hyssop of the Bible, which is believed to be a species of *Origanum*.

salt used for seasoning and is delicious sprinkled on salads or used in cheese biscuits. In dishes that need strengthening, it can replace meat stock and is excellent in soups and casseroles.

Lovage will winter well, but in colder climates the roots should have some protection.

MARJORAM (*Origanum* spp.)

This small, easily grown plant is probably one of the oldest herbs in use. "Marjoram" really covers three very different kinds of marjoram, all of which belong to the Labiatae family.

The sweet marjoram, an annual, is the most popular for flavoring, especially in sausages. It was used extensively by the Greeks, who gave it the name, which means "joy of the mountains." Its disinfectant and preserving qualities made it an invaluable culinary herb in the Middle Ages.

Pot marjoram is a perennial with a bit less flavor but more easily grown.

Wild marjoram (*Origanum vulgare*) is a wild as well as a cultivated variety with a strong flavor, the pungency varying according to where it grows. This herb, also known as oregano, is used the world over in Italian, Mexican, and Spanish dishes, and is believed to have both stimulating and medicinal properties, since it contains thymol, a powerful antiseptic when used internally or externally. The whole plant of oregano is covered with hairy oil glands. The pleasant aromatic scent, reminiscent of thyme, is very lasting—even the dead leaves and stems retain it during the winter or when dried for culinary use.

In the garden all the marjorams have a beneficial effect on nearby plants, improving both growth and flavor.

MINT (*Mentha*)

Mint is a good companion to cabbage and tomatoes, improving their health and flavor. Both mint and tomatoes are strengthened in the vicinity of stinging nettle. Mints such as apple, orange, and pineapple will thrive under English walnut trees, in part because of the filtered sunlight.

Mint

Mint deters the white cabbageworm by repelling the egg-laying butterflies. Spearmint repels ants and may help keep aphids off nearby
plants. (See also *Peppermint* and *Spearmint.*)

Mint is a repellent against clothes moths when used indoors and is
useful against black flea beetles. The leaves strewn under rabbit cages
will keep flies to a minimum, while dried leaves (or mint oils) will repel
rats and mice.

MUGWORT (*Artemisia vulgaris*)

Mugwort is one of the most useful members of the Artemisia family.
Planted in chicken yards it will help repel lice, and since the chickens like it
as a food, it is also thought to be helpful in ridding them of worms. Made
into a weak tea, it may be used as a fruit tree spray.

Mugwort is not good when it is near other garden plants because it has
a growth-retarding effect, particularly in years of heavy rainfall. The roots
and leaves exude a toxic substance. This soluble toxic, absinthin, washed
off by rain, soaks into the soil near the plant and remains active over an
indefinite period of time.

ORACH (*Atriplex hortensis*)

This beautiful annual, sometimes called French spinach, has shield-shaped,
wavy leaves of beet red. They have a mealy texture similar to its close relative, lamb's-quarters, and are also used as a potherb. While orach may be
planted in the garden, it should never be placed near potatoes, since it has
an inhibiting effect on their growth.

Old-time herbalists believed that orach had a cleansing quality either
raw or cooked and if laid upon swollen glands of the throat would cure
the condition.

OREGANO (*Origanum vulgare*)

Sow with broccoli to repel the cabbage butterfly. (See *Marjoram.*)

PENNYROYAL (*Mentha pulegium*)

Plant this with broccoli, Brussels sprouts, and cabbage against cabbage
maggot. Like tansy, it may be grown at doorways to repel ants and is also

a good mosquito repellent if rubbed on the skin. Fresh or dried sprigs have long been used as a flea repellent.

PEPPER SUBSTITUTES

Basil, summer savory, thyme, marjoram, and nasturtium can help replace pepper in cooking for those who have digestive disturbances.

PEPPERMINT (*Mentha* × *piperita*)

Of all herbs, this makes the greatest demand on the soil for humus and moisture. It will benefit from a small amount of chicken manure if well broken down.

Peppermint drives away red ants from shrubs, and planted among cabbage it will repel the white cabbage butterfly. When growing with chamomile it will have less oil, but the chamomile will benefit and have more. The oil of peppermint is increased when it is grown with stinging nettle.

Black mint is distinguished from other species by purple stems and dark green leaves. It grows about 3 feet tall and is crowned with spikes of lavender flowers in midsummer. It is widely used for medicinal and commercial purposes. (See also *Mint* and *Spearmint*.)

RUE (*Ruta graveolens*)

By now I'm sure everybody knows that rue doesn't like basil. But an authority as ancient as Pliny tells us that "rue and the fig tree are in great league and amitie together."

Rue planted near roses or raspberries will deter the Japanese beetle. It can be clipped and made into an attractive hedge, but first be sure you are not allergic to it, since the foliage can cause dermatitis as severe as that from poison ivy when the plant is coming into flower. The intensity of the eruption seems aggravated by the presence of sunlight. If you happen to be perspiring and working bare-handed with rue, you may get poisoned. If this happens, washing with brown soap or covering the exposed area with oil will help.

Rue may be grown among flowers as well as vegetables, where its good looks will add to the planting. It is protective to many trees and shrubs. It is good near manure piles and around barns for discouraging both house and stable flies.

The ancient Schola Salernitans wrote that "Rue putteth fleas to flight." However, it should be used for only dog pillows or beds, for cats do not like it. Anything rubbed with the leaves of rue will be free from cats' depredations—good to know if your house cats tend to claw the furniture.

SAGE (*Salvia officinalis*)

Sage is protective to cabbages and all their relatives against the white cabbage butterfly, and it also makes the cabbage plants more succulent and tasty.

The herb also is good to grow with carrots, protecting them against the carrot fly, and is mutually beneficial with rosemary. Do not plant sage with cucumber, which does not like aromatic herbs in general and sage in particular.

Originally sage was used medicinally in stuffing and meats to make them more digestible, but we have grown to like the flavor. It has often been made into a tea, but is now considered harmful if much is taken internally.

SALT SUBSTITUTES

The clever use of herbs can replace salt in many dishes and reduce the amount needed in others. Those on a salt-free diet can flavor their food deliciously by using such herbs as celery, summer savory, thyme, lovage, and marjoram.

SANTOLINA (*Santolina chamaecyparis sus*)

This south European plant, sometimes called lavender cotton, is a good moth repellent. The name is from *sanctum linum,* meaning "holy flax." The plants are improved by being pruned as soon as the blossoms fall.

SAVORY, SUMMER (*Satureja hortensis*)

In Germany, savory is called the bean herb because it's good to grow with beans and also to cook with them. It goes with onions as well, improving both growth and flavor.

SAVORY, WINTER (*Satureja montana*)
Winter savory is a subshrub, about 15 inches tall. Its leaves, though not as delicate as summer savory, may be used in cooking. It is useful as an insect repellent, too.

SESAME (*Sesamum indicum*)
Sesame is an herb grown in tropical countries, mainly for the oil obtained from its seeds. The oil is used in salad dressings and for cooking, while the delicious seeds are used to flavor bread, candy, biscuits, and other delicacies. Sesamin is derived from sesame oil and is used as a synergist to strengthen the effect of pyrethrum.

Sesame is very sensitive to root exudates from sorghum (*Sorghum bicolor*) and will not ripen well when grown near it.

SOUTHERNWOOD (*Artemisia abrotanum*)
Dry the leaves of southernwood, place them in nylon net bags, and hang them in the closet to prevent moths. Burned to ashes in the fireplace, they will remove any cooking odors from the house.

Southernwood has green, finely divided leaves with a lemon-mixed-with-pine scent. Grown near cabbages, it will protect them from cabbage-worm butterfly and near fruit trees will repel fruit tree moths. Among its names are old man, lad's love, and even maiden's ruin!

SPEARMINT (*Mentha spicata*)
Spearmint also is called green mint, pea mint, and lamb mint. (See *Mint.*)

SUGAR SUBSTITUTES
Three herbs that can reduce the use of sugar in cooking and sweets are lemon balm, sweet cicely, and angelica. They are particularly good in tart fruit or fruit pies made of black currant, red currant, rhubarb, gooseberry, plums, and tart apples. Not only do these herbs make it possible to use sometimes half the usual quantity of sugar but they also impart a delicious flavor. Chopped sweet cicely may be added to lightly sugared strawberries for a refreshing summer treat.

SWEET BASIL (*Ocimum basilicum*)

See *Basil*.

TARRAGON, FRENCH (*Artemisia dracunculus*)

Use potato fertilizer as a side dressing for tarragon in spring and again right after the first cutting to increase the vitality of the plant. To reset tarragon successfully, the roots must be carefully untangled. Each section of root eased apart from the clump may be reset to form another plant. This is best done every third year in March or early April. As a cooking herb tarragon is something very special, and it is particularly good for flavoring vinegar.

THYME (*Thymus vulgaris*)

Thyme has an ancient history as a medicinal and culinary herb. The oil still is used as the basis of a patent cough medicine, while thymol has anti-bacterial powers of considerable importance. But thyme is of value mainly in cooking, being very good for poultry seasoning and dressing. Lemon thyme makes a delicious herbal tea.

The herb deters the cabbageworm and is good planted anywhere in the garden, accenting the aromatic qualities of other plants and herbs.

VALERIAN (*Valeriana officinalis*)

This herb is good anywhere in the garden, particularly to give vegetables added vigor. It is rich in phosphorus and stimulates phosphorus activity where grown; it is attractive to earthworms and therefore particularly useful in the compost pile.

It is thought that the Pied Piper of Hamelin used valerian to clear the town of rats, yet many gardeners find it attracts cats, which love to nibble the leaves and roll against them. The *Universal Herbal* of 1820 states: "It is well known that cats are very fond of the roots of valerian; rats are equally partial to it—hence rat-catchers employ them to draw the vermin together."

Valerian enjoyed great prominence in colonial days as a drug plant, the strangely scented root being brewed into a sedative tea. Since the oil of valerian does have an effect on the central nervous system, the tea should not be drunk often.

Cold tea is made by this rule: Put 1 level teaspoon crushed dried valerian root to soak in 1 cup cold water. Cover and place in a cold place for
12 to 24 hours. Strain and drink about 1 hour before retiring.

WOODRUFF, SWEET (*Galium odoratum*)
Sweet woodruff is an excellent groundcover, particularly under crabapple
trees. While it will grow in the sun, the foliage is darker green and much
more abundant if it receives shade at least half of the day.

WORMWOOD (*Artemisia absinthium*)
This herb will keep animals out of the garden
when used as a border. It's good repellent for
moths, flea beetles, and cabbageworm butterfly. It discourages slugs if sprayed on the
ground. Fleas on cats and dogs may be dislodged with a bath of wormwood tea.

Many artemisias are of value as ornamentals, their cool, silvery beauty providing a fine
contrast for flowers, such as red geraniums, of
brilliant color. They do not attract honeybees,
but small wasps seem to be frequent visitors.
Keep wormwood out of the garden since most
plants growing near it do not do well, particularly anise, caraway, fennel, and sage.

Wormwood

WILD PLANTS

ALGAE

Here is a food source, not yet fully explored, that may prove of great value as populations increase, for the total amount of photosynthesis carried on by marine algae may be 10 times greater than the total of such activity in all land plants. If this is true, then we may well look to the plants of the sea as a rich, abundant, and relatively untapped source of foodstuffs.

In some parts of the world, for example Japan and China, algae have for many years been important items in the human diet. Farmers in many coastal regions cultivate brown algae on bamboo stems pushed into the ocean bottom in shallow waters. Dried preparations of these brown algae, available in many food stores in the United States, are very rich in minerals and also have moderate quantities of carbohydrates and vitamins.

Brown algae, because of their rich mineral content, also are often used as soil fertilizers (spread on fields and plowed under), or they may be dried and burned and their ash used as fertilizer. Some species are a commercial source of iodine for medicinal use.

AMARANTH
(*Amaranthus retroflexus*)

This plant, sometimes called rough pigweed and commonly found in disturbed ground everywhere, is one of the best weeds for pumping nutrients from the subsoil.

Amaranth loosens the soil for such crops as carrot, radish, and beet. It helps potatoes yield more abundantly and is good to grow with onions, corn, pepper, and eggplant—but keep it thinned. Tomatoes grown with the weed are more resistant to insect attack.

Amaranth

Euell Gibbons says in his *Stalking the Healthful Herbs* that green amaranth has a higher iron content than any other green vegetable except parsley.

A type of amaranth grown widely in rural Mexico is known as "the sacred food of the Aztecs." The small seeds, easily threshed from the large heads, can be baked with bread. Sometimes they are even popped, mixed with honey, and eaten as a confection. In India certain species are eaten as a salad or cooked like spinach, and the seed is ground into flour.

Amaranth (which is distantly related to beets) has a higher protein content than the cultivated beet, is higher in vitamin C, and has approximately the same amount of vitamin A.

ASTER (*Asteraceae*)

Many asters are soil indicators. Some like low, moist soil, so if the bushy type (*Aster domosus*) or the purple-stemmed aster (*A. puniceus*) shows up in your pastures or fields, it indicates a need for drainage. The sea aster (*A. tripolium*) grows on seasides and near salt mines and is a salt and soda collector. The poisonous woody aster (*Xylorhiza parryi*) of the West indicates an alkaline soil.

BAKED-APPLEBERRY (*Rubus chamaemorus*)

Most of us are familiar with the blackberry-dewberry family, Rubus, but one member has been neglected. The baked-appleberry, or cloudberry, is found growing in wet areas in acid peats from the Arctic regions southwest through New England. It bears a single, soft, pinkish berry, which grows on a 12-inch stem bearing a few scalloped leaves. The berries make a delicious fresh dessert and may be preserved in jams or prepared as juice.

BROOM BUSH (*Sarothamnus scoparius*)

This useful weed grows on the poorest, stoniest soils and those that are sandy and slightly to medium acid. Being rich in calcium carbonate, the plant improves the soil through decomposition of its leaves and stems. In a thin stand it will provide shelter for young tree seedlings, but it will choke them out if too many of the weeds are present.

BURDOCK (*Arctium*)

Do not allow wild burdocks to grow, since they are robbers of the soil. Particularly do not allow them to go to seed, for the burrs will adhere to the hair of sheep, horses, and dogs, even clothing, and be spread far and wide.

Burdock roots have medicinal value and are said to alleviate gout and skin diseases. An edible burdock ('Takinogawa') has been developed in Japan, where the cooked roots are greatly relished for their refreshing, pungent flavor. The Japanese also value this burdock for its reputed blood-purifying qualities and the relief it is said to give to sufferers of arthritis. (It is available from Nichols Garden Nursery; see Sources on page 442.)

CALAMUS (*Acorus calamus*)

It is said that mosquitoes are never found in swamps or other standing water in which calamus, sometimes called sweet flag or sweet root, is growing.

CHARLOCK (*Sinapis arvensis*)

Charlock or wild mustard is frequently found in grainfields. The seeds and those of wild radish can lie inert in the ground for 50 to 60 years, showing up again when the field is planted to grain, particularly oats.

CINQUEFOIL (*Potentilla* spp.)

The cinquefoils are considered a bad symptom when found on pastureland, for they indicate a very acid soil, and gradually they will choke out other grasses and clovers. They are very persistent and will last when other grasses are burned out by drought. It has been observed that the butternut or gray walnut tree (*Juglans cinerea*) and the black walnut have an inhibitory effect on the growth of the creeping cinquefoil.

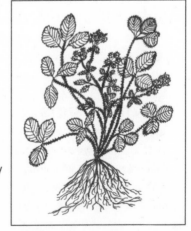

Cinquefoil

CROWFOOT (*Ranunculaceae*)

Like other members of its family, the common meadow buttercup secretes a substance in its roots that poisons the soil for clover by inhibiting the growth of nitrogen bacteria. So potent is this secretion that clover in time will disappear if buttercups begin to invade the field. Cattle will not eat the acrid, caustic plant, and children should be warned against biting the buttercup's stem and leaves, which are capable of raising blisters.

The garden monkshood (*Aconitum*) is even more dangerous, being poisonous in all its parts, while other members of this family that are more or less poisonous are delphinium, columbine, and peony. They are beautiful, but grow them with care.

CYPRESS SPURGE (*Euphorbia cyparissias*)

This funny little plant escaped from eastern gardens where it was grown as an ornamental. The milky juice once was thought to be effective against warts, and it is used in France as a laxative. Here the plant has become a weed. Do not let it grow near grapes, since it may cause the vines to become sterile, and cattle eating hay containing the spurge are made ill.

DANDELION (*Taraxacum officinale*)

Dandelion likes a good, deep soil, as do clover and alfalfa. Soil around dandelions is attractive to earthworms, for this plant is a natural humus producer.

Dandelions on your lawn may frustrate, but actually they are not in competition with the grasses because their 3-foot-deep roots take nutrients from a different level of the soil. They penetrate hardpan and bring up minerals, especially calcium, depositing them nearer the surface and thus restoring what the soil has lost by washing. When dandelions die, their root channels act like an elevator shaft for earthworms, permitting them to penetrate deeper into the soil than they might otherwise.

Dandelion

Dandelions exhale ethylene gas, which limits both the height and growth of neighboring plants. It also causes nearby flowers and fruits to mature early.

DATURA (*Datura stramonium*)

Other names for this weed are Jamestown weed, jimsonweed, apple of Peru, thorn apple, stinkweed, devil's trumpet, angel's trumpet, and dewtry. Although all parts of datura are poisonous, it is a source of valuable medicine. The weed is especially helpful to pumpkins, and it protects other plants from Japanese beetles. The smoke from dried datura leaves is calming to honeybees when opening a hive, but use it sparingly.

Datura

DEAD NETTLE (*Lamium album*)

Although classed as a nettle, this plant, sometimes called white archangel, has no relation to stinging nettle and does not sting. A similarity of the leaves may be the reason for its name. Dead nettle has a long season of showy white bloom and is one of the few herbs that will also grow in damp places in filtered sunlight. As such, it is a valuable companion for the garden and for grain crops.

DEVIL'S SHOESTRING (*Tephrosia virginiana*)

There are 19 species of this native North American weed that have valuable insecticidal properties. It is low in toxicity to animals but is reputed to contain a fish poison. Wild turkeys are fond of eating the plant.

DOCK (*Rumex crispus*)

Curled dock is both a food and a medicine. In the old days it was gathered to "thin and purify the blood" in spring of the year. There is no evidence to support this claim, but the high vitamin C content of dock undoubtedly was beneficial, particularly after a winter diet deficient in greens. The greens also are richer in vitamin A than carrots.

Dock is good to calm the pain of stinging nettle. Crush the juicy leaves and rub on the affected area.

DYER'S GREENWEED (*Genista tinctoria*)

Once this was considered a very useful plant for dyeing. Bees like it for making honey, and sheep and goats like to graze it, but it is thought that its bitter taste affects the milk of cows that eat it.

EELGRASS (*Zostera marina*)

Eelgrass has an edible grain and is a widely distributed sea plant along the coasts of North America and Eurasia. It is reportedly harvested in spring by the Seri Indians along the Gulf of California in Sonora, Mexico.

The upper stem fruits in spring, then breaks off and floats on the surface of the water. The Seri harvest the grain when large quantities of the plant are found floating loose near the shore.

Eelgrass has a content of protein and starch similar to that of grains grown on land. Once it is dry, it is separated from the seaweed with which it grows, and the grains are toasted and ground into flour. Cooked into a gruel, it may be eaten with honey. It is the only ocean plant known that has a grain used as a human food resource.

The Seri find many other uses for the eelgrass, one of which is a cure for diarrhea. It is also piled over house frames for shade and made into toys for the children. Others find it makes an excellent mulch for garden or orchard.

Fresh-water eelgrass (Hydrocharitaceae, *Vallisneria americana*), also known as tapegrass and wild celery, grows in the mud of shallow ponds, sending up ribbonlike leaves directly from the root. (See Sources.)

EUPHORBIA (*Euphorbia* spp.)

A few well-placed plants of caper spurge (*Euphorbia lathyrus*) will deter moles and mice and thus are good to plant near young fruit trees. They also are useful to repel rats and mice. Many of the spurge family like dry, light, sandy soils but will spread to cultivated land if given a chance.

Injury to stem or leaves of the euphorbias causes them to exude a milky-looking sap that is very acrid and poisonous. Great care should be taken that the sap does not touch a scratch, as it will even cause blisters on delicate skin. However, the juice of the leafy spurge (*E. elula*) and the cypress spurge have been used against warts.

Snow-on-the-mountain, one of the euphorbias, is an attractive annual plant that grows wild from Minnesota to Texas. Poinsettia, another lovely member of this family, was formerly considered highly toxic but is no longer classed as a poisonous plant.

FERN, MALE (*Dryopteris filix-mas*)

My beloved old *Pharmocopoeia of the United States of America,* 7th decennial revision, 1890, speaks of this fern eloquently, referring to its taste as "sweetish, acrid, somewhat bitter, astringent and nauseous." It does not, however, say that fern seed will make you invisible, as the Doctrine of Signatures once stated.

That ferns have medicinal value has been recognized for centuries, and they are still listed in the *Pharmacopoeia* today, the useful species including also the evergreen wood fern (*D. marginalis*). In autumn the roots are carefully dug, cleaned, and dried, and the substance oleoresin is extracted through the use of ether.

Perhaps it is because of the oleoresin that ferns and beeches (*Fagus*) have an inhibiting effect on each other. However, a compost made of ferns assists tree seeding and is useful to tree nurseries to encourage germination.

FIDDLEHEAD FERN, OSTRICH FERN
(*Matteuccia struthiopteris*)

At one time, bracken ferns were widely eaten. Today, however, there is evidence that bracken fern (*Pteridium aquilinum*) contains carcinogenic compounds, and its consumption is discouraged. Fiddleheads from the Ostrich Fern (*Matteuccia struthiopteris*), however, are safe for consumption if properly prepared. Ostrich fern grows throughout eastern North America in moist, cool, preferably partly shady areas. It will grow in the extreme cold of Zone 2 but flags in more southern areas.

Gather fiddleheads in early spring before they begin to unfurl, snapping them off by hand. They are about 1 inch in diameter and are covered with papery brown scales that are easily removed by rubbing each head through your fingers. Wrap in plastic, refrigerate immediately, and use within three days of gathering.

Prepare the ferns by washing in several changes of water, then boiling for a minimum of 10 minutes or steaming for at least 20 minutes. You may eat immediately with butter or subsequently sauté or stir-fry them.

FLEABANE (*Erigeron*)

This is one of our native plants that has spread to Europe and taken possession of stony soils. It is used for medicinal purposes, and the acrid oil as a mosquito repellent. But some people are allergic to this plant, reacting to it as they would to poison ivy.

FUNGI

Fungi are plants without chlorophyll—some are very useful and even edible, others are very troublesome.

Mushrooms have a natural affinity for plants in woods, fields, and meadows. The part of the mushroom we see is just a small portion of the entire plant. Most of it is underground, a tangled, twisted jungle of threads that form the vegetative part of the plant (mycelium). The mushroom itself is the reproductive part, the fruiting body, which grows easily under the right conditions of moisture and temperature.

My sister-in-law in Missouri once brought home some morels from her father's farm and took them outdoors to her backyard to trim. She threw the pan of trimmings under an apple tree in her small orchard and was surprised sometime later to find morels growing thriftily there. They continued to come up intermittently for several years.

The delicious morels are fairly common in the United States. The fruit body, like that of other mushrooms, grows above the ground and resembles a sponge. Because of this, they are easy to identify and safe for the collector to gather for the table. They are often found in apple orchards just about the time the trees are blossoming.

Morchella elata   *Morchella semilibera*

Many mushrooms are deadly poison and no one who is inexperienced should ever gather them for food. The deadly amanita, sometimes called the destroying angel, can cause death in less than six hours. Yet many of the poisonous mushrooms also serve a purpose in promoting healthy growth on other plants.

Truffles, another type of fungus, grow several inches below the ground and are not visible. They are rare in the United States but are quite common in Spain, England, Italy, and France, where they grow under oak and chestnut trees. They are found by dogs and pigs specially trained to locate them by smell. Should you see a French farmer following a pig on a leash, this is simply a routine way of picking truffles, which sell for fantastic prices.

Because of their close association with oak and chestnut trees, scientists believe that the truffles help the tree roots assimilate chemicals from the soil. Truffles vary from ¼ inch to 4 inches in diameter and resemble an acorn, a walnut, or a potato. The spores are borne within the tuberlike body of the fungus. They have a delicious taste and serve as a condiment. They are black and thus very attractive as a garnish for salads.

Fungus on tree roots first was reported in 1885 by the German botanist A. B. Frank. His belief that water and nutrients were entering the tree through the fungus was scoffed at, but we know today that the fungus acts as a link between the soil and the rootlets of the plant. The tree in turn helps the mycorrhizal fungus by providing root metabolites, substances that are vital to the fungus for the completion of its full life cycle. Harmless tree fungi called saprophytes help trees resist such diseases as bark canker, decay fungi, and leaf rust fungus.

Mildews are a type of fungi that can be extremely troublesome and difficult to control when they form on plants, usually due to combined moisture and humidity. They attack grapes, lettuce, tomatoes, roses, peas, tobacco, potatoes, cucumbers, and many other fruits and vegetables, usually forming a gray or white powderlike coating on the surface of the leaves. I have found that it is possible to partly control this fungus with a dusting of wettable sulfur. Sunshine and good air circulation are the best remedies.

Smut is a fungus that attacks such grains as wheat, barley, rye, corn, and oats. It looks like a large sac or tumor among the kernels when it appears on an ear of corn. The sac contains a large mass of black spores.

Smuts act on the host plant in a different way from most parasitic fungi. The mycelium that grows among the cells of the host stimulates them to produce a swelling, or gall. The spores develop as a black mass within the gall and are thrown into the air when the gall breaks.

Smuts are difficult to control as, unlike the spores of any other fungi that attach themselves to the seed, these lie dormant in the ground through winter. In spring new kinds of spores are germinated and reinfect the corn plants. Treating the seed is seldom effective in preventing corn smut. The best answer seems to be crop rotation, along with the development and use of strains of corn that are smut-resistant.

GARLIC, FIELD (*Allium vineale*)

Meadow garlic's very penetrating taste and odor give a bad taste to milk if eaten by cows, and the bulblets in wheat will spoil the flour. It's hard to eradicate because the little bulbs grow deeper and deeper into the soil with the passing years. If pastures or fields are badly infested, crop rotation is recommended.

But even wild garlic has a definite health value, and medicines derived from it are useful against high blood pressure and sclerosis.

GINGER, WILD GINGER (*Asarum canadensis*)

Ginger grows in rich woods from Virginia to Minnesota on a low plant with root cords extending from a knotty but superficial root. The leathery leaves, mottled green, are heart-shaped, growing on tough, hairy stalks. A reddish brown, cup-shaped flower blooms during April and May at ground level, emerging between two-leaf stalks. Gingerroot is best dug in October when the roots are full and are better for candied ginger at this time.

GINSENG (*Panax quinquefolius*)

Wild ginseng needs the companionship of trees to provide the filtered sunlight it requires to grow. When raised in beds for commercial use, it is covered with latticework to protect it from the heat of the sun, which would kill it.

The Chinese believe that ginseng will cure nearly every disease, yet even now Western science is not sure whether it has real value or not. Physicians regard its benefits as largely psychological, but tests in Russia and elsewhere indicate that infusions of the root actually may increase energy and resistance to infection.

GOLDENROD (*Solidago*)

There are more than 60 native species of goldenrods, some growing on dry soil low in humus, others on soil that is rich and moist. If you would eliminate the goldenrod, cut it before it goes to seed and improve your soil with organic matter and beneficial crops.

HENBANE (*Hyoscyamus niger*)

Henbane, like hemp, is an ancient narcotic that was once used to treat disease but escaped to become naturalized in some parts of the country. The drug hyoscyamine, used to dilate the pupils of the eyes, is made from black henbane.

This poisonous herb is fatal to fowl, hence its name "henbane." All parts of the plant contain poisonous alkaloids, and even hogs sometimes are killed by eating its fleshy roots.

HORSETAIL (*Equisetum arvense*)

The horsetails are the last remainder of the huge trees of the carboniferous forests. The most common is the field horsetail, which grows in sandy and gravelly soils on a high ground-water level.

Horsetail looks like a tiny Christmas tree and sometimes is called the meadow pine. The hollow-branched, jointed stems range from 1 to 3 feet tall. The plant does not produce flowers or seeds but sends up fertile spore-bearing stems resembling catkins. These are covered with powdery brown spores. After the spores drop, small green shoots emerge from the ground. The perennial horsetail has a rootstock that bears tuberous growths that store available carbohydrate against a future need.

The green shoots of horsetail contain a high percentage of silica, the controlling factor against fungus disease. If the green stems are burned in a hot but quiet flame, removing all organic parts, the white skeleton of silica that is left will show the original structure of the little stems.

Silicic acid can be found in many plants such as stinging nettle and quack or couch grass, but some, including horsetail and knotweed (*Polygonum aviculare*), are particularly rich in it. It is this silica in horsetail that has made it valuable medicinally for centuries.

Silicic acid is traditionally believed to strengthen tissue, particularly that of the lungs, and at times to add to disease resistance. It has been reported to have a good influence on inflammation of the gums, the mouth, or the skin in general. Horsetail was specially recommended for diuretic purposes by Kneipp, the German priest who was so successful in using hydrotherapy in combination with herbs.

While horsetail itself will stop external bleeding, a horsetail brew also may be used as a healing agent for abscesses, burns, cuts, and scratches, for both animals and humans. Place a good handful of the dried leaves and stems in a stewpan with just enough vinegar (5 percent acidity) to cover. Simmer for no longer than 20 minutes, cool, and strain. Keep in the refrigerator. When needed for use, add 1 part brew to 2 parts cow or goat milk. Return any not needed to the refrigerator. Horsetail brew is stingless,

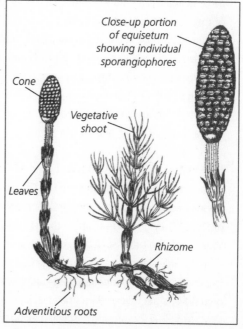

Close-up portion of equisetum showing individual sporangiophores

Cone

Vegetative shoot

Leaves

Rhizome

Adventitious roots

The ancient, primitive horsetail (called equisetum) makes a fine plant spray. The plants, containing much silica, also are used to scour pots and pans.

soothing, and gentle in its action. A plastic squeeze bottle may be used for convenient application.

A tea made of horsetail is very effective against mildew and other fungi found on fruit trees, grapevines, vegetables, and roses. It is gentle but swift in action and does not disturb soil life. Silica-rich plants are valuable on the compost pile, too.

The parts used are the dried leaves and stems of the sterile form. Boil 2 or 3 teaspoons of the crumbled herb to 1 cup of water for 20 minutes. Or soak the leaves in water for several hours, boil for 10 minutes, and steep for another 10.

Made into a spray, it is particularly useful against powdery fungus and curly leaf on peach trees. It also controls mildew on roses, vegetables, grapes, and stone fruits and has been found to have a cell-strengthening action on the plants sprayed with it.

According to Beatrice Trum Hunter's *Gardening Without Poisons,* silica in the form of an aerogel is one of our most effective insecticides. It may be used against the flour beetle, rice weevil, granary weevil, and the larva of the Mediterranean flour moth.

Horsetails have still another use: Their high silica content makes them effective as "scouring brushes" for pots and pans, and campers still find them usually available and convenient for this purpose. In German-speaking countries horsetail was called *Zinnkraut,* or "pewter plant," because the high silica content made it useful for cleaning and scrubbing copper, brass, pewter, and all fine metals.

INDIAN CUCUMBER (*Medeola virginiana*)

Indian cucumbers are found from the Great Lakes eastward and south to Florida in rich woodlands. They are good eaten raw, cubed in salad, or pickled. "They grow in sociable groups," says Grace Firth in *A Natural Year,* "and, like other sociables, are best when mildly pickled."

The single slender stems of Indian cucumbers grow out of a whorl of five or seven dark, elongate, visibly ribbed leaves. The spindly cream-colored flowers are followed by small black berries. The roots grow horizontally just below the ground surface, making the cucumbers easy to dig.

JEWELWEED
(*Impatiens capensis* or *Impatiens pallida*)

Jewelweed is an excellent remedy for poison ivy, relieving the itching almost instantly. It does not have an adverse effect on the poison ivy plants, for the two often grow side by side. In fact, wherever poison ivy grows, jewelweed is likely in the vicinity. It blooms between July and September.

Jewelweed

To make the remedy, boil a potful of jewelweed in water until the liquid is about half the original volume. The strained juice is effective both in preventing the rash after exposure and in treating the rash after it has developed. The best way to keep the extract is frozen and stored in cubes in the freezer. Pack them in small plastic bags and they will be at hand if needed.

The young, tender sprouts of jewelweed are edible and are cooked like green beans.

KNOTWEED (*Polygonum aviculare*)

These members of the Buckwheat family grow mostly on acid soils. The prostrate knotweed, frequently found in garden borders and along paths, doesn't mind being trampled on. All knotweeds are characterized by "knots" on their stems from which branches grow.

Knotweeds in pastures are thought to be troublesome to sheep. They also inhibit the growth of turnips and are very rich in silica.

LAMB'S-QUARTERS (*Chenopodium album*)

This plant, sometimes called smooth pigweed, is one of our most enduring annual weeds, producing a tremendous amount of seeds that are able to survive dormant in the soil for decades. It is among the weeds that follow human footsteps and cultivation, liking a soil with a well-fermented humus.

Lamb's-quarters is particularly stimulated when grown with potatoes, and it should be allowed to grow in the garden in moderate amounts, especially with corn. It also aids cucumber, muskmelon, pumpkin, and watermelon as well as giving additional vigor to zinnias, marigolds, peonies, and pansies.

This plant, a close relative of spinach, also is good to eat. The young shoots may be cooked and eaten like asparagus. It is richer in vitamin C than spinach, far richer in vitamin A, and, though not quite so rich in iron and potassium, is still a good source of these minerals. It is exceptionally rich in calcium.

Lamb's-quarters is a freebie that everyone should know about, for it is found in cultivated ground from north to south and east to west, and plants in the right stage for eating can usually be found from late spring until frost. It even grows in the Andes at a height of 12,000 feet and there has become an important substitute for rye and barley, which cannot survive at such an altitude.

LARKSPUR, WILD (*Delphinium tricorne*)

Wild larkspur is detrimental to cattle and too much may cause poisoning. Barley as a crop, however, is a weed deterrent and will prevent wild larkspur or poppy from establishing itself. It is thought to promote vigor in winter wheat.

LOCOWEED (*Astragalus* and *Oxytropis* spp.)

Locoweed gets its name from the Spanish word for crazy, due to the strange actions of animals poisoned by it.

Strangely enough, the poisonous effect of locoweed depends on the soil in which the plants grow, because of their ability to absorb poisonous elements from the soil. Both the green and the dry plants are poisonous, the symptoms varying somewhat in horses, cattle, and sheep.

Horses become dull, drag their legs, seldom eat, lose muscle control, become thin, and then die. Cattle react similarly, but sometimes they run about wildly, or stagger and bump into objects in their path. Sheep are less apt to be injured by the poison.

Ranchers destroy locoweed by cutting the roots about 2 inches below the surface.

LUPINE (*Lupinus perennis*)

Sometimes called old maid's bonnets, wild pea, or sundial, the plant has vivid blue flowers, sometimes pink or white, with a butterfly shape that indicates its membership in the Legume family.

Farmers once thought that lupines preyed on the fertility of their soil; hence the name derived from *lupus,* "a wolf." In fact, they help the growth of corn, as well as most other cultivated crops.

Lupines grow best on steep, gravelly banks or exposed sunny hills, liking almost worthless land where their roots can penetrate to surprising depths, in time leaving behind them fine, friable soil. They are adventurous pioneers, spreading far and wide in thrifty colonies, and are among the first plants to grow on the barren pumice after a volcanic eruption.

The lupine is one of those interesting flowers that go to sleep at night. Some fold their leaflets not only at night but also during the day when there is movement in the leaves. Sundial, a popular name for the wild lupine, refers to this peculiarity. Among the nearly 100 kinds of lupines that grow in North America, some contain poisonous alkaloids, while the seeds of others can be eaten.

MAYWEED (*Anthemis cotula*)

Sometimes this is called dog fennel or fetid chamomile because of its evil smell. Beekeepers used to rub it into their skin to repel bees. It also will repel fleas and may be rubbed into floors and walls of a granary to repel mice.

MEADOW PINK (*Lychnis flos-cuculi*)

The roots of all members of the Lychnis family contain saponin, which produces a soapy foam if stirred in water. Before the invention of soap these roots, together with those of the true saponaria, were used for washing. An interesting family member is the sleepy catchfly, so called because its flowers are closed most of the day, opening only in bright sunshine, while the gluey substance on its stems entangles flies.

MILDEW, POWDERY (*Erysiphaceae*)

Mildew is a fungus of the type called an "obligate parasite" because it feeds on living plants. When moisture conditions are just right, wind-carried spore (little seeds) resting on a plant's leaves send out germ tubes, which grow into white threads (mycelia). These branch over the leaves in a white, soft, felty coating.

This type of fungus does not grow *inside* the plant but sends its little suckers (haustoria) into the plant's sap. As chains are built up from the mycelium spore, the plant becomes covered in a few days. Eventually black fruiting bodies with the sexual or "overwintering" spores are formed.

Because it is on the surface, mildew is more easily controlled than many other fungi, and horsetail tea is an excellent spray to use (see *Horsetail* in this chapter). During the season when green plants are available, it is also good to prepare an extract by covering freshly picked plants with water. Allow them to ferment for about 10 days, then dilute the liquid and use it as a spray in the same way as the tea. Mustard seed flour or sulfur dust also may be used, while polybutenes, oil derivatives, have been used successfully to control powdery mildew on cucurbits.

MILKWEED (*Asclepias*)

All of the many milkweeds exude milky juice when their leaves or stems are punctured. Roots are considered poisonous, but Native Americans have used them for various maladies, and some say that the juice cures warts or ringworm. Cows dislike the bitter, acrid plants but may eat them if hard-pressed for food.

MISTLETOE (*Phoradendron serotinum*)

This parasite is the most legendary of plants. It was sacred to the ancient Druids, who cut it with a golden sickle (the symbol of the sun) and caught it in a cloth to prevent it from touching the ground.

Mistletoe grows on apple trees, oaks, and poplars, usually being sown by birds. Here in Oklahoma it is our state "flower" and grows profusely on hackberry trees. People used to call the mistletoe "all heal" and thought there was no illness it could not cure. It is in fact poisonous, particularly the berries. If you find it growing on your trees, remove it, for it is a parasite and eventually will weaken the tree and possibly kill it.

NETTLES
See *Stinging Nettle.*

NIGHTSHADE (*Solanaceae*)
The Nightshade family includes apple-of-Sodom, belladonna, bittersweet, capiscum, eggplant, jimsonweed (datura), petunia, potato, snakeberry, tobacco, and tomato.

Where black nightshade (*Solanum nigrum*) grows profusely, the soil is tired of growing root crops. This plant draws the Colorado potato beetle away from potatoes, since they prefer the weed though it is poisonous. The beetles eat it and die.

NUT GRASS (*Cyperus esculentus*)
The botanical name, *Cyperus esculentus,* means "edible sedge." Nut grass is related to the tules and bulrushes and is almost as old as civilization. The ancient Egyptians developed cultivated strains more than 5,000 years ago. The "nut," which is really a tuber, may be made into many tasty and unusual dishes. But consider well before growing this plant, for it can become a fearful weed. Having battled against it in my garden for many years, I have come to the conclusion that this is a plant only Euell Gibbons could love.

If the native species of nut grass plagues you, it can be discouraged by growing a heavy cover crop of cowpeas on the plot for several summers. Sow the cowpeas thickly to form a dense mat that shades the ground. Plow them under in late fall, in October or November, and they will add nitrogen to the soil. St. Augustine grass will choke out nut grass on a lawn.

OPUNTIA (*Opuntia humifusa*)
In the Southwest and Mexico, people eat the flat-leaved joints on opuntia cactus, boiled or fried, and make the flowers into salads. The juicy fruits of opuntia are eaten raw or cooked; the seeds are ground into a

Opuntia

meal and made into cakes. The bushy cacti grow in hedges around houses; where little else will survive, they serve as windbreaks and groundcover.

In some regions in times of drought ranchers "burn off" the cactus spines so that cattle may eat the plants. Plant breeder Luther Burbank developed a spineless cactus that proved to be a useful source of food for both people and animals, and in some sections of California the opuntia type is a commercial crop. It is grown on sandy loam, fertilized with chicken manure, and needs no insecticides. The crop of young leaves, which measure about 8 inches long at 1 to 2 months, are hand-picked to be sold diced, shredded, and spiced or pickled.

Almost all useful species of cactus will grow well with each other, requiring similar soil and cultural practices. Many have beautiful delicate blossoms and are grown as ornamentals, and there is a subzero cactus of the opuntia type that will grow in the North. (See Sources on page 442.)

OX-EYE DAISY (*Chrysanthemum leucanthemum*)
Ox-eye daisy seeds are beneficial in small quantity mixed (1 to 100) with wheat grains, but in larger quantity the daisy will overwhelm the wheat.

PANSY, WILD (*Viola tricolor*)
The wild form of the cultivated pansy is Shakespeare's "heartsease" and once was listed in the *U.S. Pharmacopoeia* as a medicine. Many species of viola were candied as a sweet and were thought to be soothing and therapeutic to the heart.

Rye helps the wild pansy germinate and is itself seemingly improved by a few pansies. But the pansy has an inhibiting effect on wheat.

PAWPAW TREES (*Asimina triloba*)

These quick-growing, rangy trees rarely exceed 20 feet but bear large, out-size leaves. The fruit usually falls to the ground before it is good to eat but will ripen if held at room temperature. Mashed pawpaws may be made into a bread similar to banana bread.

PENNYCRESS (*Thlaspi arvense*)

Like shepherd's purse, this is often abundant where grain is grown. The seeds are 20 percent oil and if accidentally ground with grain will spoil the flour. Mountain pennycress (*Thlaspi montanum* var. *montanum*) likes soils containing zinc.

PERUVIAN GROUND CHERRY (*Physalis peruviana*)

When planted in quantity near a barn or in a stableyard, it will repel flies, and it also is effective against whitefly.

PIGWEED

See *Amaranth* in this chapter.

PLANTAIN (*Plantago*)

Of the many different plantain species, the large-bracted plantain (*P. aristata*) is the most prolific, one plant producing well over 3,000 seeds.

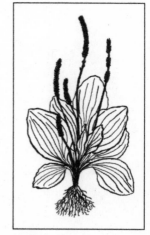

The narrow-leaved (*P. lanceolata*) has been used as a home remedy for treating bruises and strained joints. It also has a cooling and astringent effect if a few leaves are squeezed over a bee sting.

Plantain and red clover frequently are found growing together, this because the plantains occur as impurities in grass and clover seeds. If plantains appear on the lawn, it is best to dig them out.

Plantain

POISON IVY (*Toxicodendron radicans*)
Jewelweed will relieve the itching of poison ivy (see *Jewelweed*). If your land has poison ivy, the best thing you can do is eradicate it. Mow close to the ground in midsummer and follow this with plowing and harrowing, grubbing out small patches. Under trees or along a fence where mowing might be difficult, try smothering it with heavy cardboard or tar paper. A deep mulch of hay or straw may work as well.

Poison ivy

Vines growing in trees may be cut near the ground and then pulled down a few days later. Be sure to wear gloves and protective clothing, washing well afterward, preferably with yellow soap.

POKEWEED (*Phytolacca americana*)
I find that pokeweed grows well under my figs, Scotch pines, black walnuts, and other trees. Pokeberries and roots are poisonous, but the tiny, pinkish green, asparagus-like shoots are simply delicious. Poke is one of the first greens to come up in early spring, and these shoots should be cooked lightly in several changes of water. The berries and roots contain phytolaccin, a cathartic and slightly narcotic substance used for treating rheumatism. Pokeweed is an excellent food plant for backyard birds.

Poke should never be confused with the completely unrelated Indian poke, or white hellebore. This latter plant, also poisonous, grows in wet places and comes up very early in the spring before the edible poke starts growing.

POTHERBS
There is a wealth of wild greens available in April. Most of these belong to the Mustard family, Cruciferae, so called because their

Pokeweed

flower petals form a cross. The mustard cousins, Brassica, together with strong-growing young shoots of peppergrass (*Lepidium*), horseradish (*Armoracia rusticana*), pennycress (*Thlaspi arvense*), searocket (*Cakile edentula*), scurvy grass (*Cochlearia*), and other cress greens are a gift of spring to those who love to search for wild greens. Cook them as you would turnip greens or spinach. If their turnipy-horseradishy flavor is a bit much for you, combine them with milder greens such as young amaranth (*Amaranthus*), Japanese knotweed (*Polygonum cuspidatum*), or dayflowers (*Commelina communis*). Shepherd's purse (*Capsella bursa-pastoris*) also combines well. Gather potherbs when the shoots are very young. You may wish to cook them through several waters.

PUFFBALL (*Fungi*)
Sometimes these "smoke balls" or "devil's snuffboxes" grow to be more than 2 feet across. A cut that is bleeding profusely may be covered with the powderlike spores from those that produce a puff or "smoke" when disturbed, and it will stop the bleeding. Puffball powder is very explosive, so if you store it, keep the container closed and away from fire.

PURSLANE (*Portulaca oleracea*)
Purslane has a liking for good cultivated soil and is frequently found in gardens. But it is not altogether unwelcome, for though often considered a weed, it is cultivated in both England and Holland. It is a refreshing green with a slightly acid taste, and it may be cooked like spinach. About 100 grams of purslane contains 3.5 milligrams of iron, and this is all the more remarkable because the plant is 92.5 percent water.

PYRETHRUM (*Chrysanthemum cinerariifolium*)
Pyrethrum is absolutely bug-proof and will keep pests from plants close by. Few ticks are ever found where pyrethrum or sage forms a groundcover. Pyrethrum powder, used in days gone by as an insecticide, is made from the dried flowers. It has a very short residual action, breaking down rapidly in sunlight. Because of this, it was used as a preharvest spray.

Records show that pyrethrum may have been used nearly 2,000 years ago in China. As an insect repellent, it became popular again in the 19th century, when it was the "secret ingredient" in Persian insect powder. In 1828 this powder was produced on a commercial scale and introduced into Europe by an Armenian trader. By 1860 it was becoming well known in the United States.

The active principles in pyrethrum are the esters pyrethrin and cinerin. Certain nontoxic plant products such as asarinin (from the bark of southern prickly ash), sesamin (from sesame oil), and peperine (from black pepper) were added to pyrethrum to strengthen its effect.

RATTLEBOX (*Crotalaria*)

This weed is valuable for its soil-improving qualities, but one variety, C. sagittalis, found on bottomland in the Missouri and Mississippi basin, is very poisonous to cattle and horses and should be eradicated. Following its cutting or plowing under, plant a crop like cotton or corn that needs repeated cultivation.

ST.-JOHN'S-WORT (*Hypericum perforatum*)

This common pasture plant contains a red oil sometimes used as a home remedy for bronchitis and chest colds. It also is astringent and has been used against diarrhea and dysentery. The leaves have oily cells and a strong, peculiar smell. They look perforated if held against the light. It was once believed that if the plant was collected during St. John's Night (June 24), it would afford protection against witches and evil spirits.

SHEPHERD'S PURSE (*Capsella bursa-pastoris*)

Shepherd's purse is very rich in minerals. Along with mustard, it absorbs excessive salts in the soil and returns them in organic form. If grown on a salty marsh and plowed under while still green, it will both sweeten the soil and discourage the weeds ordinarily growing on such soil. It has medicinal qualities and has been used as a styptic.

SOW THISTLE *(Sonchus arvensis)*

This plant has creeping, deep-growing roots, contains a milky, yellow-tinged juice, and grows on moist soil. It aids watermelon, muskmelon, pumpkin, and cucumber and in moderate amounts onions, tomatoes, and corn.

A cousin called blessed thistle (*Cnicus benedictus*) has medicinal and industrial uses and is a basic ingredient of the Benedictine liqueur as well as certain bitter tonics.

SPURGE

See *Euphorbia* in this chapter.

STINGING NETTLE *(Urtica dioica)*

Stinging nettle has many helpful qualities. It makes neighboring plants more insect-resistant. It also helps plants withstand lice, slugs, and snails during wet weather; strengthens growth of mint and tomatoes; and gives greater aromatic quality to herbs such as valerian, angelica, marjoram, sage, and peppermint. The nettle protects fruit from mold and thus enables it to keep better. Fruit packed in nettle hay ripens more quickly.

Stinging nettle helps stimulate fermentation in compost or manure piles, according to British author M. E. Bruce, who advises making a crushed nettle solution. Good results can be had with less trouble by using the nettle in its original form, placing it in layers in the compost before the nettle seeds ripen. The plant is said to contain carbonic acid and ammonia, and these may be the factors that activate the compost. If you have the space, you might try raising a crop of nettle—somewhere away from the garden, for the plant spreads quickly.

Euell Gibbons reported that stinging nettles combined with young horseradish leaves are delicious as spring greens. They also combine well with lettuce or spinach. Since nettles are rich in

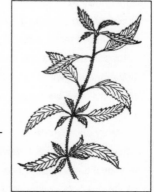

Stinging nettle

vitamins and iron, they are a good remedy for anemia, while aiding blood circulation and acting as a stimulant.

The plant leaves, as rich in protein as cottonseed meal, are good for animals, too, though they will touch them only when the nettles are mowed and dried. Horses improve in health and cows will give more and richer milk. When powdered nettle leaves are added to their mash, hens will lay more eggs and the eggs will have a higher food value, chicks will grow faster, and turkeys will fatten. Even the manure from nettle-fed animals is better than that from others.

Be sure to wear gloves when picking your nettles, for the fine hairs on the leaves and stems contain formic acid, which irritates the skin when touched. Nettle rash can be relieved with the juice of the nettle plant itself, or by rubbing the skin with jewelweed, with rhubarb, or with any member of the Sorrel family.

TANSY (*Tanacetum vulgare*)

Tansy, once used as a medicinal tea, is now considered dangerous if taken as an infusion (see the chapter on Poisonous Plants). It is also thought to be poisonous to cattle. But planted under fruit trees, particularly peach, it repels borers and is a good companion to roses, raspberries, blackberries, grapes, and other cane fruits. It deters flying insects, Japanese beetles, striped cucumber beetles, and squash bugs and helps repel flies and ants. The dried leaves are useful for storing woolens and furs. Because of its concentration of potassium, tansy is useful on the compost pile.

Tansy also goes by the names of bitter buttons, ginger plant, hind heal, and scented fern. Its fernlike, attractive foliage is topped by composite heads of buttonlike flowers, and their scent is delightful. Though the roots have been used, it is the tops that are of primary importance.

Tansy once was used as a culinary herb in place of pepper, was widely used in churches in medieval times as a "strewing herb," and is one of the plants associated with the Virgin Mary.

Tansy

THISTLE (*Cirsium*)

All thistles are rich in potassium and thus are useful in compost; but their prickly leaves make them unpopular in pasture, and certain types rob grainfields of food and moisture. To kill out thistles, be careful not to cut them before the blossoms are open or many more will grow from the rootstocks. If you cut just the blossom heads after the blossoms are pollinated, the plant will bleed to death.

Blessed thistle (*Cnicus benedictus*) has medicinal and industrial uses, as well as being an insect repellent, while constituting a basic ingredient of the Benedictine liqueur and stomach bitters.

THORN APPLE (*Datura stramonium*)

See *Datura* in this chapter.

TILLANDSIA (*Tillandsia usneoides*)

Tillandsia (also called Spanish moss, though not a true moss) occurs naturally from Virginia to Florida and Texas and southward to Argentina and Chile. Though it may be seen on other trees, its favorite host is the live oak. It is not a parasite, sapping the life of the tree, but a lodger that finds its own food supply through slender, grayish stems that look like hair hanging from the trees. It is related to the air plants of the Bromelia or Pineapple family, which largely draw their sustenance from the air itself. It may be used for mulching or for compost. The dried stems also are used industrially to stuff uphostery.

VENUS FLYTRAP (*Dionaea muscipula*)

This hungry little plant, which captures its own meals, must be grown in high humidity indoors or out. The insect traps on young plants, which develop in 3 to 4 weeks, consist of two leaves hinged in the center when open, and when closed forming a pouch in which the trapped insects are digested. When a leaf has caught several insects it withers and dies, but new ones take its place. Venus flytrap grows naturally in bogs where the soil lacks available nitrogen, and the insects supply this nutrient in the plant's diet.

WATER HYACINTH (*Eichhornia crassipes*)

This tropical American plant has "escaped" and is now growing profusely in the southern states, sometimes choking ponds and streams with its growth of floating leaves. The roots hang down in the water and receive the spawn of fish. The lovely violet flowers are large and showy.

If you live in an area where water hyacinth has gone wild, you will benefit both yourself and the waterway by dredging it out and using it for compost.

WATER LILY (*Nymphaea odorata*)

The American water lily, related to the lotus, sends long, stout leaf and flowerstalks from the mud bottom of clear, shallow water. The beautiful flowers usually rise above the water on long stalks and may be as large as 1 foot across. In cultivation, water lilies grow well in a mixed planting of other water plants. In the wild state they, like water hyacinth, sometimes threaten to choke shallow pools of water and should be dredged out and composted.

WEEDS

Someone once said, "A weed is a plant out of place," but I am inclined to go along with Ralph Waldo Emerson, who believed that "a weed is a plant whose virtues have not yet been discovered." Weeds, wisely used, are some of our most important companion plants. Of course, they never should be allowed to overwhelm the food plants, but a few left here and there may surprise you by the influence they exert.

The extensive root growth of weeds penetrates the subsoil, breaking it up and making it easier for the roots of crop plants to go farther than usual as they search for water and nourishment. A few weeds are useful in shading the ground to keep seedling vegetables from drying out in the sun's heat. Meanwhile, moisture from the subsoil will travel by capillary action up the outside of the weed roots to a level where the young vegetables can use it.

Deep divers such as pigweed, lamb's-quarters, and thistles bring up minerals from the lower soil by way of their stalks and leaves. When these weeds are turned under, the minerals become available to shallower-rooting crops. Minerals, including trace minerals that may be leached away or become exhausted under a succession of crops, are thus retained.

Another interesting fact is that weeds seem to accumulate the nutrients in which a particular soil is deficient. Such weeds as sheep sorrel and plantain, which thrive best in acid soil, are rich in alkalinizing minerals such as calcium and magnesium. Bracken, which grows best in phosphorus-poor soil, is high in phosphorus. Turning these weeds under will release these minerals into the topsoil, again making them available to food plants.

Weeds also benefit the soil by conditioning it. Their extensive root systems leave fibrous organic matter, which decays, adding humus to both topsoil and subsoil. Not only this, but they also leave channels for drainage and aeration. When decomposed, the root systems of dandelions provide subterranean channels for earthworms, which, in turn, enrich the soil with their castings. Soil texture is vastly improved and soil-inhabiting bacteria will multiply enormously.

Learning to read weeds can be very useful, for they are excellent indicators of the type of soil they select to grow on.

Weeds that delight in acid soil, and also indicate increasing acidity, are the docks, fingerleaf weeds, lady's thumb, and sorrels. Horsetail indicates slightly acid soil, as do hawkweed and knapweed.

Weeds that indicate a crust formation and hardpan are pennycress, morning glory, horse nettle, field mustard, chamomiles, quack grass, and pineapple weed.

Weeds most likely to occur on cultivated land are chickweed, buttercup, dandelion, lamb's-quarters, plantain, nettle, prostrate knotweed, prickly lettuce, field speedwell, common horehound, celandine, mallows, rough pigweed, and carpetweed.

Sandy soils are favored by arrow-leaved wild lettuce, yellow toadflax, onions, partridge pea, broom bush, flowered aster, and most goldenrods.

On alkaline soils we are apt to find sagebrush and woody aster, while limestone soils grow field peppergrass, hare's ear mustard, wormseed, Canada bluegrass, cornelian cherry, pennycress, Barnaby's thistle, mountain bluet, yellow chamomile, and field madder.

If a plot of land grows healthy weeds, it will probably grow good vegetable crops, too. Let the weeds reach full growth but cut them before they go to seed. Let them wilt a few days and then plow them under for green manure.

You may even find it helpful to your compost pile to bring in extra weeds such as those cut by the highway department along public roadways. This largesse often includes such items as nettles, sunflowers, yarrows, and sweet clover. These should be thoroughly composted to kill their seeds before being placed on the garden.

Weeds are not necessarily our enemies. With good management, they may well become friends and co-workers.

WILD CARROT (*Daucus carota*)

Wild carrot does not always indicate bad soil, for its deep taproot implies a deep soil capable of good cultivation. A rich stand indicates a soil worth improving for crops. But it can become a pest, so prevent it from seeding by cutting the plant close to the ground shortly after pollination. Do not cut too early or many plants will spread out from the root.

WILD MORNING GLORY (*Convolvulus* and *Ipomoea* spp.)

We have it from Native Americans that wild morning glory is beneficial to corn, but if allowed to go to seed it can become a great pest, coming up for years afterward. It may be killed out by spraying a little white vinegar into the center of each vine.

WILD PARSNIP (*Pastinaca sativa*)

Wild parsnip is a nourishing food plant that will give a good yield even on poor soils, but it soon becomes a weed and is hard to eradicate. The cow parsnip (*Heracleum lanatum*) is poisonous.

Wild morning glory

WILD RADISH (*Raphanus raphanistrum*)

Wild radish spreads quickly in soils worn out from growing too many grain crops and depleted in nitrogen. It flourishes well, especially in wet years, where manure is scarce and potassium fertilizer abundant. Nevertheless, cattle are very fond of it, and it produces a good honey as well as an oil from the seed.

WILD ROSE (*Rosa eglanteria*)

When this pretty weed migrates from hedgerows to pastures, it indicates the pasture has not been grazed adequately and needs mowing and harrowing. The prickly canes are troublesome to sheep and cattle but do not particularly bother goats, which love all kinds of rosebushes. To eradicate, cut the canes while they are still soft.

WILD STRAWBERRY (*Fragaria virginiana* and *F. vesca*)

Wild strawberries are small but have a delicious flavor quite unlike any other. Their presence in pastureland is an indicator of increasing acidity.

YARROW (*Achillea millefolium*)

Yarrow is a plant of both mystery and history. For centuries the Chinese mystic has cast yarrow stalks when consulting the *I Ching*.

According to the Bio-Dynamic book *Companion Plants,* yarrow has a definite effect on the quality of neighboring plants, increasing not so much their size as their resistance to adverse conditions, and thereby improving their health. It is a good companion for medicinal herbs, enchancing their essential oils and increasing their vitality. It is also said to help cuts to heal.

Yarrow also gives nearby plants resistance to insects, perhaps because of its acrid, bitterly pungent odor.

Yarrow tea or yarrow hay is helpful to sheep, and I have given it to milk goats after kidding. It will grow almost anywhere and under any conditions and does not mind being walked or trampled on. Where it grows in lawns and is cut by the mower, it simply spreads out in a low growth pattern.

Yarrow

GRASSES, GRAINS, AND FIELD CROPS

Sugarcane and bamboo are both giant grasses, while the great cereals of the world, grasses too, are wheat, corn, oats, rye, barley, and rice. Wild grasses include bluegrass, esparto, reed, sandbur, Sudan grass, and wild barley, as well as many others.

Many grains and grasses grow well planted together, the growth of both being enhanced and increased.

COTTON (*Gossypium*)

Alfalfa planted before cotton will put nitrogen in the soil, to the cotton's benefit, and alfalfa planted with it will discourage root rot.

Cotton growers try to keep the pink bollworm under control by isolating infected fields, sterilizing seeds and cotton, and using machines that chop up leaves and other trash from the cotton. Shredding stalks in late summer and plowing them under helps control the worm and the boll weevil.

Farmers now protect their crops against the diseases of older cotton plants by growing varieties that are bred to resist such diseases as wilt and blight.

CRABGRASS (*Digitaria sanguinalis*)

Crabgrass is one of the most troublesome lawn pests. Hand-pulling is recommended on lawns (before the plants form a mat), and mulch and frequent cultivation in gardens. As with Bermuda grass, dry, hot weather will wilt the roots if they are brought to the surface. (See also *Grass, Quack*, in this chapter)

ESPARSETTE (*Onobrychis viciifolia*)

This perennial forage legume of Eurasian origin also is called medick, sainfoin, lucifer, snail clover, and great trefoil. The plant was introduced into England by the Romans, but it is not much grown today, possibly because it takes several years to reach fruition. It produces three crops a year, however, and grows again from the same roots once it reaches maturity. Oddly enough, in England the spikelike flowers are violet but in China they are yellow.

Esparsette is a good food for cattle and is equally nourishing for humans. It is valuable for those suffering from weight loss yet reduces weight in those who are too heavy. It is considered a tonic for both the brain and spinal cord. The roots of a variety called black medick make a good tooth powder.

Esparsette is recommended as a border plant for small grains or vegetables, and in a thin stand it aids growth of corn. It also may be grown as a lightly scattered stand with small grains. The seed retains viability for up to 3 years, and grows well in limestone soils.

FLAX (*Linum usitatissimum*)

Flax is a good companion to both carrots and potatoes, improving both their growth and flavor. Flax planted near potatoes will protect against the Colorado potato beetle. (It is, however, poisonous; see the chapter on Poisonous Plants.)

Leaf extracts of the false flax (*Camelina sativa* and *C. microcarpa*), frequently found growing in flax fields, have an inhibiting effect on flax itself.

GRASS, KOREAN (*Zoysia japonica*)

This group of eastern Asiatic, perennial creeping grasses is widely used in the Southwest. Some are important horticulturally as lawn grasses, others as ornamentals. In hot, dry climates Korean grass is strikingly effective planted with such succulents as the porcelain-like aeonium, the texture contrast enhancing both plants. It grows well with *Cotyledon undulata* and the delicately beautiful rosettes of *Graptopetalum paraguayense*.

GRASS, MOLASSES (*Melinis minutiflora*)

See *Molasses Grass* in the Pest Control chapter.

GRASS, PAMPAS (*Cortaderia*)

This ornamental grass, best grown in the South, produces beautiful flower plumes that, if cut when fully developed, are useful for decorative purposes indoors during the winter. It is increased by root division, and it grows well as a specimen plant in the lawn.

GRASS, QUACK (*Agropyron repens*)

Quack grass indicates a crust formation and/or a hardpan in the soil. Choke it out by sowing millet, soybeans, or cowpeas, making sure that the land first is thoroughly cultivated and the weather hot and dry. Two successive crops of rye also will choke it out.

A concentrated brine of common salt (sodium chloride) will kill it out, too, if used after grass is freshly cut and applied in dry weather several times. Dry weather will wilt the roots of quack grass if they are brought to the surface. Hand-pulling is recommended if there are but a few plants.

Like so many other things, quack grass isn't all bad. It is a good cattle feed, and because of its persistence, it makes a useful covering for gullies and road banks where live soil has been cut open and few other plants will grow. Though hard to get rid of once it is started, it does prepare the soil for better things. Oddly, it is wheat's nearest relative.

GRASSES, LAWN

Bermuda grass (*Cynodon dactylon*) is an excellent lawn grass for the southern states. Bermuda withstands both heat and drought and will grow reasonably well even on poor soils. It may be started by seeding or sodding.

Never allow Bermuda to get started in the garden or flowerbeds, for it spreads quickly on cultivated soil, competing with flowers or vegetables for moisture. It may be killed out in the summer by hoeing and exposing the rhizomes to hot sunlight.

Kentucky bluegrass (*Poa pratensis*) is an excellent grass for the North and East. It needs a quick-germinating and quick-growing grass, such as redtop, planted with it to provide a rapid groundcover that will help crowd out weeds during its early development. After it gets a good start, the bluegrass will crowd out the nurse grass.

St. Augustine (*Stenotaphrum secundatum*) also thrives in the South, being particularly good under trees or in other shady areas where Bermuda will not do well. It forms a thick mat and smothers weeds.

Zoysia (*Zoysia matrella*) in cultivated species forms dense turf and is very valuable for planting on sandy soils, especially in the South. It is propagated vegetatively by means of small pieces of turf called plugs. Zoysia will choke out crabgrass and weeds.

GRASSES, PASTURE

With the development of "beefalo" hybrid cattle by D. C. Basolo of Tracy, California, rich pasture grasses suitable for grazing livestock take on even greater importance. The Beefalo, which is three-eighths buffalo, three-eighths Charolais, and one-quarter Hereford, can be produced more economically than other breeds because it gains weight faster and can finish out on grass rather than grain.

Another old-time breed is the Texas Longhorn, which, though not as tasty as the Beefalo, also will finish out on grass.

Bermuda (*Cynodon dactylon*), a very persistent and nutritious grass of the southern United States, is useful for both pasture and lawn.

Buffalo grass (*Buchloe dactyloides*), which grows on the western range where bison used to graze, still serves as food for herds of cattle. It is also useful for binding the soil, preventing erosion.

Grama grass (*Bouteloua* spp.), occurring mostly in the Great Plains area, is excellent forage for livestock. It is widely used also in conservation to prevent soil erosion.

Johnson grass (*Sorghum halepense*) grows wild all over the Southwest and is often a great pest in gardens, yet in pastures it is very nutritious for cattle.

Ryegrass (*Lolium* spp.) is often grown in nut orchards and serves as food for cattle that are allowed to graze and fertilize the land, from which they are removed at harvest time. Orchard grass growing under fruit trees can suppress the root growth of pears and apples.

Teosinte (*Zea mexicana*), an annual grass often used as fodder for livestock, is considered the nearest relative to maize or Indian corn of all the wild grasses. It grows wild in moist soil from Connecticut west to Kansas and south to Florida and Texas.

NURSE GRASS

A quick-germinating and quick-growing grass such as redtop frequently is used in lawn seed mixtures to provide a rapid groundcover that helps crowd out weeds during the early development of the more permanent grasses, such as Kentucky bluegrass. The bluegrass may take two or three years to reach full development, but once it attains this under favorable conditions, it will crowd out the nurse grass.

OATS (*Avena sativa*)

A cover crop of mixed clover and oats following sod and before corn is planted will lessen the white grubs that infest corn. Oats and vetch do well planted together.

Oats sometimes can be grown effectively as a trap crop to lure red-winged blackbirds away from other grains. The stand should be grown at some distance from the birds' roosting places.

PASTURE WEEDS

Do not let sneezeweed (*Helenium autumnale*), sometimes called dogtooth daisy, grow in pastures. Most cows respect the bitter leaves, but many a pail of milk has been spoiled by a mouthful of helenium among the herbage. If you are wondering why this plant is called sneezeweed, take a whiff of snuff made from the dried and powdered leaves.

Wild larkspur (*Delphinium tricorne*) is poisonous to cattle.

Other plants that should never be allowed to grow in pastures are field garlic (*Allium vineale*) and meadow garlic (*A. canadense*). Just a few minutes after a cow has eaten some field garlic her entire body is penetrated, and after half an hour the milk is flavored with it and remains so for several hours. To avoid damage to the milk, it may be necessary to keep the cows off the pasture or let them graze for only a short time after milking, then remove them to another pasture.

POPPY (*Papaver* spp.)

Poppy and wild larkspur (*Delphinium tricorne*) like
to grow with winter wheat but dislike barley.
Wheat fields heavily infested with poppy yield a
poor harvest of lightweight seeds.

Poppies are grown for both seed and oil but
they rob the soil of nutrients, causing it to need
rest and reinforcement afterward. This factor may
be used to advantage, however, to choke out
weeds that cannot be gotten rid of by any other
means.

Poppy seeds may lie dormant in the ground for
years and then show up again with a grain crop,
particularly winter wheat.

*Poppies can become too
much of a good thing,
especially in plantings of
barley, which they inhibit.*

RAPE (*Brassica napus*)

Rape is an annual plant cultivated for its leaves and used as temporary
pasture crop for livestock. Because of its deep taproot, it loosens soil and
improves drainage, leaving the land friable and ready to grow a more
useful crop. It helps heal soils injured by overdoses of mineral fertilizer.

The succulent rape grows fast, producing best under cool, moist conditions. It also resists rather severe frosts and is best seeded in fall in the
southern states and in spring in the northern ones.

Do not grow rape near hedge or field mustard since both will inhibit
its growth.

RICE, WILD (*Zizania aquatica*)

This aquatic grass is not really rice at all, nor is it related to rice. It grows
from 4 to 8 feet tall in the shallow lakes of Minnesota, Wisconsin, and
central Canada. There it has traditionally been harvested by Indians, who
bend the heads of the plant over the edge of boats or canoes, beating the
grains loose with two sticks.

Wild rice may be cultivated, growing best in quiet, pure water from
1 to 6 feet deep, along the margins of streams, ponds, and lakes or the

floodplains of rivers with rich mud bottoms. It likes a slow current and will not grow in stagnant lakes or pools. A fairly shallow farm pond fed by streams can provide a good supply of this vitamin B–rich delicacy.

Do not try to plant the product you find in the grocery store, for only unhulled seed will sprout. Rice for planting must be sacked and kept wet. The seed may be planted by scattering over the surface of the water at the rate of 1 bushel per acre. The good seed will sink rapidly. If your area is small, use a large handful to a 6- by 6-foot space. The best planting time is just before ice forms in late fall.

RYE (*Secale cereale*)

Rye is an excellent crop to choke out chickweed and other low-growing weeds that survive the winter. Planted twice in succession it even will choke out quack grass. A cover crop following sod will reduce black spot on strawberries and pink root on onions.

Rye will be benefited by cornflowers in the ratio of 100 to 1. A few pansies in the field will aid it, and the wild pansy (*Viola*) will germinate almost 100 percent if grown nearby. Rye has an inhibiting effect on field poppy, retarding both the germination of the seed and its growth.

Rye flour sprinkled over cabbage plants while they are wet with morning dew will dehydrate cabbageworms and moths. Refined diatomaceous earth is useful as an insecticide for stored rye, but it is not injurious to warm-blooded animals.

SORGHUM
(*Sorghum bicolor, S. vulgare,* or *S. halepense*)

Several insect-resistant strains of sorghum have been developed: 'Atlas' is resistant to the chinch bug, while 'Milo' is susceptible; 'Sudan' is resistant to the corn leaf aphid, while 'White Martin' is susceptible.

The sweet sorghums or sorgos are grown especially for the production of sorghum syrup, which is made by pressing the juice out of the stems. For the gardener who would like to be self-sufficient, here is a source of sweetening for his other foods. To get the maximum amount of sugar in the juice, sorghum should be seeded on soils that are not too fertile. Large vigorous stalks usually are lower in sugar than those grown more slowly and not over ½ inch in diameter.

Root exudates of sorghum apparently are poisonous to sesame and wheat. Stored sorghum grain can be kept free of insects by refined diatomaceous earth used as a desiccant dust.

Sudan grass (*Sorghum vulgare sudanense*) is a tall annual sorghum whose thin stalks grow quickly and may reach a height of 10 feet. It serves as excellent summer pasturage and grows well with soybeans if sufficient moisture is present.

Johnson grass (*S. halepense*), a perennial sorghum, grows as a weed in the southern United States. It resembles Sudan grass but spreads by creeping rootstocks in gardens or on land needed for cotton or other row crops to become a pest, but it makes excellent hay for cattle feed.

SOYA BEAN-SOYBEAN (*Glycine max*)

Soya beans, native to China, are so rich in protein they have been called the "meat without a bone." They are perhaps the world's oldest food crop, and they have meant meat, milk, cheese, bread, and oil to the Asiatic peoples for centuries. Like all legumes, they loosen and en-rich poor soil and are an excellent crop to grow preceding others that need nitrogen. They grow faster and thicker than weeds and will choke them out.

Soybeans planted near corn protect it against chinch bugs and Japanese beetles. They grow well with black-eyed peas and will choke out weeds be-cause they grow so rapidly.

SUGAR BEET (*Beta vulgaris*)

Grain can be partially replaced as stock feed by sugar beets, which are liked by all animals and are good for increasing the milk flow of cows.

Cheat grass is often a despised weed but has the ability to quickly form a groundcover over denuded soil, preventing erosion. At the same time it replaces

Soya bean

plants that are host to beet leafhopper, making it of considerable importance to sugar beet growers.

TIMOTHY (*Phleum pratense*)

Timothy, a valuable, cool-season grass perennial, sometimes called herd's grass and by the English cat's-tail, has slender stems bearing round spikes of tiny, tightly packed flowers. Farmers in both Canada and the United States often sow timothy in rotation with oats and other grains. It does not last long when cattle or other animals graze on it continually and is not considered a satisfactory pasture grass unless mixed with hardier types.

Timothy and other small grains are benefited by planting them with legumes such as alfalfa and sweet clover as a protection against white grubs.

TRITICALE (*Triticale*)

The International Wheat and Maize Center of Mexico produced a new grain, triticale, by crossing wheat and rye, gaining the high yield of wheat and the disease and drought resistance of rye. This was accomplished by the tedious process of cross-fertilization among different species. By nurturing the resulting embryo and chemically causing its chromosomes to duplicate themselves, the scientists succeeded in producing fertile plants bearing the characteristics of both parents.

The name triticale (pronounced *trit-i-kay-lee*) derives from the scientific names for wheat and rye, *triticum* and *secale*. The cross has a higher protein content and efficiency ratio than either wheat or corn—comparable to soy concentrate—and is also higher than wheat in lysine and methionine, two of the life-sustaining amino acids.

A delicious bread made from triticale is now obtainable in many grocery stores throughout the Southwest. The flour's baking qualities are better than rye's.

Triticale has shown the ability to produce two or three times as much per acre as either wheat or rye and can be grown

Triticale

anywhere in the world where wheat is found. The grain is being improved constantly as new strains are developed at various experiment stations throughout the country.

VETCH (*Vicia*)

Vetch, a relatively slow-growing perennial, is a good companion for oats and rye. Plant fast-growing rye or oats as a "nurse crop" to provide shade and check competitive growth. However, if this is done, the vetch should be planted more thinly than ordinarily or the annual nurse crop may be choked out by the sturdy perennial. Fall-planted vetch is one of our most valuable green manure crops. Being a legume, it enriches the soil with both nitrogen and humus.

WHEAT (*Triticum* spp.)

There are two stories about the origin of wheat, both extremely interesting. The first is that bread wheat appeared around 8000 B.C.E. (B.C.) when wild wheat by accidental cross-pollination apparently formed a hybrid with a type of "goat grass," resulting in much plumper grains. This new plant, called emmer, again crossed with goat grass, forming an even more luxuriant hybrid. Because the husk of this grain was so tight that the whole grain would not scatter to the wind as other grass seeds do, the continued existence of "wheat" depended on humans, and thus bread wheat came into being.

The Theosophists, however, believe that mankind at a certain stage in his development was assisted by some high initiates coming from the planet Venus. They believe that these advanced beings not only gave moral and social guidance to humans but also brought with them wheat grains to supply a better cereal, bees to produce honey and fertilize flowers, and ants. Rye, they think, was produced by people in imitation of wheat by selective breeding. Oats and barley are thought to be hybrids brought about by crossing with earthly grasses.

In some regions, poppies spring up and become a weed in wheat fields. They should not be allowed to spread, for they check the wheat's growth. On the other hand, chamomile is beneficial when permitted to grow with wheat in a very small ratio (1 to 100), while in larger amounts it is harmful. Wheat will be increased by the presence of corn.

The growth of wheat is adversely affected by cherry, dogwood, pine, and tulip, as well as by proximity to the roots of sorghum. Canada thistle and field bindweed are harmful to both wheat and flax.

I have grown a good stand of winter wheat by sowing it in fall on my Bermuda grass lawn. In our mild climate it grows intermittently all winter, heading about the last of June. After it is harvested, the Bermuda grass takes over again and you would never know the wheat had been there. I don't know whether this would work with other lawn grasses.

If you do make a sowing of winter wheat, avoid the Hessian fly by planting it late, timed according to when this fly appears in your area.

FIRST STEPS FOR HOME FRUIT GROWING

For the gardener on a small lot, the site of the home orchard may be limited by necessity, the placement of trees being to a large extent dependent on the overall landscape design. The homesteader, who has several acres, has at least a modest choice.

Since fruiting plants are more permanent than vegetables, their placement in the landscape design becomes more important. And often their usefulness may be doubled by considering also their ornamental and shade values.

Apples, plums, peaches, and pears are such beautiful flowering trees that they may be used for the same design scheme as crabapples, dogwoods, and redbuds. Pecans and walnuts (as well as apples and pears) make fine shade trees, too.

In areas where they grow well, blueberries will fit in nicely with other flowering shrubs such as forsythias, hydrangeas, and spireas. A trellis or arbor becomes both useful and beautiful if bunch grapes or muscadines are planted to grow on it. Unsightly fences may be covered or a patio comfortably shaded if a few grape plants are placed thereon.

Pollination, as applied to fruit and nut trees, vines, and bramble fruits, really is a matter of "companion planting," yet we seldom hear it called this.

Fruit and nut trees almost always do better if at least two of each kind are planted. For some varieties the need is imperative—they will bear scarcely at all without pollination help.

Few home gardens can accommodate more than two or three different kinds of fruit. To grow them successfully, it is very important to consider varieties known to be self-fertile (also called self-fruitful), or known to be good pollinators for the other types you wish to grow. If your home site is not large enough for many trees, check around the neighborhood and list the fruit and nut trees you find there. Some of them may be good pollinators for trees you would like to plant.

In this limited space I cannot possibly list every variety of each fruit that will ensure pollination, but there are a few general rules to follow for good results, and I have tried to include them in the chapters that follow. But remember that pollination, important as it is, is only one factor in success.

While I won't go into the details of cultural practices here, it should be said that trees in a healthy growing condition will naturally derive more benefit from correct companion plantings. Healthy trees produce more pollen. And this applies to all trees, whether standard or dwarf types.

Getting Started

If you have room to set aside a definite orchard area, the first year you should do subsoiling, plowing, disking, and grading well in advance of planting. If possible, choose a gently sloping site with good air and good soil drainage. There is nothing a tree dislikes more than hardpan and wet feet.

Soil that absorbs water readily is the best, and you can test this by digging a 10-inch-deep hole and filling it with water. If the hole drains completely within about eight hours, drainage may be considered satisfactory. However, if the water remains much longer, drainage is poor. To prevent root rot, work crushed rock, gravel, or peat moss into the soil. Mixing compost with the soil will help in more ways than one.

Grow a nourishing cover crop such as rye, vetch, or soya beans, and disk this in after well-rotted manure or compost has been spread. Allow time for its decomposition, for the trees do not like raw manure or organic matter around their roots. In a natural forest setting, raw organic matter remains on top and only decomposed humus touches the roots.

For shrubs or bramble fruits, the materials should be worked into the soil at least 1 foot deeper than planting depth. For trees, mix the additions to the soil about 2 or 3 feet deeper than the intended planting hole.

The actual planting of the trees comes after the soil has settled. Planting trees in early spring is the generally accepted practice, but in the South or Southwest it is often possible to plant with good results in fall or early winter.

Shortly before planting, fill the hole with water and allow it to drain completely. This will prevent the surrounding soil from absorbing most of the water applied to the freshly planted shrub or tree.

Maintaining a layer of mulch around new plantings helps their growth, since it preserves moisture and in time becomes compost, providing plant nutrients.

Culture

Dr. Ehrenfried E. Pfeiffer, author of *The Biodynamic Treatment of Fruit Trees, Berries and Shrubs,* believed that a mixed culture in the orchard as well as in the garden helped keep down insect pests. He advocated growing nasturtiums between fruit trees as a means of transmitting a "flavor" to the tree that made it disagreeable to insects. He considered it particularly effective when the flowers were grown under apple trees to repel woolly aphids. Washing down the trees with nasturtium juice was recommended, if planting them was not possible.

Dr. Pfeiffer also suggested for orchard use stinging nettle, chives, garlic (against borers), tansy, horseradish, and southernwood. Permanent covers considered beneficial are clovers, alfalfa, and pasture grasses. Temporary crops to turn under for green manure are such biennial clovers as mammoth clover, red clovers, and incarnate clover. He believed buckwheat useful on a light, sandy soil.

Though a mixture of red clover and mustards is considered ideal, Dr. Pfeiffer cautioned that mustard, while it sweetens the soil, can become a rapidly spreading weed and for this reason should not be allowed to go to seed. Alfalfa hay, particularly if shredded or chopped, was thought to have special benefit as a mulch.

Dr. Pfeiffer also recommended a paste for all fruit trees consisting of equal parts of cow manure, diatomaceous earth, and clay, to which horsetail tea is added. This mixture is applied with a whitewash brush or with spraying equipment in the larger orchard.

A number of excellent preparations for fruit trees are obtainable from the Bio-Dynamic Farming and Gardening Association (see Sources on page 442).

More Hints for Fruit Growers

Here are some other helpful suggestions concerning fruiting plants:

- *Marigolds* planted near apple trees or between rows of nursery stock will benefit the trees used in grafting and budding.

- *Wild mustard* is beneficial to grapevines and fruit trees, but cut it before it goes to seed.

- *Dandelions* near fruits and flowers will stimulate them to ripen quickly.

- *Chives* improve the health of apple trees and will prevent apple scab. Use chive tea as a spray against apple scab and for powdery and downy mildew on gooseberries.

- *Pollination* is accomplished mainly by bees and other insects, so no sprays of any kind should be used at blossoming time.

- *Ripening apples* give off small amounts of ethylene gas, which sometimes limits the height of nearby plants but causes their flowers or fruit to mature earlier than normal.

- *Oats* may have an inhibitory effect on the growth of young apricot trees.

- If you must replace a *young fruit tree* on the same spot where an old one has been removed, choose a different variety.

- *Garlic juice* or the powdered extract contains a powerful antibacterial agent effective against diseases that damage stone fruits.

- Do not place *apples* near carrots in a root cellar, as they may cause the carrots to take on a bitter flavor. If apples and potatoes are stored near each other, both will develop an "off" flavor.

- *Nut trees* usually take a little longer to bear than fruit trees. While you are waiting for them to grow, interplant with peanuts (legumes). They will improve the soil and give you a crop as well.

- *Nut trees* are good to plant in pastures and near stables or manure and compost piles, to repel flies on cattle.

The Fungus Connection

One gardener reported that her unthrifty young peach trees apparently were assisted by moldy oats from the cleanings of the oat bin when 1 bushel was applied to each tree. After several weeks all her slow-growing trees were putting out new, healthy leaves.

A possible reason for the good growth of the peach trees is in the mold, rather than the oats, for almost all trees have a symbiotic relationship with some fungi (including molds). The fungi grow around the plant roots and furnish vitamins and other natural compounds necessary for a fast-growing and healthy tree.

This brings us back to the soil again. Because of this relationship, it's a good idea to have some of the original soil packed around the roots when transplanting a shrub or tree. Quite likely there will be fungi in the soil beneficial to the plant.

You may even do this: If you have a tree that isn't doing well after being set the first time, take some soil from another tree of the same variety that *is* growing well and dig it in around your problem tree. There's a good chance that your tree will perk up and grow.

This will work well not only with trees but also with other ornamentals and even with houseplants. If possible, investigate the original, invigorating habitat of such plants, remove some of the soil, and see if nature doesn't have a cure for the ailing plant far better than any commercial fertilizer you could buy.

Small Fruits

BLACKBERRIES (*Rubus* spp.)

Some self-unfruitful varieties of blackberries require cross-pollination. Others, even though self-fruitful, may benefit from the pollen-distributing visits of insects.

The flowers of blackberries are very attractive to their primary pollinators, honeybees. If a variety of blackberry is known to require cross-pollination, ensure a sufficient supply of pollinators in large acreages by placing colonies of bees in or near the field.

Do not grow blackberries near raspberries. Plant them in moderately acid soil, 5.0 to 5.7 pH.

Mulberries, chokecherries, and elderberries may be used to attract birds away from valued blackberry crops. Blackberries themselves are strong vital plants that help prepare the soil to support the growth of trees.

BLUEBERRIES (*Vaccinium* spp.)

Have at least two different varieties—any two—in a blueberry planting.

Blueberries like very acid (4.0 to 5.0 pH) and open, porous soils, such as a mixture of sand and peat with loam. The water table should be 14 to 30 inches below the surface.

BOYSENBERRY (*Rubus ursinus* 'Boysen')

Boysenberries are sometimes called trailing or semitrailing blackberries.

GRAPES (*Vitis* spp.)

Bunch grapes such as 'Concord', 'Fredonia', and 'Niagara' are self-fertile, and one vine will give an abundance of grapes even if planted alone. Grapes like a moderately acid soil of 5.0 to 5.7 pH. Perhaps more than any other fruiting plant, they need good air circulation to prevent fungus disease such as mildew. This is particularly impor-tant in moist, humid climates.

Grapes

Grapes in their natural environment swing high in the trees, doing especially well if the tree happens to be an elm or a mulberry. Such grapes are seldom troubled by either brown rot or mildew. Since growing grapes in trees is impractical for most, the best solution is perhaps a terraced hillside unsuitable for other crops.

Try planting hyssop with your grapes for an increased yield, or use legumes as an intercrop. Cypress spurge is unfriendly, so do not let it grow nearby. To discourage the rose chafer, keep grass out of the vineyard, since its larvae feed on grass roots.

(See also *Muscadine* below.)

MUSCADINE

As natives of the southeastern United States, muscadines do well under the high temperature and humidity found in this area, but they also are re-sistant to drought conditions and disease. Under favorable conditions they will live many years, but are not hardy in the northern United States. Some varieties are self-pollinating, while others require a pollenizer.

RASPBERRIES (*Rubus* spp.)

Raspberries, which like a near-neutral soil (6.5 to 7.0 pH), are self-fertile.

Because of virus disease, black and purple raspberries should be planted no closer than 600 feet from red varieties. Do not grow raspberries and blackberries near each other, either. Do not plant any raspberries near potatoes, since they make the potatoes more susceptible to blight.

STRAWBERRY (*Fragaria* × *ananassa*)

Almost all strawberries now sold, both June-bearing and everbearing, are self-fruitful. The "best" varieties vary from one area to another.

A cover crop of rye following sod will reduce the incidence of black rot on strawberries. They do well in combination with bush beans and spinach.

Strawberries will benefit if a few plants of borage, also a good attractant for honeybees, are grown near the bed. Lettuce is good used as a border. Pyrethrum, planted alongside, serves well as a pest preventative. A spruce hedge also is protective. White hellebore will control sawfly, and marigolds are useful, too, if you suspect the presence of nematodes.

Pine needles alone or mixed with straw make a fine mulch, said to make the berries taste more like the wild variety. Spruce needles also may be used as a mulch, but my personal preference is chopped alfalfa hay.

In some areas, growers plant strawberries as an intercrop in peach, apple, fig, orange, or other tree-fruit orchards. When the orchard is first planted, strawberries may be set out and grown for several years before the trees need all the ground. The strawberries furnish some income from the land, or at least pay the expense of caring for the orchard. The intensive cultivation given strawberries is especially good for young orchards. Also, because strawberries do not bear well unless moisture conditions are good, they will prove a good indicator of the orchard conditions.

Tree Fruits

APPLES (*Malus* spp.)

Only a few apples will bear well if grown alone, producing a good crop from self-pollination. Most should not be planted alone or be depended on for pollination in a combination; they are either low or lacking in viable pollen. Suppliers of fruit trees have information available on which combinations work best, and you should be sure you understand the needs of any tree you plan to buy.

If you have room only for one tree, there is still a way that you can have your favorite apple and pollinate it too. Graft a branch of a good pollinator somewhere on the host tree and this will serve your purpose. Apples like a near-neutral soil with a pH of 6.5 to 7.0.

APRICOTS (*Prunus* spp.)

All apricots are self-fertile, but they will benefit from cross-pollination to bear more heavily.

CHERRIES (*Prunus* spp.)

All sour pie cherries are self-fruitful and have no pollination problems. A single tree may be planted and expected to produce well from its own pollen.

Sweet cherries all are self-unfruitful and will require another variety nearby to enable them to set fruit. To further complicate things, there are even instances of pollen incompatibility among this group. A good nursery will give you information on pollination needs of the trees you are interested in.

Wheat is suppressed by the roots of cherry trees, and potatoes grown in the vicinity are less resistant to blight.

CITRUS (*Citrus* spp.)

Lime, lemon, orange, and grapefruit trees grow better in the area of guava, live oak, or rubber trees, which apparently exert a protective influence.

CRAB APPLES (*Malus* spp.)

Crab apple trees are often planted simply for their beauty. But they are self-fertile, and a good variety such as 'Dolgo' will provide both beauty and fruit.

FIG (*Ficus carica*)

Many people consider figs a tropical fruit, but there are varieties that will do well elsewhere. The fruit of the fig tree is peculiar in that the flowers form inside the fruit's skin. Pokeweed grows well as a fig's companion.

Most figs offered by general nurseries are self-fertile, but some varieties
will not mature their fruits unless the
tiny female flowers are fertilized by
pollen from a special kind of fig tree
called a caprifig. Other varieties bear
larger fruit if they are subjected to
this process, which is known as caprification. The pollen is transferred by a
tiny wasp that spends part of its life in

Fig

the fruits of the caprifig. In regions where Smyrna and other figs requiring caprification are grown, caprifigs are planted also.

MULBERRIES (*Morus* spp.)

Mulberry trees have rather insipid-tasting fruits but can be very useful to
lure birds away from cherries and berry plants. The birds seem actually to
prefer mulberries.

The Russian mulberry (*Morus alba* 'Tatarica'), a rapid-growing tree,
bears an abundant crop resembling blackberries, which may be made up
into pies and jams.

NECTARINE (*Prunus persica* var. *nucipersica*)

Nectarines are self-fruitful. They also will pollinate peaches, and peaches
nearby will help the nectarine to set a larger crop.

PEACH (*Prunus persica*)

Most peaches are self-fruitful, but a few require a pollinator (which can be
any other variety of peach). Peaches like a near-neutral soil with a pH from
6.5 to 7.0.

Never plant a young peach tree where an old one has been removed—
plant a different fruit tree.

If peach leaf curl appears and only a few leaves are affected, pull them
off by hand. Feeding the tree with well-rotted manure or compost high in
nitrogen will help the tree back to health. Garlic planted close to the trunk
will protect against borers.

PEARS (*Pyrus* spp.)

Almost all pears require other varieties nearby for a good fruit set, the exceptions under most conditions being 'Duchess' and 'Kieffer', which are self-fruitful. 'Bartlett' and 'Seckel' are not compatible, and 'Kieffer' is not always a good pollinator for 'Bartlett'.

If you live in an area where fire blight prevails, plant resistant varieties.

Some orchardists believe that pears are suppressed by the root excretions of grass, but a successful pear grower in California, believing the opposite, lets a variety of grasses and weeds grow in his orchard. He also sprays against codling moth and leaf roller, using ryania because it kills only chewing insects. He uses chicken manure to provide nitrogen, plus other animal manures, cottonseed meal, compost, and dried blood.

PERSIMMONS (*Diospyros* spp.)

There are two species of importance. The first, American persimmon (*Diospyros virginiana*), is native to a large part of the United States, and the second, the Oriental or Japanese persimmon (*D. kaki*), is a native of China and Korea. American and Japanese trees are not interfruitful. Persimmons come in dozens of cultivated varieties, which are considered superior to the wild type.

The common persimmon is a small, low-growing tree perfectly adapted for the homeowner with limited space, since it ordinarily attains a height no greater than 40 or 50 feet. The inconspicuous, greenish yellow, urn-shaped male and female flowers are borne on separate trees. A number of excellent grafted persimmon varieties are offered by the Louis Gerardi Nursery. (See Sources on page 442.)

Persimmon

PLUMS (*Prunus* spp.)

Almost all plums require pollination, though there are a few that will fruit alone. Plums like a moderately acid soil of 5.0 to 5.7 pH.

QUINCE (*Cydonia oblonga*)

Quince trees are self-fruitful.

NUTS

ALMOND (*Prunus dulcis*)
The almond is not a true nut but belongs to the Rose family. All varieties produce better if pollen from another tree is available. Peaches and almonds, being of the same family, will pollinate each other.

BUTTERNUT (*Juglans cinerea*)
Butternut has an inhibitory effect on plants within its immediate vicinity, but to a lesser degree than the black walnut. (See *Walnut* entries in this chapter.)

Butternuts

CASHEW (*Anacardium occidentale*)
This native of Brazil has become naturalized in many tropical countries and will grow on sandy soils in Florida. Cross-pollination is not necessary.

CHESTNUT, AMERICAN (*Castanea dentata*)
The chestnut blight has just about wiped out the native American species. However, much work has been done to develop disease-resistant varieties, and a revival of this tree is in progress.

CHESTNUT, CHINESE (*Castanea mollissima*)
Plant two or more varieties for cross-pollination.

FILBERTS AND HAZELS (*Corylus* spp.)
In ancient times many believed that a forked hazel twig had supernatural powers. Such twigs are mentioned in the Bible, while the Romans also describe the magical quality of the branches and told of hazel divining rods being used to find water and precious minerals underground.

   Hazels furnish valuable cover and food for wildlife. Homeowners also plant them as ornamentals or to shelter other plants. In some forests, hazels form such dense thickets that tree seedlings cannot grow and heavy machinery is needed to uproot them so more valuable timber can be planted.

Hazel trees and bushes are beneficial in pastures and elsewhere against flies. Cows like to nibble on the leaves, which increase the butterfat in their milk, while the tannic acid also acts as a cleansing agent for their digestive systems.

Hazelnuts

It is recommended that two varieties be planted for cross-pollination and better crops.

HICKORIES (*Carya* spp.)

The hickories, like the walnuts, have male and female flowers growing separately on the same shoot of the current season's growth. Many varieties appear to be self-unfruitful, so it is good practice to plant several varieties together to ensure cross-pollination.

PECAN (*Carya illinoinensis*)

These trees in all their varieties give no evidence of cross-incompatibility, and all will bear larger crops if two or more varieties are planted together.

Pecan trees like plenty of nitrogen. In the orchard, plant a winter and spring cover crop such as clover, which harbors nitrogen-fixing bacteria. For a lawn specimen, let a dense mat of grass grow near the trunk to conserve soil moisture and prevent sunscald of the roots. It is good to mulch with grass clippings, too.

The casebearer and hickory shuckworm, the most serious pecan pests,

are best foiled by releasing trichogramma wasps in the orchard. Do not store pecan meats near onions or oranges.

WALNUT, BLACK (*Juglans nigra*)

Grafted varieties of these self-fruitful trees usually produce each year, while wild trees generally produce well only in alternate years, some only every third year.

Pecans

Black walnut trees are known to produce a substance called juglone, which is washed from the leaves to the soil, inhibiting the growth of many plants within the area where the trees grow. Cultivated plants not compatible with black walnuts are apples, alfalfa, potatoes, tomatoes, blackberries, azaleas, rhododendrons, and heathers. The

Walnuts

butternut also seems to have this quality, but plants near it are less severely affected. (See also *Sycamore* in the Ornamental Trees and Shrubs chapter.)

Toxicity is contained in the roots of black walnuts as well as in the leaves, and because of this many plants will not grow near the tree. But not all are discouraged. Right at the drip line of a black walnut I have a bed of rainbow-colored iris, interplanted with daylilies, grape hyacinths, and daffodils, none of which appear to be in the least affected.

Drawbacks aside, the black walnut is prized for its valuable wood and delicious nuts. In addition, the tree's leaves scattered around the house or put in the dog kennel will repel fleas.

A Russian remedy to prevent sunburn is to rub freshly ground walnut leaves on the skin. The dark juice of walnut hulls applied to ringworm is said to heal the scalp. (See also *Walnut, English,* below.)

WALNUT, ENGLISH (*Juglans regia*)

Unless you live in a favorable climate, you will probably be more successful with a tree of the Carpathian type, which will do well farther north.

Walnuts are monoecious—that is, the male and female blossoms are separate on the same tree. They are self-fertile but produce better in plantings of several nearby.

English walnuts do not have the level of detrimental leaf and root excretions found in black walnuts, but their shade makes it difficult to grow some plants nearby. Many of the fruit mints, such as apple, orange, pineapple, and spearmint, will do well, however, as will angelica, sweet anise, and other herbs that like filtered sunlight.

ORNAMENTAL TREES AND SHRUBS

ALDER (*Alnus*)

Closely related to the hornbeams and birches is this small, water-loving tree that grows very rapidly and serves definite, special uses. The genus *Alnus* includes 20 species, nine of which grow in North America and six of which reach the height of trees. Alders may be planted in hedges along the borders of streams where their closely interlacing roots hold the banks from crumbling and keep the current clear in midstream. Like willows, alders are of assistance in draining wet soils.

In America the black alder (*Alnus glutinosa*) is often found in horticultural varieties. The daintiest are the cut-leaved forms, of which 'Imperialis', with leaves fingered like a white oak, is a good example. The root nodules add nitrogen to the soil, the black alder being the only nonleguminous plant that is able to perform this function. (See *Legumes* in the Soil Improvement chapter.)

AZALEA (*Rhododendron*)

Azaleas, holly, pieris, and rhododendrons are good companions for a landscape planting because all like humusy, acid soil. Do not plant azaleas or rhododendrons near black walnut trees. The substance called juglone present in the leaves, roots, and nuts of black walnuts is detrimental to them.

BEECH (*Fagus*)

Beech trees and ferns often grow together, and scilla bulbs do well under the trees. Beech trees in their infancy do well under the shade of other trees, so each fruiting tree is the mother of many young ones.

BIRCH (*Betula*)

It is believed that birch roots excrete substances that encourage fermentation and make the trees useful to plant around manure and compost piles. Dr. Ehrenfried Pfeiffer, one of the early advocates of the biodynamic method of farming and gardening, observed that composts fermented in the vicinity of the gray birch derived benefit from it and suffered no losses of nutrients, even if the roots actually penetrated the heap. It is considered best, though, to maintain a distance of at least 6 feet from the tree when building a compost pile.

CONIFERS

Turpentine substances washing from the leaves of conifers such as pine trees will inhibit the fermentation process of compost piles. Interplanting onions with conifers will help prevent damage by squirrels, which eat the buds of Scotch, white, and red pines. Winter-hardy Egyptian onions are the best kind.

Pine needles make an attractive mulch and will increase the stem strength, flavor, and productiveness of strawberries. In general, conifers have an adverse effect on the growth of wheat, since rain washing over them picks up substances that inhibit the germination of seeds.

Squirrel damage in pines and other trees can be limited by planting onions nearby.

ELDERBERRIES (*Sambucus nigra* and *Sambucus canadensis*)

Elderberries, having a liking for moist soil, are helpful near compost yards that are difficult to drain, and will also assist in the fermentation of the compost. Elderberries are noted for their ability to produce very fine humus soil about their roots.

ELM (*Ulmus*)

Grapevines that climb trees, swinging high in the air, are greatly benefited by good air circulation and sunlight. It is the sunlight on their leaves rather than on the grapes that causes them to ripen to perfection. Elm trees are particularly beneficial as supports for grapevines.

The slippery elm (*Ulmus fulva,* or *U. rubra*) is also known as the red elm, because its wood is red, and the moose elm, because moose are fond of browsing its young shoots. When the bark is stripped from this valuable tree, it is possible to scrape from its inner surface the thick, fragrant, mucilaginous cambium—a delectable substance that allays both hunger and thirst. The inner bark, dried, ground, and mixed with milk, is a valuable food for invalids. Fevers and acute inflammatory disorders have been treated with the bark, and poultices of the bark also relieve throat and chest ailments.

HEDGES

Hedges used as windbreaks are of particular value in dry, windy areas. Blueberries make a delightful hedge where they can be grown. *Rosa rugosa* makes an almost impenetrable hedge and also affords a harvest of vitamin-rich rose hips. 'Cardinal' autumn olive (*Elaeagnus umbellata* 'Cardinal') and dwarf burning bush (*Euonymus alatus* Compactus) are beloved by birds, as is red-leaf barberry (*Berberis thunbergii* Atropurpurea). For brilliant color there are 'Golden Prince' euonymus (*Euonymus fortunei*) and 'Goldflame' spiraea (*Spiraea japonica* 'Goldflame').

LOCUST (*Robinia*)

Sweet pea-type blossoms on a tree, or pods like the pea's swinging from the twigs, mean that it's a member of the pod-bearing Leguminosae family, to which both herbaceous and woody plants belong.

The black locusts (*Robinia pseudoacacia*) have nectar-laden, white flowers of "butterfly form," which honeybees (leading a host of other insects)

Being a leguminous tree, the black locust is a good companion to lima beans. There are toxins dangerous to humans in the leaves, bark, and roots, however.

swarm about as long as a flower remains to offer its sweet nectar. Cross-fertilization is the advantage the tree gains from all it gives.

Locust, good to plant as a border, has leaves, roots, and bark that are poisonous if eaten, but the pods of honey locust (*Gleditsia triacanthos*) contain a sweetish pulp used as cattle feed and occasionally eaten by small boys, who brave the tree's thorns to get them.

MAPLE (*Acer*)

The single genus *Acer* includes from 60 to 70 species widely distributed over North America. *Acer saccharum,* the sugar maple, is the best known and economically the most important for both its beautiful wood and its sap, which yields maple syrup. The black maple (*A. saccharum* spp. *nigrum*) is the sugar maple of South Dakota and Iowa. Red maple (*A. rubrum*), perhaps the most beautiful of all, is a swamp lover but will thrive on hillsides if the soil is moist. It is widely planted in parks and along streets.

Maples have shallow, spreading root systems, and it is difficult to get other plants to grow near them. They may also excrete substances that inhibit the growth of some plants, particularly wheat. Maple leaves laid in layers between apples, carrots, potatoes, and other root vegetables have a preservative effect.

MULBERRIES (*Morus alba, Morus rubra,* and *Morus nigra*)

White mulberry is the chosen food of silkworms and no substitute has ever topped this tree's preeminence. The berries of the red mulberry (*Morus rubra*) do not compare with the cultivated type, but are of value in poultry yards and hog pastures, where they are eagerly devoured. The black mulberry (*M. nigra*), believed a native of Persia, has large, dark red, juicy fruits but is hardy only in the southern and Pacific Coast states, where it is a desirable tree because it is so attractive to birds.

Mulberry trees are particularly good as a support for grapes. Tree-grown grapes are more difficult to pick than trellised grapes, but they will be relatively free of fungus diseases due to better circulation of air around them. Worms in horses may be repelled by mulberry leaves, and Russian mulberry is sometimes used as a trap crop to protect cherries and strawberries.

NURSE TREES

As abandoned fields again become covered with vegetation, the brushland is gradually reforested. The first trees are quick-growing, short-lived types that provide conditions suitable for the slower-growing, longer-lived trees. Looking at the forest floor, you will see very few pine seedlings. Other seedlings—young oaks, black cherries, and hickories—do better. Gradually, the pines will die off and the young hardwoods grow up and take their place. Should a forest fire occur, the whole process will start over again.

NITROGEN-FIXING TREES

Pod-bearing (leguminous) trees have the power to take nitrogen out of the air and store it in their roots and stems. The decay of these parts restores to the soil the plant food that is most often lacking and most expensive to replace. These trees and shrubs include black locust (*Robinia pseudoacacia*); bristly locust (*Robinia hispida*), sometimes called rose-acacia; clammy locust (*Tobinia viscosa*); Scotch broom (*Cytisus scoparius*); honey locust (*Gleditsia triacanthos*); Kentucky coffee tree (*Gymnocladus dioicus*); redbud (*Cercis canadensis*), sometimes called Judas tree; yellow-wood (*Cladrastis lutea*); woadwaxen (*Genista tinctoria*), sometimes called dyer's greenweed; indigo bush (*Amorpha fruticosa*), sometimes called false indigo; mesquite (*Prosopis juliflora*); screwbean (*Prosopis pubescens*), a slender-trunked mesquite, sometimes called screw-pod; Palo Verde (*Cercidium floridum*); Jamaica dogwood (*Piscidia piscipula*); horse bean (*Parkinsonia aculeata*); Texas ebony (*Pithecellobium flexicaule*); and frijolito (*Sophora secundiflora*)

Black alder (*Alnus glutinosa*) also adds nitrogen to the soil. It is the only known nonleguminous shrub with root nodules that can do this. (See *Alder* in this chapter.)

OAK (*Quercus*)

Oaks grown with American chestnuts seem to give them some resistance to chestnut blight. During their growth, oaks accumulate a large amount of calcium in their bark, yet amazingly the most calcium has been found in the ash of oak trees that grew in calcium-deficient soil.

A mulch of oak leaves serves to control radish and turnip maggots as well as repelling slugs, cutworms, and grubs of June bugs, but some gardeners believe the leaves have an inhibiting effect on certain vegetables. Therefore they should be fully composted before being spread on the garden.

In Germany it has long been a practice to control greenhouse pests such as ants, aphids, and small mites with the smoke from oak leaves. The smoke is not considered poisonous and will not kill bacteria in the soil or leave harmful residues.

Live oaks are believed to exert a protective influence on citrus trees.

The trichogramma wasp, whose larvae feed on moth eggs, helps keep oak trees green by controlling gypsy moths. *Bacillus thuringiensis* (see the Pest Control chapter) will also control and kill various caterpillars on the trees.

OSAGE ORANGE (*Maclura pomifera*)

This thorny tree is native from Arkansas to Texas and is hardy as far north as New England and central New York. It is valued for windbreaks or to grow in poor soils and is an excellent hedge plant, being almost impenetrable when fully mature. It was widely planted by the pioneers as a living fence around their homes before barbed wire came into use. The name refers to the Osage Indians and to the yellow fruit, which looks like an orange but is inedible.

PINE (*Pinus*)

Pine boughs are good to lay over peonies in winter for protection. Remove them in spring before growth starts. Pine needles make a good mulch for azaleas, rhododendrons, and other acid-loving plants and will increase vigor and flavor in strawberries.

Pine needles contain terpene, which, washed down by rain, has an inhibiting effect on seed germination. It is not good to place a compost heap near pine trees.

POPLAR (*Populus*)

The quick-growing, short-lived poplar often fulfills the function of "nurse tree." When a fire sweeps through the forest, it is likely to be the first tree to grow again on the bare land. The poplar's abundant seed, much like willow's, is wind-sown far and wide. Lombardy poplars, which look like exclamation points, are often planted to shelter other plants from the wind.

In Canada, very good stock feed has been made by boiling poplar wood under pressure.

Rosa rugosa

ROSA RUGOSA (*Rosa rugosa*)

This "hippy" rose has become so famous that it deserves to be mentioned all by itself. Grown in a mass, it makes a charming windbreak as well as an almost impenetrable barrier for animals. It grows better with purslane, parsley, and mignonette around it; is protected from rose bugs by alliums or onions nearby. Keep boxwood away. It blooms prolifically and is an excellent source for berries (hips) rich in vitamin C, containing more than oranges. The hips are used for making teas, jams, soups, and other dishes.

ROSE (*Rosa*)

All the alliums—garlic, onions, chives, and shallots—are beneficial to roses, protecting them against black spot, mildew, and aphids. For a recipe to overcome black spot in roses, see *Tomatoes* in the Vegetables chapter.

Garlic and onions are particularly beneficial to roses. In Bulgaria, where attar of roses is produced for perfumes, it is a common practice to interplant them with roses since they cause the roses to produce a stronger perfume in larger quantities.

Roses also are aided by parsley against rose beetles, by onions to repel rose chafers, by mignonette as a groundcover, and by lupines to increase soil nitrogen and attract earthworms. Marigolds are helpful against nematodes, and geraniums or milky spore disease against Japanese beetle. (See *Milky Spore Disease* in the Pest Control chapter.)

A carpet of low-growing weeds from the Purslane family will improve the spongy soil around the roots of rosebushes. An infusion of elderberry leaves in lukewarm water sprinkled over roses is thought to control cater-pillar damage and is also recommended for blight.

Do not plant roses with other plants that have woody, out-spreading roots that will compete with the roses for soil nutrients.

SASSAFRAS (*Sassafras albidum*)

Sassafras is sometimes called the mitten tree from its peculiar leaves, which grow in three different shapes: the simple ovate leaf, a larger blade (oval in form but with one side extended and lobed to form a thumb), and third, a symmetrical three-lobed leaf, the pattern of a narrow mitten with a thumb on each side.

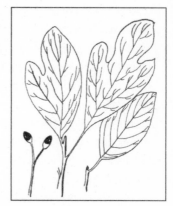

Sassafras will repel mosquitoes. The pun-gent oil has antiseptic properties, and the bark mixed with dried fruit wards off insects.

Sassafras

A tea made from the bark of young sassa-fras roots has been used for digestive disturbances. The dried leaves, called file, were formerly much used in the southern states as an ingredient in soups. However, sassafras is now regarded as unsafe for internal use.

SPRUCE (*Picea*)

Three species of woodpeckers were credited with controlling a serious infestation of spruce beetles in Colorado in 1947. Naturally occurring *Bacillus thuringiensis* (see the Pest Control chapter) has been found to give good control of this beetle in some forests.

SYCAMORE (*Platanus occidentalis*)

Studies conducted by American and Iraqi scientists show that sycamores inhibit the growth of other herbaceous plant species and the decaying leaves cause significant reduction in seed germination and seedling growth. Organic compounds leached from the leaves often are allelopathic to plants, and virtually no herbaceous plants will grow under the trees.

Sycamore bark has value, however. Boiled in water and made into a poultice, it is good to use for poison ivy.

WILD CHERRY (*Prunus pensylvanica*)

The wild bird, pin, or red cherry grows from Newfoundland to Georgia and west to the Rocky Mountains in rocky woods, forming thickets that are valuable as nurse trees. Wild cherry often springs up in burned-over districts where its bird-sown pits take root, the young trees sheltering new pines and hardwoods. It provides berries for birds and nectar-laden flowers for bees, so it can scarcely be called worthless, even though it is a short-lived tree.

The wild black cherry (*P. serotina*) is sometimes called the rum cherry. A tonic is derived from its bark, roots, and fruit, and brandies and cordials are made from its heavy-clustered fruits, which hang until late summer, turning black and losing their astringency when fully ripe. The wild black cherry makes an attractive shade and park tree, too.

The wild black cherry and the chokecherry (*P. virginiana*) are both of value to attract birds. Unfortunately, the tent caterpillar favors them to lay its eggs, making the trees unpopular with farmers. The egg rings in the outer smaller branches are easily seen and removed.

WILLOW (*Salix*)

The tough and fibrous roots of willow are useful in binding the banks of streams that may erode. Nature seems to have designed them specifically for this purpose, for wherever a twig lies upon the ground, it will strike root at every joint if the soil is sufficiently moist. The wind often breaks off twigs and the water carries them downstream where they lodge on banks and sandbars, which soon become covered with billows of green.

For thousands of years the bark and leaves of the willow have yielded resins and juices that eased the aches and pains of rheumatism and neuralgia or alleviated the distress of fevers. In the 1820s, salicin, the active principle of willow bark, was isolated, and in 1897 a synthetic derivative gave the world aspirin.

WINDBREAKS

Before planting a windbreak, study your land carefully and plan to put it where it will do the most good. Consider prevailing wind directions and the location and relationship of your buildings to the area you want protected. Most often windbreaks are planted across the west and north sides of a property, but of course there are exceptions to this rule, depending on the configuration of the land and the winds.

Do not plant your screen too close to the garden, for if the windbreak is to consist of trees and shrubs, they will rob the soil of moisture and nutrients. If you have sufficient land, plant the windbreak at least 50 feet from field crops. Very possibly you do not have this much room, but be as generous as you can.

The protective factor of a windbreak is 20 times its height. Thus a 10-foot screen would give you protection up to 200 feet downwind from it. You will also receive protection for several feet in front of the tree belt because it causes the air to back up and act as an invisible wall before it hits the planting of trees. Not the least of its uses is to hold down soil against heavy winds and to keep snow from drifting over walks and driveways. It may even help you reduce fuel bills.

In the prairie regions in particular, shelterbelt plantings have a marked influence on local climate, especially if they are placed at right angles to prevailing winds. A chain of such belts checks movement of the air, slowing down the wind velocity even before the windbreak is reached, and starts up a whole series of favorable climatic influences. These influences, such as a reduction of evaporation by increasing the humidity of the air, improve the yield of crops grown under their protection. (See *Hedges* in this chapter, and *Vertical Gardening* in Garden Techniques.)

WITCH HAZEL (*Hamamelis virginiana*)

The witch hazel is a stout, many-stemmed shrub or small tree, characteristically an undergrowth of larger trees. Native Americans were the first to use the bark of the witch hazel for curing inflammations. An infusion of the twigs and roots is made by boiling them for 24 hours in water to which alcohol then is added. The extract distilled from this mixture is used for bruises and sprains and to allay the pain of burns.

Perhaps the alcohol is the effective agent, for chemists have failed to discover any medicinal properties in either bark or leaf—yet who knows, they may still find it.

The tree has the peculiar property of throwing its seeds, particularly in dry, frosty weather. This does for the parent tree what the winged seeds of other, taller trees accomplish.

Witch hazel gets its name from the fact that superstitious English miners once used the forked twigs as divining rods.

GARDEN TECHNIQUES

BORDER PLANTS

Castor beans planted around the perimeter of the garden will repel moles, while borders of daffodil, narcissus, scilla, and grape hyacinth around flowerbeds will discourage mice. If used in small amounts, dead nettle (henbit), sainfoin, esparsette, hyssop, lemon balm, and valerian are helpful to all vegetables. Yarrow is a good plant in paths, as well as borders, as it will grow well even if walked on. Planted as a border to the herb garden, it enhances the growth of essential oils in the herbs.

CATCH CROPPING

This simply means growing a quick-to-mature crop of some vegetable in ground you've reserved for a planting of a later or slower-growing crop such as tomatoes, or a member of the Cabbage family such as broccoli or cauliflower. While you are waiting, put in radishes, lettuce, or spinach as a catch crop.

CLIMATE

Since climates vary greatly throughout the world, where you live should always be taken into account when you plan your garden. Maximum summer and minimum winter temperature should be considered, as well as annual rainfall.

For best success, try plants recommended for your area, making these your garden basics. This determined, you can then have fun experimenting each year with a few borderline plants those that do best in either a warmer or colder climate. Often, by providing shelter or otherwise creating a "mini" climate you may grow these successfully. Winter protection will help in the North, shade or a windbreak in the South. Some natural feature of your land, such as a pond, may enable you to grow something that your neighbor a few miles away cannot. (See *Microclimate* in this chapter.)

Mulching to keep the ground cool may be helpful for certain plants. Improving soil with humus often makes it possible to grow vegetables or plants that formerly were unsuccessful.

DAMPING-OFF

This is a disease caused by fungi, apparently present in the soil, that kills many young plants. It is characterized by collapse of the stems, or the seedlings falling over. It may occur before the seeds germinate or after the seedlings emerge. To avoid this, you can start seeds in a commercial soil-less medium; but if you make your own potting soil in which seeds are to be planted, it should be treated to kill the fungi by steam-heating to 180°F for half an hour or more.

A simple method for the home gardener is dry-heating the soil in an oven. Place the soil 4 to 5 inches deep in a pan and bury a small potato about 1½ inches in diameter in it. Bake in 200°F oven until the potato is done and the soil is sterilized and ready to use.

FRENCH INTENSIVE GARDENING

This type of gardening, which stresses maximum use of the soil and first became popular in the 1800s, is largely accomplished by using raised beds. These may be any length but narrow enough to permit easy handling from either side. Raised beds have the advantage of improved drainage and better aeration. The soil does not become waterlogged in winter and as a result it warms up faster in spring and produces earlier crops.

Prepare the soil by loosening to a depth of 12 inches and removing all weeds. Add compost or well-decomposed manure as well as any other or-ganic amendments (agricultural lime, gypsum, bonemeal, phosphate rock, etc.) that a soil test may indicate. Double digging is then done. This means that where the first spade-depth of soil is removed, a second spade-depth of soil is loosened before soil from the top layer is replaced.

If you are working with extremely poor soil, the bottom spade-depth may need to have additional incorporations of sand, compost, and loamy soil. All this sounds like a lot of work, and it is, but as the soil is improved each year, the work gets easier.

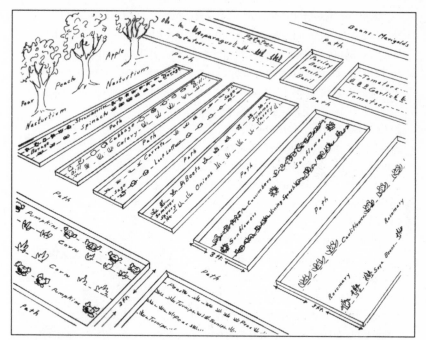

French intensive gardening is an ideal way to save space and to use companion planting effectively.

The benefit derived from this intensive gardening method is the increased number of plants that may be grown in a very small area. Perhaps in no other form of gardening is companion planting so important, since herbs and vegetables are so closely crowded together.

In general, the smaller vegetables and salad greens are best suited to this type of culture, but there is no law that says you can't grow corn and pumpkins and sunflowers and cucumbers this way if you want to!

FROST

Vegetables frequently are classified according to their ability to survive frosts. The USDA defines the differences as follows:

Hardy or cool-season crops will survive medium to heavy frosts. Seed from this group (peas, beets, kale, etc.) can be planted as soon as the soil can be prepared in spring, or in midsummer for a late-fall crop.

Semihardy vegetables will survive a light frost. Seed will germinate at relatively low temperatures and can be planted 2 to 3 weeks before the last frost date. This will vary in different sections of the country.

Tender or warm-season crops (tomatoes, eggplant, bell pepper, etc.) are injured or killed by frost, and their seeds seldom germinate in cold soil.

HONEYBEE (*Apis mellifera*)

Both in the garden and in the orchard, honeybees are an important agent of pollination. They are particularly attracted to the often inconspicuous flowers of herb plants.

A hive of bees is a good weather indicator, for if drones are forced out of the hive during fair weather, it is a sign that cold, wet weather is imminent. When hiving a new swarm of bees, rub the hive's inside with lemon balm, which the bees like. A smoke from jimsonweed (datura) calms the bees when a hive is opened.

INTERCROPPING

This is really the heart of companion planting, for the idea is to have two or more different vegetables growing on the same piece of ground, or in the same row, providing diversification. And this idea need not be confined to vegetables. Flowers and herbs can happily bump shoulders with each other. In fact, they *should*.

If your garden is small and you don't want to have empty spaces between your peas and bean rows, intercrop with broccoli, Brussels sprouts, cabbages, cauliflower, kale, or even radishes or carrots. After the early peas and beans are out, the slower-growing vegetables have all the space to themselves, and you have room to walk again. This may make things a bit cramped at times, but if you have little space and a short growing season, it's well worth trying. Of course, to be successful, you must keep up the fertility of your soil.

I like to keep a sort of "floating crop game" going in my own garden, making small plantings of quick-growing vegetables that sprout readily from seed, such as lettuce, radish, spinach, celery, cabbage, kale, chard, collards, and other greens. Staggered plantings mean fresh supplies coming on all through the season.

Plants that will help each other are put together as often as possible, either in the same row (for instance, marigolds with bush beans) or in adjacent rows. Lettuce and onions do well together, so I pop in a lettuce plant each time I pull a green onion for the table. I plant onions close together and pull every other one, letting the remaining onions mature for dry onions. In my climate even eggplant and green peppers benefit from a bit of shade, so I plant these together in a row next to okra.

Many vine crops, such as squash, cucumbers, and pumpkins, grow well with corn and may even protect it from raccoons. The corn is helpful in protecting vine crops from wilt. Many early crops do well following spinach, which is rich in saponin. (See *Saponin* in the Soil Improvement chapter.) Early spinach also may be intercropped with strawberries.

Many vegetables are pretty enough to put in the flowerbeds. Parsley between bulbs provides an attractive background in the spring. Tomatoes can grow with roses and at the same time protect them against black spot.

Chive clumps are another attractive planting for the rose garden (see *Chive* in the Herbs chapter for benefits). They grow larger each year, late spring bringing a pincushion of lavender blossoms that last for many days.

Hardy amaryllis, a member of the Lily family, sends out long strap leaves in early spring. After the leaves ripen and die, the ground is bare until August, when a sturdy stem emerges to grow quickly and bear fragrant pink lilies. A lettuce planting between the bulbs contrasts with the flower, making it far more beautiful.

Some vegetables have been especially hybridized for ornamentation in a flower planting. Flowering cabbage and kale come in colorful shades of red, white, and green, yet have excellent flavor. Plant them with the same herbs (mints, thyme, rosemary, sage, hyssop) as the garden varieties.

In garden intercropping, try not to put a plant that needs light where other, taller-growing plants will shade it, or a moisture-loving plant with another that is greedy for water.

Just follow the general rules: asparagus with tomatoes; beans with carrots or summer savory; beets with onion or kohlrabi; members of the Cabbage family with aromatic plants or potatoes or celery; leeks with onions, celery, or carrots; turnips with peas.

Remember also the dislikes, and do not plant beans with onions, garlic, or gladiolus; beets with pole beans; the Cabbage family with strawberries, tomatoes, or pole beans; potatoes with pumpkin, squash, cucumber, sunflower, tomato, or raspberry.

MICROCLIMATE

A microclimate—a small area with special growing conditions—may result from an unusual natural feature on your land, such as a pond that moderates air temperatures. Or you can create a microclimate yourself by varying your plantings, adding a hedge, or covering a fence with vines. A hedge makes a permanent windbreak, but rows of tall corn will grow quickly and serve the same purpose for a season: to shade, protect, and limit air circulation for tender plants. So will vine plantings, such as grape, and also cucumbers (though these must be kept well watered during the summer, particularly if they take the western sun).

MULCH

Mulch can be almost anything that retards loss of moisture from the soil, but organic mulches, many of which also add nutrients, are considered the most helpful. These include chopped bark, buckwheat hulls, cocoa shells, coffee grounds, corncobs, cottonseed hulls, cranberry vines, evergreen boughs, grass clippings, hay, hops, leaves (particularly oak leaves, which repel slugs, cutworms, and grubs of June bugs), manure, peanut hulls, peat moss, pine needles (great to increase stem strength and flavor of strawberries), poultry litter, salt hay, sawdust, seaweed, stinging nettle, straw, sugarcane residue, tobacco stems, and wood chips and shavings. Particularly comprehensive information is found in *The Mulch Book* by Stu Campbell. (See Suggested Reading.)

MULCH, SAWDUST

There is much to be said about a sawdust mulch, both for and against. Mulches like sawdust are particularly susceptible to spontaneous combustion, fresh sawdust can cause a depletion of soil nitrogen, and it is not good to use in summer because earthworms will avoid it.

On the good side, it is claimed by many authorities that blueberries mulched with sawdust will develop a larger, more fibrous root system and as an end result have a far higher yield. It is considered good mulch for raspberries and should be put on immediately after transplanting. Mixed with animal manures or poultry litter, it makes an acceptable mulch for many plants and shrubs where either one alone would not work well. Shavings or sawdust used for animal bedding makes an excellent mulch.

The type of tree from which the sawdust comes also has a bearing on the situation. Unweathered pine sawdust will decompose very slowly, so give it a bit of time to weather and turn gray before using. Sawdust from hardwood trees will rot much more rapidly than pine, spruce, or cedar, especially if weathered before using. Studies now show that the tannins and terpenes in sawdust that gardeners often fear really do little if any harm to the soil. (See *The Mulch Book* in Suggested Reading.)

pH

Experienced garden writers take it for granted that everybody knows what pH is all about. If you don't, relax; it isn't scary at all.

The pH of anything simply indicates its active acidity or alkalinity, expressed in units. The term is generally used in horticultural science to indicate a condition of the soil, and it's important to know, because many plants thrive only when the pH value of the soil closely approximates the optimum for their particular kind.

Soil acidity may be of two kinds, active and potential. It is a state in which the concentration of hydrogen ions (H) exceeds that of hydroxyl ions (OH-). When you have an exact balance of H and OH- ions, you have neutrality. When the OH- ions exceed the H ions, you have alkalinity.

Acitve soil acidity represents the excess of H ions over the OH ions present in the soil solution. It is expressed in pH units on the pH scale. On this scale, 7.0 represents neutrality; higher readings indicate alkalinity and lower ones acidity. It is rare to find a soil with greater acidity than 3.5, or with greater alkalinity than 8.0. You should note, however, that the relationship between the figures is geometric. Acidity at pH 5.0 is 10 times as great as at 6.0, and at pH 4.0, 100 times.

What can you do about it if a soil test shows too much one way or the other? To neutralize acidity, the gardener adds lime, preferably the agricultural type. Gypsum or sulfur can be used to correct an alkaline condition. In my opinion all soils, but particularly alkaline ones, benefit from the use of compost or humus in the form of decomposed organic matter. A green manure crop plowed under also helps.

SHADE

Shade is sometimes the decisive factor in companion planting. Nature does not arrange plants in long, straight rows, as we often do in our gardens. Try radishes in a foot-wide bed with no thinning. Put fast-growing lettuce such as 'Buttercrunch', 'Simpson', or 'Oakleaf' between cabbages, broccoli, Brussels sprouts, or even tomatoes, which will shade the young plants while they are growing. The lettuce will be up and out of the way when the slower-maturing plants need the room. You'll have a double crop on half the ground and with half the work, and you will also find that the taller plants give the lettuce just enough shade to keep it coming on crisp and sweet right into hot weather.

If you interplant early beets with late potatoes, the shade of the growing potatoes will benefit the beets, keeping them tender and succulent right into warm weather.

Plant melons between your onion rows, and by the time the onions are harvested the melons will be taking over the ground. While the vines are growing, the onions will protect them from insects.

After you harvest your early corn, let the stalks remain a while to shade a planting of fall cabbage, beans, peas, and turnips. When the fall garden is well established and the sun less warm, remove the cornstalks and use them for mulch right on the ground where they grew.

Many of the mints take kindly to shade and may be grown under trees. Sweet woodruff also likes shade and makes an excellent groundcover, while retaining moisture for other plants that give it protection from the sun. Tarragon and chervil like partial shade, too.

SUCCESSION PLANTING

This technique will enable you to make the most of a supply of compost or fertilizer. Heavy feeders such as broccoli, Brussels sprouts, cabbage, cauliflower, celeriac, celery, chard, cucumber, endive, kohlrabi, leek, lettuce, spinach, squash, sweet corn, and tomato should be planted in soil newly fertilized with well-decomposed manure.

Follow these heavy feeders with light feeders such as beet, carrot, radish, rutabaga, and turnip, which also like finely pulverized raw rocks and compost.

Legumes, the third group in succession planting, include broad and lima beans, bush and pole beans, peas, and soybeans. These soil improvers collect nitrogen on their roots and restore it to the soil.

SUICIDE IN PLANTS

Why do most annual plants die in autumn? It is possible that seeds inside mature fruits such as soybean pods send out hormones that cause plants to yellow and die even before nights cold enough for freezing cut them down.

Gardeners for years have known that if faded flowers are picked before they form seeds, the plants will continue to produce more flowers. Pansies are a good example. Among the vegetables, okra will continue from early spring to frost if the pods are kept picked before they harden.

This idea has been tested on soybeans. Growing pods were plucked from one side of the plant only and allowed to remain on the other. The side with the mature pods and seeds turned yellow and died, while the other remained healthy.

TWO-LEVEL PLANTING

Vegetables that occupy different soil strata often make good companions. Among these are asparagus with parsley and tomatoes, beets with kohlrabi, beets with onions, leeks with vine plants, garlic with tomatoes, carrots with peas, and also strawberries with bush beans.

Many combinations like this are possible, enabling the gardener with little space to virtually double the garden's yield, and at the same time improve the health and flavor of the vegetables planted together.

Do not put together plants that are competing for the same space and light, such as sunflowers and pole beans, or plants whose root excretions react unfavorably on each other, such as carrots and dill.

TWO-SEASON PLANTING

Gardeners in areas that have a long growing season may find both a spring and fall garden possible. In fall here in Oklahoma I can grow cauliflower, broccoli, Brussels sprouts, cabbage, collards, lettuce, radishes, and English peas, and they are practically insect-free. Some of the vegetables that require a long growing season to head up are a complete failure for me if planted in spring, because of the hot midsummer conditions. I cannot give you exact dates; this has to be worked out by area according to where you live, but is well worth trying experimentally.

In many places, early-planted squash is more likely to withstand borers, which lay their eggs in July. Where I live, we have a rule: "Plant squash and cucumbers the first day of May before the sun comes up and they will be free of beetles." I find that squash planted in midsummer for a fall crop often escapes insects as well.

Radishes and cabbage may escape root maggots by careful timing. The Hessian fly's attacks on winter wheat may be avoided if the wheat is sown after the first week in October, when the fly is no longer active.

South Texas cotton farmers have found they can control the pink bollworm without insecticides by carefully establishing deadline dates for both planting cotton and destroying the stalks after harvest.

Observe when insect infestations are worst on certain crops, and plant either earlier or later than you have in the past.

VERTICAL GARDENING

If you have a fenced garden, here is an opportunity for both beauty and increased productiveness. Many plants take kindly to climbing. Cucumbers (such as Burpee 'Burpless') grow longer and straighter when trained on a fence. Scarlet runner beans climb rapidly and make a beautiful as well as a tasty display. I plant these with my chayotes, which bear good-tasting and attractive fruit in September.

Morning glories and pole beans do well together, and rambling roses are happy with gourds. When the roses are gone, the gourds will bear attractive blossoms and fruits without damage to the roses. If you grow the birdhouse type of gourd, these will be a bonus for your garden, dried and hung the following season to attract birds.

You might follow Oriental practice to relieve the somber dark green of pines by allowing clematis to grow into the trees, particularly the white-blooming type that forms huge panicles of scented flowers in late fall.

If you don't have a fence, try a tepee or wigwam made of four or more poles fastened together near the top and with soft wire or twine tied from pole to pole. The growing plants are trained to the poles by tying loosely. When they reach the top, pinch out the growing point of each plant, causing them to produce side shoots. This system is very good for vining squash. Soon the wigwam will be covered with a mass of attractive flowers, bright green leaves, and squashes.

SOIL IMPROVEMENT

ALFALFA (*Medicago sativa*)

This is one of the most powerful nitrogen-fixes of all legumes. A good stand can take 250 pounds of nitrogen per acre from the air each year. Alfalfa needs a deep soil without hardpan or an underlying rock layer, because it sends its roots down deep. Researchers have traced them for well over 100 feet, and 20 to 30 feet is average.

Alfalfa's deep-rooting ability is the source of its great nutritional power, feeding as it does from mineral-rich subsoil that has not been worn out and depleted. Alfalfa is strong in iron and is also a good source of phosphorus, potassium, magnesium, and trace minerals.

You can easily sprout alfalfa seeds in the kitchen, and you may even want to grow some in your garden for highly nutritious alfalfa greens, or use its leaves for tea. Alfalfa used as a meal is a great compost stimulant and activator, particularly good for composting household garbage.

Alfalfa will make good growth wherever dandelions grow. Dandelions themselves are deep divers, their presence indicating that the subsoil is easy to penetrate. Alfalfa grown in pastures will give protective shelter for shallower-rooting grasses, keeping other plants alive longer during spells of dry weather. As a trap crop, alfalfa will draw lygus bugs away from cotton. Just 2 percent alfalfa provides sufficient control, but it should be planted about a month before the cotton.

BUCKWHEAT (*Fagopyrum esculentum*)

Buckwheat is valuable as a soil builder, and it will grow on very poor soils while collecting lots of calcium. Used in this manner it will take the light away from low-growing weeds, choking them out. If plowed under as green manure, it will sweeten the soil and make it more suitable for growing other crops. Buckwheat does not like winter wheat.

CALCIUM

Peas, beans, cabbages, and turnips revel in soil containing lime, but a few plants—notably those belonging to the Heath family, such as erica, azalea, and rhododendron—actually dislike it. Potatoes and a few cereals are not at their best if lime is applied to the ground immediately before they are planted or sown.

Land in need of lime does not respond to cultivation and manuring as it should, and often coarse weeds such as sheep sorrel flourish. Sometimes a green scum grows over the surface. A soil test that reveals excessive acidity indicates the need for liming.

Buckwheat accumulates calcium, and when composted or plowed under as green manure enriches the soil. Lupine (*Lupinus*) has roots that penetrate to surprising depths even on steep, gravelly banks or exposed sunny hills. It adds calcium to the soil, too, and is of value to grow on poor, sandy soils worthless for other purposes.

Scotch broom, a member of the Legume family, also accumulates calcium but may become a weed unless kept in check. Melon leaves are rich in calcium and should be added to the compost heap when the plants are spent.

CLOVER (*Trifolium* spp., *Melilotus* spp.)

Planting clover between rows of grapes will add nitrogen to the soil. This also works well in orchards or with companion grasses. Clover dislikes henbane and also members of the Buttercup family, which secrete a substance in their roots that inhibits the growth of nitrogen bacteria and poisons the soil for clover. This poisoning is so effective that clover will disappear in a field if buttercups are increasing. Clover has a stimulating effect on the growth of black nightshade (*Solanum nigrum*).

COMPOST

Compost is largely composed of decayed organic matter that has heated sufficiently to kill weed seeds and then has thoroughly decomposed. Plant preparations may be used to influence or speed up the fermentation process, and these—even when added in small amounts—can influence the entire operation. Once conditions are right, earthworms enter the compost pile and assist the other microorganisms in the breaking-down process.

Certain plants such as stinging nettle may be used to speed up or assist fermentation in the compost heap or manure pile. This plant is an excellent soil builder and, like comfrey, has a carbon-nitrogen ratio similar to barnyard manure. Nettles also contain iron.

Several other herbs are particularly well endowed with minerals and can be of value when incorporated into compost. These include dandelion, which absorbs between two and three times as much iron from the soil as other weeds; salad burnet, with its rich magnesium content; sheep sorrel, which takes up phosphorus; and chicory, goosegrass, and bulbous buttercup, which accumulate potassium. Horsetail shares honors with ribwort and bush vetch for a capacity to store cobalt. Thistles contain copper as a trace element.

Compost is the best fertilizer for herbs as well as garden vegetables and is particularly rich if weeds are put in the pile instead of being destroyed. Use all herb refuse obtained in the garden, too, and naturally all the kitchen waste, particularly in a household where cooking with herbs is a frequent feature.

Any organic material can be added, but refuse of a woody nature will decompose more rapidly if it is crushed or chopped first and used as a base mixed with materials such as grass clippings. Leaves picked up in fall make good compost, but since they are apt to pack, they should be mixed with other material. Bulky materials form a solid mass of rotting vegetation, lack oxygen, and quickly become sour and slimy.

When making a pile, the site first should be well dug to allow for a quick entry of earthworms. Incorporate sods into the heap. Build your pile up to a height of about 6 feet with straight sides and a slightly concave top. The contents must be kept moist and allowed to heat so that weed seeds are killed. The quicker the heating, the earlier the heap should be turned and restacked.

Turpentine substance, washing from the leaves of conifers, will retard fermentation. Birch trees in the vicinity assist fermentation, even if their roots penetrate the heap. It is best, however, to have them at least 6 feet away.

FERTILIZERS, NATURE'S OWN

We are all familiar with the ability of legumes to draw nitrogen from the air and "fix" it to their roots. Actually, there are many plants that get only about 5 percent of their nourishment directly from the soil.

Have you ever noticed how plants, particularly grass, look greener after a thunderstorm? This is not an optical illusion. They really *are* greener as a result of the electrically charged air, which frees its 78 percent nitrogen content in a water-soluble form.

Rain and lightning are fertilizing agents. Each time lightning strikes the earth, large amounts of nitrogen are charged into the ground. One authority states that 250,000 tons of natural nitrogen is produced every day in the 1,800 thunderstorms taking place somewhere on the earth. In some places this may amount to more than 100 pounds per acre per year. Rain also brings nitrogen—in some areas as much as 20 pounds per acre annually.

Sulfur comes down with the rain, possibly producing as much as 40 pounds per acre per year. Rainwater also contains carbonic acid, forming carbon dioxide in the soil where it is needed for the plant-feeding process. Millions of tons fall yearly, and when we consider that nearly half the makeup of a plant is carbon we realize how important this is. Evidence also seems to show that rare minerals such as selenium and molybdenum are washed down in rain.

Snow, which furnishes not only nitrogen but also phosphorus and other minerals, yields an extra bonus denied warm-climate areas. Snow contains 40 percent less heavy water, or deuterium oxide, than normal water. Deuterium is a heavy isotope, a form of hydrogen but a little different. Combined with water it does not form H_2O, the water molecule, but D_2O instead. Heavy water, according to the Russian scientists who observed this, slows down some chemical and biological processes of growing plants. When the heavy-water molecules are removed, plants seem to grow faster. Thus crops are aided in short-season, snowy climates such as that of Alaska. Even fog contributes to the soil's fertility, especially along the seacoast, where it brings in large quantities of iodine, nitrogen, and chlorine.

Dust, though sometimes disagreeable, has its good points too, containing minerals, organic matter, and beneficial organisms often in substantial quantities essential to plant growth. Dust may be carried for thousands of miles, even being held suspended for long periods in the upper atmosphere to be washed down eventually by rain. Many believe that dust is one of the most significant factors in restoring minerals to the exhausted soil and that it also contains bacteria important to healthy soil life.

Here's a way you can obtain some of the benefits of electroculture without waiting for a thunderstorm. Tie your tomato plants to metal poles or trellises (ours are concrete-reinforcing wire bent in an inverted V-shape, with a row of tomato plants on each side), using nylon strips cut from discarded panty hose. These sturdy supports also attract static electricity. A friend who tried this reported that she harvested an abundant crop of extremely large tomatoes, and her plants continued producing right up to frost.

GREEN MANURES

Green manures are cover crops, usually achieved by planting low-priced seed. If fall-planted, the cover protects the soil surfaces of fields or garden plots from erosion from winter winds, snowstorms, and quick thaws. It acts as an insulating blanket, keeping the soil warmer in winter and cooler in summer. This encourages soil life activity in general and earthworms in particular. The

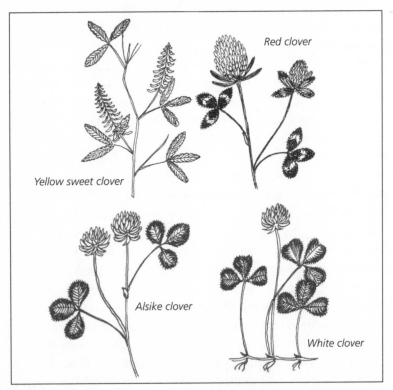

Red clover

Yellow sweet clover

Alsike clover

White clover

The clovers make fine green manure crops. Clover's growth is limited by Henbane and Buttercup family members.

more earthworms in the soil, the more channels they'll burrow deep in the subsoil, bringing to the surface useful minerals and nutrients that will increase the health and insect-resistance of food plants.

The roots of many green manure crops themselves reach deeply into the subsoil, where they absorb and bring up valuable nutrients. These revitalize the soil when the plants are plowed under to decompose.

Certain green manure crops, the legumes, have the ability to capture and fix large amounts of nitrogen from the air, also adding this important nutrient to the soil. Alfalfa is one of the best of these, and it is also high in terms of protein, which breaks down into usable nitrate fertilizer.

Other useful green manure cover crops are barley, bromegrass, buckwheat, cheat grass, alsike clover, cowpeas, lespedeza, millet, rape, spring rye, Italian ryegrass, winter rye, sorghum, soybeans, Sudan grass, sunflowers, common clovers, hairy vetch, and winter wheat.

For the home garden, kale makes a good, thick, nonleguminous green manure cover crop for the winter months. It is an easily grown, tasty food with a flavor that is improved by frost, and you can actually dig it out from under the snow in the dead of winter. When the kale starts growing again in spring, it can be tilled under to add green manure to the soil for the spring garden.

INOCULANTS

Sold under various names such as Legume Aid (Burpee) and Nitragen (Farmer), these preparations are aids to peas, beans, soybeans, sweet peas, etc., for better blooms and increased yields. Treating legume seeds with the proper inoculants helps them develop root nodules, which convert free nitrogen into plant food.

LEGUMES

Though nitrogen makes up 80 percent of the volume of the atmosphere, it is almost useless to most plants, for it must be changed into a compound before it can be used. Lightning combines or fixes small amounts of this nitrogen and oxygen in the air, thus forming oxides of nitrogen that are washed out of the atmosphere by rain or snow to reach the soil. (See *Fertilizers, Nature's Own* in this chapter.)

Nitrogen-fixing bacteria living in nodules on the roots of legumes—alfalfa, beans, clover, esparsette, kudzu, lespedeza, peas, peanuts, soybeans, winter (hairy) vetch, and others—can change atmospheric nitrogen into nitrogen compounds useful to themselves and other plants. Farmers for centuries have rotated their crops to take advantage of this increased soil fertility produced by legumes.

Clover is particularly beneficial if used as a green manure crop and plowed under before planting a crop of wheat or corn the following season. The decaying legume, rich in fixed nitrogen, increases the nitrogen content of the soil without the need for commercial fertilizers. A green manure crop of alfalfa will benefit a crop of cotton. Red clover may be used on soils too acid and too poorly aerated for alfalfa. The optimum pH for red clover is between 5.8 and 6.8, but it can stand a pH below 6.0 and still do reasonably well.

The steps in the nitrogen cycle can be traced in the growth and use of clover, as follows:

1. Atmospheric nitrogen is changed into proteins by the action of nitrogen-fixing bacteria growing in nodules on the clover roots.

2. After plowing under, the clover proteins are changed into ammonia by ammonifying bacteria.

3. Ammonia is then changed into nitrates by nitrifying bacteria.

4. Both ammonia and nitrates are used by other plants to form plant proteins.

The nitrogen-fixing bacteria of legumes benefit not only themselves but also other plants nearby. Peas and beans, for instance, benefit potatoes, carrots, cucumbers, cauliflower, cabbage, summer savory, turnips, radishes, corn, and most other herbs and vegetables.

Peanuts, another legume, can be used as a second crop after an earlier one is out, provided that you have a long growing season.

On larger acreages, lespedeza, kudzu, and esparsette may be used to aerate the soil and put nitrogen into it, while winter vetch and soybeans make excellent cover crops. Legumes sown with a small amount of mustard are helpful to grapevines and fruit trees. Peanuts are excellent to grow in an orchard of newly set nut trees.

NITROGEN-FIXING PLANTS
See *Legumes,* above.

OATS
See the chapter on Grasses, Grains, and Field Crops.

POTASSIUM
Certain weeds indicate a soil rich in potassium. These are marsh mallow, knapweed, wormwood, opium poppy, fumitory, Russian thistle, tansy, and sunflower. Red clover, however, is a good indicator of potassium deficiency.

Tobacco takes up potassium in its leaves and stalks and is thus a good plant on the compost pile if it has not been sprayed with chemicals. Vegetables that like potassium are celeriac and leek.

RHIZOBIUM
These are the bacteria that live in nodules on leguminous plants and turn atmospheric nitrogen into a useful form for building plant proteins. Unfortunately, they withhold their talents from such useful plants as corn and wheat. (See *Legumes* in this chapter.)

SAPONIN
Soapberry or chinaberry is the name of a group of genus of trees and shrubs that bear fruit containing a soapy substance called saponin. Bouncing Bet is probably our best-known saponin-rich plant. Other unrelated plants such as primroses and carnations also contain this substance, as well as many legumes, cyclamen tubers, camellia, viola, mints, horse chestnut, orach, pokeweed, runner beans, tomatoes, mullein, potatoes, and spinach. These plants are important because their decomposing remains create a favorable environment for the plants that come after them.

Commercial saponin is used as a foam producer in beverages and fire extinguishers and also as a detergent useful in washing delicate fabrics.

SEAWEED
Kelp, a type of brown alga, provides fertilizer and is a source of the chemical element iodine, particularly the giant kelp of the Pacific. Chemists also

extract from kelp large amounts of algin, which is useful in such commercial products as ice cream and salad dressings by virtue of its ability to hold several different liquids together.

Kelp is about 20 to 25 percent potassium chloride, and it also contains common salt, sodium carbonate, boron, iodine, and other trace elements. For gardeners who live on or near the seacoast, seaweeds are a natural and usually wasted resource that can be utilized for mulching and in compost piles. They are especially good materials to put around fruit trees. Another advantage is that decomposing seaweed is less attractive to mice than is straw.

Chopping seaweed may be advantageous if only for cosmetic reasons. It may be also advisable to rinse off the salt, but it is not necessary to be too fussy about this. The small amount left clinging will do no harm. Eelgrass, which dries into a light "hay" and doesn't pack down, is a perfect non-smothering material for many plants. (See *Eelgrass* in the Wild Plants chapter.)

Another use for "sea power" to promote fertility is liquefied seaweed, which may be applied as a foliar fertilizer directly to the leaves of plants. It is particularly helpful to trees.

It has been found that seaweed as a fertilizer helps promote frost resistance in tomatoes and citrus fruits, as well as increasing the sweetness of some fruits and giving better resistance to pests and diseases. Beets and parsnips respond badly to boron shortages in the soil, so chopped kelp makes an excellent mulch for them.

Seaweed helps to break down certain insoluble elements in the soil, making them available to plants. This quickens seed germination and further aids development of blossoms and fruits, resulting in increased yields.

For those who cannot readily obtain seaweed, there are seaweed preparations and fish emulsion, available at most garden centers and through gardening catalogs.

Eelgrass makes an enriching mulch and is fine for composting.

Another idea for gardeners who, like myself, live far inland is to use water plants. Though not as rich in nutrients as seaweed, they make good mulch if chopped and are good additions to the compost. Water hyacinth, a plant of tropical America, has escaped into the wild and has become a pest, choking many small streams. Dredged from waterways, it is valuable on the mulch pile. Here where I live, water lilies grow profusely in shallow pools and are easily obtainable. My husband, Carl, and I usually bring back a tubful of water weeds of some kind every time we go fishing.

SWEET CLOVER (*Melilotus alba* and *Melilotus indica*)

This is not a true clover, although it is a legume. White and yellow sweet clovers live for 2 years, and their large roots penetrate deeply into the soil. At the end of the second season, they decay to enrich the soil with nitrogen and decomposing vegetable material.

Sour clover, a kind of sweet clover used almost entirely to improve the soil, is often called melilot.

Spoiled sweet clover hay or poorly preserved silage never should be fed to animals. The hay contains coumarin, an anticoagulant, which develops toxicity as the clover decomposes and may cause both internal and external bleeding.

WEEDS AS SOIL BUILDERS

Many weeds seem to have a mysterious capacity for enriching soils. Jimsonweed (datura) grown near pumpkins will promote their health and vigor; the best watermelons come from the weediest part of the patch; onions grown with weeds (but not allowed to be overwhelmed) are apt to be larger than those in clean-cultivated rows.

This is particularly true if the weeds are the so-called deep divers that break up the subsoil, allowing the roots of the vegetable plants to have a larger-than-usual feeding zone. Deep divers sometimes bring up from below the hardpan mineral elements that the roots of food plants cannot reach. The high mineral content of weeds is another reason for adding them to the compost pile. (See *Compost* in this chapter.)

PEST CONTROL

BACILLUS THURINGIENSIS (BT)

This is a selective bacterial disease, highly effective against caterpillars and the larvae of moths. During spore production, *Bacillus thuringiensis* produces crystals that act as a stomach poison on the insects eating the treated plants. This substance is not toxic to plants, people, or animals, and can be applied even up to the day of harvest.

BT controls the fruit tree leaf roller, as well as tent caterpillars—the disease attacking the pest in the caterpillar stage after they have come out of the tent. The disease is widely used to protect commercial crops of celery, lettuce, cabbage, broccoli, cauliflower, mustard, kale, collards, and turnip greens. It is also effective against the tobacco budworm and the bollworm.

Special strains of BT have been developed to kill other pests such as mosquito larvae and Colorado potato beetles. BT products are widely available, under various brand names, from garden centers and mail-order catalogs.

BEAN AND POTATO COMBINATION

Bush beans planted with potatoes protect them against the Colorado potato beetle. In return, the potatoes protect the bush beans from the Mexican bean beetle. It is considered best to plant the beans and potatoes in alternate rows.

BIRDS

Birds around the garden are generally recognized as one of the best controls against insect pests. Particularly useful are the purple martins, which must catch flying insects almost constantly in order to live. Well-made martin houses set up away from large trees will attract martin colonies. Feeders and bird waterers will also encourage other birds to visit and nest near your garden.

Birds are particularly attracted to hackberry, chokecherry, elderberry, Tartarian honeysuckle, mulberry, dogwood, Japanese barberry, red and black raspberries, viburnum, Hansen's bush cherry, Russian olive, hawthorn, and sunflowers. Evergreen trees and thornbushes are attractive for nest-building. Certain forms of cactus are attractive for this purpose in the Southwest.

Birds, however, can be too much of a good thing at times and may become very destructive to the food plants we grow for ourselves.

The purple martins rid orchards and garden areas of incalculable numbers of injurious insects, devouring them in the winged stage.

From the Chinese we have this suggestion: When fruits start to ripen, hang sliced onions in the trees—the birds dislike the scent and will avoid the fruit. Hanging empty milk cartons in fruit trees from a string so they twirl in the breeze will deter many birds. Change their position occasionally, and perhaps also use bright, fluttering ribbons or strips of cloth.

If crows are a problem for a corn or watermelon patch, put up several stakes and string white twine around the patch, crisscrossing it through the center. To the birds this will look like a trap and they will avoid the patch.

BLACKFLY (*Aphididae*)

This insect is particularly detrimental to broad beans, and it is advisable to use a fermented extract of nettle to keep it under control. Its natural enemy, the lady beetle, will help. Intercropping with garlic, or placing an occasional plant of nasturtium, spearmint, or southernwood in the rows of beans, also is a good plan.

BORERS

Nasturtiums planted around fruit trees repel borers, while garlic and other alliums such as onions and chives also are good.

CASTOR BEAN
(Ricinus communis)

Castor bean

Experiments have shown that castor beans planted around a garden will repel moles, and they also are good to repel mosquitoes.

This is an agricultural crop in some areas where it is grown for the oil in the seed, yet all parts of the plants are poisonous to livestock and humans, particularly the seed. Two or three seeds eaten by a child can be fatal and as few as six can cause the death of an adult. The beans also carry an allergen that causes severe reactions in some people when handling castor pomace. This danger can be eliminated if the seed heads are clipped off or destroyed before they mature.

To form an effective mole repellent, plant castor beans every 5 or 6 feet around the perimeter of the garden. Use them also as a companion crop: Plant several pole beans close to their bases and let them climb the tall-growing plants. The largest variety of castor bean, 'Zansibarensis', grows 8 feet tall, and has large leaves and beautiful variegated seeds of various colors. Sanguineus grows 7 feet tall and 'Bronze King' 5 feet.

COCKROACH *(Blatella)*

Extracts of chinaberry have been found useful against cockroaches and termites. The little-known cockroach plant (*Haplophyton cimicidum*) from Mexico also is valuable in controlling this pest.

CUTWORM *(Noctuidae)*

A 3-inch cardboard collar around young plants extended 1 inch into the ground and 2 inches above will foil cutworms. Use cardboard cut from toilet paper or paper towel rolls, or cut off the top and bottom from a quart-size milk carton and cut the remainder into three collars.

A used matchstick, toothpick, small twig, or nail set against the plant stem will keep the cutworm from wrapping itself around the plant and cutting it. Oak leaf mulch will repel cutworms, too.

DIATOMACEOUS EARTH

This effective remedy against many insects is made from the finely ground skeletons of small, fossilized, one-celled creatures called diatoms, which existed in the oceans and constructed tiny shells about themselves out of the silica they extracted from the waters. The microscopic shells, deposited on the floor of the ancient seas, collected into deposits sometimes thousands of feet deep.

This earth contains microscopic needles of silica, which do their work by puncturing the bodies of insects, allowing vital moisture to escape from them. The insects die from dehydration. This earth is so finely milled that it poses no threat to either humans or animals, but these particles, when taken internally by insects, interfere with breathing, digestion, and reproductive processes.

Diatomaceous earth will not harm earthworms, which are structurally different from the insects. The earthworm's outside mucus protection, coupled with its unique digestive system, enables it to move through soil treated with diatomaceous earth without harm.

Many gardeners use diatomaceous earth as a dusting agent to give effective control against gypsy moths, codling moths, pink boll weevils, lygus bugs, twig borers, thrips, mites, earwigs, cockroaches, slugs, adult mosquitoes, snails, nematodes, all species of flies, corn worms, tomato hornworms, mildew, and so on. For field crops and in orchards, the diatomite particles are best applied with an electrostatic charger, which gives the particles a negative charge, causing them to stick to plant surfaces.

EELWORM

Eelworms or nematodes are tiny, sightless, eel-shaped organisms that pierce the roots of plants to feed or to lay their eggs inside, causing knots to form. When organic matter is incorporated into the soil, it encourages beneficial fungi and other nematodes that feed on the plant-parasitic variety of eelworms. These beneficial fungi grow in decomposing vegetable matter and kill the nematodes. It is possible to reduce wireworm and nematode attack, too, by placing a heavy dressing of barnyard manure on grassland before plowing.

These microscopic nuisances are discouraged by marigolds, scarlet sage (*Salvia*), or dahlias (*Dahlia*). Asparagus is a natural nematocide. Tomatoes grown near asparagus thus are protected, while in turn the tomatoes protect the asparagus from the asparagus beetle.

FLEA BEETLE (*Epitrix*)

Flying insects such as flea beetles are known to dislike moisture. Very often they can be discouraged by watering the garden in full sunlight. I find them annoying but not particularly destructive to eggplant, tomato, turnip, and radish plants. The damage is mostly cosmetic, and strong plants quickly grow out of it, the plant becoming less attractive to the beetles as the leaves enlarge and toughen a little. Light cultivation and the addition of organic matter to the soil both discourage the beetles and help the plants.

Bruised elderberry leaves laid over the rows of plants are a deterrent to the beetles, and they also are repelled by mint and wormwood. Beetles attracted to radish or kohlrabi may be controlled by interplantings of lettuce.

FUMIGATION

Greenhouse gardeners, who frequently have difficulty controlling aphids, ants, and termites as well as that all-time pest, whitefly, find smoke from oak leaves effective. The leaves are not poisonous, do not kill soil bacteria, and leave no harmful residue. Smoke the leaves for about a half hour, keeping the greenhouse door tightly closed.

GARLIC

See the Herbs chapter.

GOPHER (*Geomyidae*)

These burrowing rodents may be repelled by plantings of scilla bulbs. The scillas, sometimes called squills, are flowering, bulblike ornamentals that have grasslike leaves and clusters of flowers at the top of long stems. They are easy to grow and bloom in early spring. They may be grown among vegetables as well as in flowerbeds. But be cautious—the bulbs should never be eaten.

GRASSHOPPER (*Tettigoniidae* and *Locustidae*)

Grasshoppers are very difficult to control, especially where they come in from surrounding fields, but this spray will help: Grind together two to four hot peppers, one mild green pepper, and one small onion, and add to 1 quart water. Let stand 24 hours and strain. This mixture also is good against aphids.

During times of grasshopper infestation, an after-harvest plowing will discourage egg-laying, while spring tilling before seeding will prevent grasshoppers from emerging from the eggs still present. Extracts of chinaberry have proved useful, while sabadilla dust or extract is effective against these and many other insects. (See *Insecticides, Botanical* in this chapter.)

Grasshoppers can be attracted by this bait: Fill several 2-quart mason jars with a 10 percent molasses solution and place around the area where the infestation is the worst.

Chickens are of great value in an orchard, for they eat grasshoppers and other insects and at the same time add their manure to aid the fertility of the soil. The coop may be moved every few days to a new location.

Wild birds attracted to the garden eat a great many grasshoppers and surprisingly cats will kill and eat them, too. I think this is partly for the fun they get out of the chase.

Grasshoppers will eat almost anything except horehound, but grasshopper-resistant varieties of corn and wheat have been developed.

ICHNEUMONID WASP (*Ichneumonidae*)

This wasp has been found by the Brownsville Experiment Station in Texas to parasitize at least 27 destructive species of moths and butterflies, but it *prefers* to deposit its eggs in bollworms and tobacco budworms.

INSECT ATTRACTANTS, BOTANICAL

Insects are largely attracted by scent and may be lured away from certain crops by other plants placed nearby. Plants such as nasturtium and mus-

tard, both of which contain mustard oil, frequently are used for this purpose. These are called trap crops. Insects feed on and lay their eggs in trap
crops, which should be destroyed before the eggs hatch.

INSECT REPELLENTS, BOTANICAL
Insect repellents may be prepared from crushed leaves, infusions, or essential oils of such botanicals as citronella, eucalyptus, pennyroyal, bergamot,
cedarwood, clove, rose geranium, thyme, wintergreen, lavender, cassia,
anise, sassafras, bay laurel, and pine tar. Ginkgo, elder, pyrethrum, and
lavender repel ticks or other insects. Other plants such as cedar, quassia,
and teakwood themselves are immune to insects.

See also the *Insect Control through Companion Planting* chart, on page
140, in this chapter.

INSECTICIDES, BOTANICAL
Several plant derivatives are available commercially as insecticides. (See
Sources on page 442.) These materials break down more quickly than
man-made insecticides and (except for nicotine) are relatively nontoxic to
humans when used as directed. They kill bees and other beneficial insects as well as pests, however, and can be toxic to other life-forms such
as fish. As with any pest-control products, read the label and follow the
directions exactly.

Pyrethrum is prepared from the flowers of *Chrysanthemum cinerari-
afolium* or *C. roseum*. It is used alone or in combination with other botanicals against a wide variety of pests, including aphids, leafhoppers, spider
mites, harlequin bugs, and imported cabbageworms.

Rotenone comes from the roots of tropical plants and is a powerful
stomach poison for chewing insects.

Nicotine. Tobacco and its main alkaloid, nicotine, have been used as an
insecticide since the late 17th century. Nicotine, a contact poison, is toxic
to mammals as well as insects and should be handled very carefully.

Ryania, from a Latin American shrub native to Trinidad, does not always
kill insects outright but rather makes them ill, causing them to stop feeding.

Sabadilla. When heated or treated with an alkali, the seeds of this plant become toxic to many insects. Oddly enough, this quality increases with age during storage. The extract is effective against a large group of insects such as grasshoppers, corn borers, codling moths, webworms, aphids, cabbage loopers, and squash bugs. Be cautious in handling the dust and do not breathe it in.

Neem oil is made from the seeds of the neem tree, native to India. It is useful against Japanese beetles and the adults of many other chewing insects, acting as a repellent and as an appetite suppressant, and also stops the development of larvae.

INSECT-RESISTANT VEGETABLES AND GRAINS

Every garden has both "good" and "bad" bugs, yet from a gardener's point of view only a few are relatively destructive. Some vegetables seem to have a natural, built-in resistance: carrots, beets, endive (including escarole and witloof chicory), chives, okra, Egyptian onions, parsley, peppers, and rhubarb. Under good growing conditions, lettuce might be added to this list, too.

Numerous vegetables and herbs listed in this book help other vegetables resist insects when grown with or near them. Scientists have done a great deal of research, also, on why certain other plants are attractive to insects. They have come up with something that organic gardeners knew all along: Insects prefer to eat plants having high concentrations of free amino acids, such concentrations being enhanced if the plants are improperly nourished. Organically grown vegetables produced on balanced, healthy soils have significantly lower levels of free amino acids in their tissues than plants grown where chemical fertilizers have destroyed the balance. Such "organic" vegetables are less tasty to insects.

In addition, vegetable varieties have been bred for resistance to specific pests. Your county's Cooperative Extension Service may have recommendations on good choices for local conditions.

Insect Control through Companion Planting

Legumes planted in a rotation will protect grain crops and grasses from white grubs and corn rootworm. Chinch bug on corn and flea beetles are controlled by growing soybeans to shade the bases of the plants. The herbs in this chart may be planted as specific controls.

Herb	Deters
Basil	Flies and mosquitoes
Borage	Tomato worms
Castor bean	Moles and plant lice
Catnip	Flea beetles
Datura	Japanese beetles
Dead nettle	Potato bugs
Flax	Potato bugs
Garlic	Japanese beetles, aphids, weevils, fruit tree borers, spider mites
Henbit	General insect repellent
Horseradish	Potato bugs (plant at corners of plot)
Hyssop	Cabbage moths
Lavender	Clothes moths (dry and place in garments)
Marigold	Mexican bean beetles, nematodes, and many other insects
Mint	White cabbage moths, dried against clothes moths
Mole plant	Moles and mice (mole plant is a species of euphorbia)

LADYBUGS

These famous eaters of aphids may be purchased for garden release. (See Sources on page 442.) The problem, especially in small gardens, is to keep them there. If there is a real need for their services and there is plenty of ladybug food around, they are more likely to stay.

The way you release them will make a difference. Do not strew them about like grains of corn but rather place them by handfuls—carefully—around the base of infested plants. Their natural instinct is to climb the nearest plant and start hunting for food. Do your "seeding" gently, since rough handling, especially in warm weather, may excite them to flight. Early morning or evening is the best time.

Herb	Deters
Nasturtium	Aphids, squash bugs, striped pumpkin beetles, woolly aphids
Pennyroyal	Ants and plant lice
Peppermint	White cabbage butterflies, ants
Petunia	Beetles
Pot marigold	Asparagus beetles, tomato worms, and many other insects
Rose geranium	Oil or crushed leaves as insect repellents
Rosemary	Cabbage moths, bean beetles, carrot flies, and malaria mosquitoes
Rue	Japanese beetles
Sage	Cabbage moths, carrot flies, ticks
Santolina	Moths
Sassafras	Plant lice
Southernwood	Cabbage moths, malaria mosquitoes
Spearmint	Ants, aphids
Stinging nettle	Aphids, blackflies
Summer savory	Bean beetles
Tansy	Flying insects, Japanese beetles, striped cucumber beetles, squash bugs, and ants
Thyme	Cabbageworms
White geranium	Japanese beetles
Wormwood	Animal intruders, cabbageworm butterflies, black flea beetles, malaria mosquitoes

Lamb's-quarters, which sometimes harbor the leaf miner, also may play host to the beneficial ladybug. Newly placed ladybugs depend on their hosts (which also may be Chinese celery, cabbage, or other plants), and they must find aphids in sufficient quantity to keep them in the vicinity and ensure reproduction. After one mating a female will produce from 200 to 1,000 offspring.

In spring you can often find the eggs of ladybugs on the undersides of leaves, near their early food supply, aphids. You will see them standing on end in clusters of 5 to 50, generally yellow or orange in color. The alligator-shaped larvae are blue-black and orange-spotted.

MICE

Mice and rats are repelled by fresh or dried leaves and the oils of mints, by camphor, and by pitch pine. Mothballs repel rabbits as well as mice but should not be used where food crops are grown or where children can pick them up.

Sea onions, white lavender, wormwood, and spurge repel mice, while everlasting pea is useful against field mice, and leaves of dwarf elder protect against mice in granaries.

If mice like your garden too much, repel them by planting of daffodils, narcissuses, scillas, or grape hyacinths.

MILK

Cows and goats give more and richer milk when fed on stinging nettle hay or members of the Umbelliferae family. We always saved discarded carrots and carrot tops when we kept milk goats, which also liked prunings from rosebushes if the thorns were not too prominent.

Skim milk may be used as a spray on tobacco and other plants subject to mosaic virus, and on peppers and tomatoes grown in the greenhouse. Pickers in commercial pepper and tomato plots find it useful to dip their hands in skim milk to avoid transmitting the mosaic virus, while whey proteins are effective, too.

A milk and blood spray has been used in orchards to control fungi, and a milk and coal tar mixture against chinch bug.

Sour milk or buttermilk may be sprinkled over cabbages against cabbageworms.

MILKY SPORE DISEASE (*Bacillus popillae*)

This widely used bacterial organism gives protection against the Japanese beetle by producing a fatal disease in the grub. Because it brings about an abnormal white coloring in the insect, it is called "milky."

It was developed in 1933 when a field survey in central New Jersey discovered a few abnormally white Japanese beetle grubs, which examination showed were teeming with bacterial spore.

This disease was studied by the Bureau of Entomology and Plant Quarantine, where attempts were made to culture it for release. Treatment with the milky spore on experimental plots showed a more than 90 percent mortality in 2 months. In areas where the disease had been established naturally, equally high kills were found. This disease, occurring naturally in Japan, has kept the beetle from becoming a serious problem there.

The spore ordinarily needs only one application and then continues and spreads itself. It can be applied to the soil at any time except on a windy day or when the ground is frozen, but it is best to treat mowed or cropped areas. Apply 1 teaspoonful of the spore disease powder on grass or sod spots 3 to 4 feet apart and in rows the same distance apart. The beetle grub feeding in the soil will then take in the bacteria spores.

When a healthy grub becomes infected, the spores give rise to slender vegetative rods, which multiply in the blood by repeated divisions. In a short time the normally translucent blood of the grub will become milky in appearance, and eventually the grub dies. The spores stored in the body cavity then are released in the soil and are taken up by other grubs that in turn become infected. As the cycle continues, the spores increase in number and, since more grubs are killed, fewer and fewer adult beetles emerge to feed on crops. The disease is cumulative and self-perpetuating.

Milky spore disease is sold under several trade names. (See Sources on page 442.)

Neem oil is being increasingly used as a Japanese beetle control with considerable success. (See *Insecticides, Botanical* in this chapter.)

Less effective but still useful methods of combating the Japanese beetle include companion plantings of geraniums among roses and grapes to drive the beetles away. Larkspur eaten by the beetles will kill them, while soybeans work as a trap crop. The beetles are rarely destructive to cabbage, carrots, cauliflower, eggplant, onions, lettuce, parsley, peas, potatoes, radishes, spinach, squash, sweet potatoes, tomatoes, and turnips.

MOLASSES GRASS (*Melinis minutiflora*)

This is one of nature's very own insect traps. The covering of glandular hairs exudes a viscous oil capable of trapping small pests such as ticks. It does not kill them, but stops them from crawling upward to come in contact with an animal. Cattle pastured on this grass in Guatemala were found to be free of ticks within a year, and it has been planted in Florida with good results. At the same time it provides good forage for cattle, which like to graze on it. It also repels mosquitoes and the tsetse fly.

MOLES (*Scalopus*)

Moles generally are considered a nuisance, and they do consume beneficial creatures such as earthworms, but they also feed on Japanese beetle grubs. They are deterred by a few plants of caper spurge (*Euphorbia lathyris*) strategically placed, by daffodil bulbs, and by castor bean plants. Thorny twigs of raspberry, rose, hawthorn, or mesquite pushed into burrows will scratch the moles and cause them to bleed to death.

MOSQUITOES

Garlic-based oil (see *Garlic* in the Herbs chapter) is effective in killing mosquito larvae in ponds. Myristicin, a synthesized compound found in parsnips, also is effective as a selective insecticide against the larvae, as is *Bacillus thuringiensis*.

The leaves of molasses grass (*Melinis minutiflora*) and sassafras are mosquito repellents, and I find that castor beans planted around my garden make it more pleasant to work there in the long, cool western twilights when I do most of my garden work.

Euell Gibbons says that American pennyroyal (*Hedeoma pulegioides*), sometimes called squaw mint and a native American plant not to be confused with the European pennyroyal (*Mentha pulegium*), is a natural insect repellent. A handful crushed and rubbed on the skin will not only emit a pleasant smell but also repel mosquitoes and gnats.

MOTHS AND MILLERS

If moths and millers are troublesome in fruit trees, place 1 cup molasses in 1½ cups water and hang in small buckets or cans from the limbs. Remove the insects occasionally or make new solutions. This remedy is particularly helpful if used in peach trees.

NEMATODES

See *Eelworms* in this chapter.

OIL SPRAYS

Used properly and at the right time, dormant oil sprays are effective, particularly in orchards, against many chewing and sucking insects. The oils make a tight film over insect eggs, causing suffocation.

Apply heavy dormant oil sprays only in early spring over leafless trees. However, lighter, more refined oils have been developed in recent years and can be used at other stages of growth.

PEPPER, HOT (*Capsicum frutescens, C. annuum*)

Hot red peppers are among the most useful plants in the garden as well as the most flavorsome.

Grind hot peppers and mix with water and a little powdered real soap to make an infusion for spraying plants infested with aphids. Dry cayenne pepper may be dusted on tomato plants attacked by caterpillars.

However, if you note long green hornworms on tomatoes, do not be too quick to spray. Watch to see if any of the tiny parasitic wasps that build noticeable white cocoons all over tomato hornworms are present. Do not harm these predators with sprays—they may be doing your work for you.

Ground red peppers placed around eggplants and rubbed on the leaves will help repel eggplant pests, and the dried pods, pulverized and sprinkled on corn silk, will give protection against raccoons.

Another all-purpose spray may be made of ground pepper pods, onions, and a bulb of garlic. Cover this mash with water, let stand 24 hours, and strain. Add enough water to make 1 gallon of spray. Use several times daily on roses, azaleas, chrysanthemums, or beans to hold down serious infestations. Do not throw away the mash, but bury the residue among the plants where insects occur.

Protect your hands with gloves when you work with hot peppers, and be very careful not to get the juice in your eyes.

PRAYING MANTIS
(*Mantidae*)

Despite its ferocious appearance, this is one of our most beneficial insects, and it will not harm any vegetation in the garden, dining only on other insects.

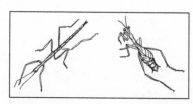

Walking stick   *Praying mantis*

When young it eats the soft-bodied insects, the cutting and sucking aphids and leafhoppers; when fully mature it kills and eats chinch bugs, crickets, locusts, bees, wasps, beetles, flies, spiders, tent caterpillars, and many others.

Though mantids usually are found in warm countries, the common European mantid can live in the northern United States. A full-grown large mantid varies from 2 to 5 inches in length, depending on the kind, yet it will easily escape notice because in form and color it closely resembles the plants on which it rests.

Female mantids lay their eggs in masses, gluing them to trees and shrubs with a sticky substance from their bodies. The eggs remain there all winter, and if you carefully examine thorny bushes, brush, hedges, and berry bushes in autumn after leaves have fallen, you may find the hardened froth mass egg cases. You can collect some from marshes and waste areas for your own garden, but never strip the area clean.

To place them properly, allow one egg case for each major shrub or about four for each quarter acre (without shrubbery). Select a sheltered spot and tie or tape the cases securely about 2 to 4 feet above the ground.

Early in spring the mantis will start to aid you, emerging in bright sunshine usually from early May to late June, just at the time when a large variety of insect fare is likely to be available.

Each creature is fastened to the egg by a tenuous cord that it must break. After doing so it will drop and then climb to surrounding foliage. Mantids are poor flyers and slow movers and usually remain in the area as long as they continue to find food. And once introduced, they are likely to multiply and extend control. Though many of the young may not live, still enough survive to perpetuate themselves. You can obtain mantid egg cases through the mail, usually between November and early May. (See Sources on page 442.)

RABBIT (*Oryctolagus cuniculus*)

Onions are repellent to rabbits and may be interplanted with cabbage and lettuce. An old garden hose cut in lengths of a few feet each and arranged to look like snakes will frighten rabbits away.

Animal fat can be painted on young fruit trees, and a thin line of dried blood or blood meal sprinkled around the garden often acts as a repellent.

Try a dusting of aloes on young plants, or shake wood ashes, ground limestone, or cayenne peppers on plants while they are wet with dew.

RACCOON

(*Procyon lotor*)

Farmers have been planting pumpkins and corn together for centuries to discourage raccoons. Put the pumpkin seeds about 4 feet apart. As the corn approaches maturity, the big, wide pumpkin leaves will grow around the stalks. It is believed that the coons will not come into the corn

Repellents for the corn-loving raccoon include nearby cucumber, melon, pumpkin, or squash vines. Red or black pepper on the corn silk also may help.

rows because they like to be able to stand up and look around while they eat, and the big leaves make that impossible.

Other methods to control raccoons include sprinkling black or red pepper on the corn silks. This does not affect the taste of the corn.

For a small planting, a wire "corn cage" or an electric fence is sure protection. If a 6- or 12-volt battery is used, the fence will be harmless. It need be turned on only at night, since raccoons sleep during the day. Another solution is to use a small transistor radio (enclosed in a plastic bag to protect it from rain or dew), placing it in the patch at ripening time. Turn it on at night.

ROOT KNOT

If the fresh tops of asparagus are crushed and steeped in water, the solution may offer protection to vegetables where root knot, stubby root, and meadow nematodes are suspected. Marigolds are also useful against meadow nematodes. (See *Eelworms* in this chapter.)

ROOT MAGGOT

Root maggots may be repelled by a mulch of oak leaves.

SALT, COMMON (*Sodium chloride*)

Salt has a damaging effect on most plants and should not be used except where the soil is to be rested for a time afterward. It is useful to kill Canada thistle or quack grass and will give best results if put on after the

weeds are freshly cut. Apply several times, taking care to do so only in dry weather. Salt put on slugs will dissolve them.

SKUNK

The skunk's protection, butyl mercaptan, can be neutralized with tomato juice. Skunks have some benefit in keeping large insects and mice in check.

Skunks will visit yards at night to feed on beetle grubs.

SLUGS AND SNAILS (*Agriolimax campestris* and *Helicidae*)

These beasties cause most of their damage at night in places where the ground is damp. They are topers: They love beer and will drown in saucers of it. Even empty beer cans placed in the garden attract them by their odor. They also love honey and will drown in saucers of that. What a way to go!

Since ordinary table salt will dissolve slugs, I find it handy to carry a small salt shaker in my pocket to use when I spot them. I feel no compunction, for these slimy creatures will destroy my choicest cabbage and lettuce heads if they get a chance.

Tobacco stem meal discourages slugs, and white hellebore controls them on grapes and cherries. They dislike tanbark or oak leaves, and wormwood will repel them by its bitterness. Snails are reluctant to cross lines of ashes or hydrated lime.

SOAP

In using most sprays made from botanicals it is a good idea to add a small amount of real soap to help the materials adhere. Washing plants down with just plain soapy water is an excellent practice in many instances, for soap appears to have antiseptic qualities useful against many plant diseases. Insecticidal soaps, which can be useful against a variety of insects, are available from garden centers and by mail order (see Sources on page 442).

SPIDER (*Arachnida*)

Some mites and spiders are natural predators, our valuable allies that dine on many destructive insects. One type of predaceous mite is used to control plant-feeding insects on avocado and citrus crops.

The garden spider is one of the best helpers to the vegetable grower.

SPIDER MITE (*Acarina*)

Spider mites, also called red spiders, are apt to show up suddenly in hot weather. They sometimes can be removed from plants by spraying forcibly with a stream of plain water, and once dislodged from a plant they seldom return. A 3 percent oil spray, pyrethrum dust, or a spray of onions and hot peppers may be used, and I have found that garlic will repel the mites on tomato plants, while ladybugs are their natural enemies.

TEA LEAVES

Mix tea leaves with radish and carrot seed to prevent maggots.

TERMITE (*Termitidae*)

Since most termites cannot live without water, cutting off the source of moisture usually kills them. Silica aerogel is effective, and extracts of chinaberry will kill about 98 percent of them.

TOAD (*Bufo*)

Both toads and frogs consume many insects, one toad being able to eat up to 10,000 insects in three months' time—and many of these will be cutworms. Other pests eaten by toads include crickets, grubs, rose chafers,

Horned toads, like true toads, are great friends of the gardener, devouring great numbers of insects. Help them find shelter in your garden area.

rose beetles, caterpillars, ants, squash bugs, sow (pill) bugs, potato beetles, moths, mosquitoes, flies, slugs, and even moles.

Toads do *not* cause warts and are not poisonous to humans, though they exude a slime distasteful to their enemies. If you want to secure some toads for your garden, look around the edges of swamps and ponds in spring. Once secured, they need shelter. A clay flowerpot with a small hole broken out of the side will serve for housing if buried a few inches in the ground in a shady place. They also need a shallow pan of water and, if your garden is not fenced, protection from dogs and other creatures.

In the Southwest, we use the horned toads (really lizards) in the same way. They can withstand more heat and thrive in my garden, apparently living on dew and insects. They bury themselves in the ground during winter, emerging when the weather warms up.

TOBACCO

See *Nicotine* under *Insecticides, Botanical,* in this chapter.

TRAP CROPS

See *Insect Attractants, Botanical,* in this chapter.

WHITEFLY (*Trialeurodes vaporariorum*)

Whitefly is one of the insects known to thrive on certain shortages of minerals in the soil, and experiments have shown that greenhouse whiteflies attack tomatoes only when phosphorus or magnesium is deficient.

Botanical controls include nasturtium, particularly good to grow in greenhouses with tomatoes. Oak leaves burned in a greenhouse for a half-hour period are helpful, and nicotine may be used as a spray.

Whitefly can be controlled biologically by a small parasite called *Encarsia formosa* (see Sources on page 442). Ladybugs also are a control, as are aphis lions (the larvae of lacewing flies).

WHITE HELLEBORE (*Veratrum album*)

White hellebore is a member of the Lily family whose roots and rhizomes contain insecticidal substances. It becomes less toxic on exposure to light and air and has little residual effect, making it less poisonous to use than more persistent materials.

The use of white hellebore is centuries old. The Greeks mixed it with milk to kill flies, and it was a favorite remedy of the Romans against mice and rats. Today it is used to control many leaf-eating insects such as sawflies, which attack ripening fruit, and also for slugs and cabbageworms.

WOOD ASHES

Wood ashes sprinkled around the base of cauliflower and onion plants are a popular remedy to control maggots. They are also used against clubroot, red spider, bean beetles, and scab on beets and turnips, as well as aphids on peas and lettuce. They are good, too, around such plants as corn, which need to develop strong stalks.

A thin paste made of wood ashes and water and painted on the trunks is used to control tree borers. A handful each of wood ashes and hydrated lime, diluted with 2 gallons of water, makes a spray for the upper and lower sides of cucurbit leaves to control cucumber beetles.

WOODCHUCK (*Marmota monax*)

If woodchucks are a problem, spray the plants they are nibbling with a solution of water and hot pepper.

POISONOUS PLANTS

Humans perhaps were given dominion over the birds and beasts and the lilies of the field (including thorn apple and castor bean) because they have (or should have) the ability to distinguish the good from the bad.

Since the beginning of time, people the world over have lived close to hundreds of plants that can cause irritation, illness, or even death. A few are seriously poisonous; a far greater number are moderately so, producing varying degrees of illness or irritation. Some cause dermatitis, hay fever, or other illnesses as a result of the allergic sensitivity rather than the direct toxicity of the plant.

It would be virtually impossible to ban every plant that is inedible, drug-producing, or even irritating to the skin. Moreover, many such plants are the source of valuable medicines or serve us well as natural insecticides when wisely and carefully used in our gardens.

Children should be taught which plants are harmful and which may be used or eaten or touched with safety. Caution the child old enough to understand about eating wild berries, fruits, and nuts or chewing on bark, branches, or stems. Watch younger children just as you keep them from running into the street or keep dangerous household preparations out of reach. For poisonous plants, whether present in your own garden or in fields nearby, *will* be encountered.

Dozens of these plants are attractive garden flowers or shrubs, greatly prized for their beauty or usefulness in landscaping, so many adults and most children do not realize their poisonous characteristics. Who would think, for instance, that daffodils are dangerous? Nevertheless, the bulbs can be fatal if eaten. The beautifully marked castor beans are

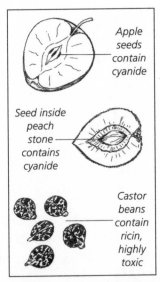

Apple seeds contain cyanide

Seed inside peach stone contains cyanide

Castor beans contain ricin, highly toxic

Apples, peaches, and castor beans contain highly poisonous parts.

a natural plaything for a small child, and little children often put things into their mouths. Yet for this poison there is no available antidote.

Even older children have been known to try the "nuts" inside a peach stone. How do I know? Because as a child I tried one of these seeds. They resemble almonds, which I dearly loved. I didn't know that they contain cyanic acid and that a few of them could kill me. Fortunately, the bitter flavor stopped me after the first bite. Parents can't think of everything to tell children not to do, so it is indeed fortunate that many dangerous plants do have an unpleasant taste.

But if my mother didn't know about the peach stones (and I kept my experiment a secret), there wasn't much else she didn't know about. She was a skilled herbalist, and I grew up with the salves and tonics she prepared and brewed from the plants she gathered.

She often told me of her adventures in the Indiana woods where she went to gather ginseng. Later, when she and my father were married, she transferred her herb-gathering activities to Kentucky, and still later, when the family came out to Oklahoma, then Indian territory, she literally found new fields to conquer. She learned much herbal lore from Native Americans, who gave her the name Hill-lea-tah-ha, meaning "one-who-thinks-good." In those days when there were few doctors and most people made their own medicines, her knowledge was greatly respected. And she was fortunate to be completely unaffected by poison ivy, an immunity that I too am happy to have.

To this day I still grow many of the plants my mother brought in from the woodland to use for their therapeutic or insecticidal properties. Until recently I did not think of them as "companion plants," though I was clearly aware of their effect when grown in proximity to flowers, fruits, and vegetables.

While I have known since childhood most of the plants likely to be encountered in garden or woodland, there are many things about plants that I am still learning. For example, I now know that some plants require high soil concentrations of selenium in order to live, drawing the selenium into their leaves and stems as they grow. When they die, the selenium is left on the surface of the soil, where it is absorbed by other plants that grow there. These plants, which do not require selenium to grow, nevertheless absorb enough of the substance to make them toxic to grazing animals.

I have no degree in botany, nor am I an expert on poisonous plants. Neither shall I try to cover *all* of them in this brief chapter. Excellent books are available for those who are interested in pursuing the subject in greater detail. Here are only those plants that are most likely to be found in flower or vegetable gardens, or that may arrive, blown in by the wind or other means, as weeds. I have noted with a **D** those plants likely to cause dermatitis, a susceptibility that of course varies from person to person.

Poisonous Plants

Scientific Name	Common Name	Poisonous Part of Plant
Abrus precatorius	Rosary pea	One rosary pea seed causes death
Acokanthera spp.	Bushman's poison	All parts very poisonous
Aconitum spp.	Monkshood	All parts, especially roots and seeds, very poisonous; characterized by digestive upset and nervous excitement
Adonis aestivalis	Summer adonis	Leaves and stem
Aesculus spp.	Horse chestnut, buckeye	Leaves and fruit
Ailanthus altissima	Tree of heaven	Leaves, flowers; **D**
Alstroemeria spp.	Peruvian lily, lily of the Incas	All parts
Amaryllis belladonna	Belladonna lily, naked lady	Bulbs
Anemone patens	Pasqueflower	Young plants and flowers
Arisaema triphyllum	Jack-in-the-pulpit	All parts, especially roots (like dumb cane), contain small, needle-like crystals of calcium oxalate that cause intense irritation and burning of the mouth and tongue
Asclepias spp.	Milkweed	Leaves and stems
Asparagus officinalis	Asparagus	Spears contain mercaptan, which may cause kidney irritation if eaten in large amounts; young stems; **D**
Baileya multiradiata	Desert marigold	Whole plant
Buxus sempervirens	Common box	Leaves; **D**

Scientific Name	Common Name	Poisonous Part of Plant
Cephalanthus occidentalis	Buttonbush	Leaves
Cestrum spp.	Cestrum, night-blooming jessamine	Leafy shoots
Cicuta	Water hemlock	All parts fatal, causing violent and painful convulsions
Clematis vitalba	Traveler's joy	Leaves
Colchicum autumnale	Autumn crocus, meadow saffron	Leaves and bulbs are very poisonous
Conium maculatum	Poison hemlock	All parts
Convallaria majalis	Lily of the valley	Leaves and flowers very poisonous
Corynocarpus laevigatus	Karaka nut	Seeds
Crinum asiaticum	Crinum lily, poison bulb	Bulbs
Crotalaria agatiflora	Canary bird bush	Seeds
Cypripedium spp.	Lady's slipper orchid	Hairy stems and leaves; **D**
Cytisus scoparius	Scotch broom	Seeds, leaves, twigs (stems)
Daphne spp.	Daphne	Bark, leaves, and fruit fatal; a few berries can kill a child
Datura spp.	Angel's trumpet, thorn apple, jimsonweed	All parts cause abnormal thirst, distorted sight, delirum, incoherence, and coma; has proved fatal
Delphinium spp.	Larkspur, delphinium	Young plants and seeds cause digestive upset; may be fatal
Dicentra spp.	Dutchman's-breeches, bleeding heart	Leaves and tubers may be poisonous in large amounts; fatal to cattle

The handsome lily of the valley has poisons in both leaves and flowers.

The digitalis in foxglove is a powerful heart stimulant. It may be fatal if the leaves are eaten.

Scientific Name	Common Name	Poisonous Part of Plant
Dieffenbachia spp.	Dumb cane	Stems and leaves cause intense burning and irritation of mouth and tongue; death can occur if base of tongue swells enough to block the throat
Digitalis purpurea	Foxglove	Leaves cause digestive upset and mental confusion; may be fatal in large amounts
Duranta repens	Golden dewdrop	Fruits and leaves
Eichium vulgare	Blue weed	Leaves and stems; **D**
Euonymus spp.	Burning bush, spindle tree, strawberry bush	All parts
Eupatorium rugosum	White snakeroot	Leaves and stems
Euphorbia spp.	Euphorbia, snow-on-the-mountain, poinsettia	Milky sap; **D**
Ficus spp.	Fig	Milky sap; **D**
Gelsemium sempervirens	Yellow jessamine	Flowers and leaves; roots; **D**
Ginkgo biloba	Ginkgo, maidenhair tree	Fruit juice; **D**
Glechoma hederacea	Ground ivy	Leaves and stems
Gloriosa spp.	Climbing lily	All parts
Hedera helix	English ivy	Leaves and berries
Helenium spp.	Sneezeweed	Whole plant
Helleborus niger	Christmas rose	Rootstocks and leaves; **D**
Heracleum lanatum	Cow parsnip	Leaves and root slightly poisonous; dangerous to cattle
Heteromeles arbutifolia	Toyon, Christmas berry	Leaves

Scientific Name	Common Name	Poisonous Part of Plant
Hyacinthus	Hyacinth	Bulb causes nausea, vomiting; may be fatal
Hydrangea macrophylla	Hydrangea	Leaves
Hymenocallis littoralis	Spider lily	Bulbs
Hypericum perforatum	St.-John's-wort	All parts when eaten; **D**
Ilex aquifolium	English holly	Berries
Impatiens spp.	Impatiens	Young stems and leaves
Ipomoea alba	Moonflower	Seeds
Iris spp.	Iris	Rhizomes; **D**; if eaten causes digestive upset but not usually serious
Jasminum	Jessamine	Berries may be fatal
Juglans spp.	Walnut	Green hull juice; **D**
Kalmia latifolia	Mountain laurel	Leaves
Laburnum anagyroides	Golden chain	Leaves and seeds cause severe poisoning; may be fatal
Lantana spp.	Lantana	Foliage and green berries may be fatal
Laurus	Laurel	All parts fatal
Leucothoe fontanesiana	Drooping leucothoe, fetterbush	Leaves and nectar from flowers
Ligustrum spp.	Privet	Leaves and berries
Linum usitatissimum	Flax	Whole plant, especially immature seedpods
Lobelia spp.	Lobelia	Leaves, stems, and fruit; **D**
Lupinus spp.	Lupine	Leaves, pods, and especially seeds
Lycium halimifolium	Matrimony vine	Leaves and young shoots
Macadamia ternifolia	Queensland nut	Young leaves
Maclura pomifera	Osage orange	Milky sap; **D**
Malus	Apple	Seeds contain cyanic acid
Melia azedarach	Chinaberry	Fruit, flowers, and bark
Menispermum	Moonseed	Berries (resemble small wild grapes) may be fatal if eaten
Myoporum laetum	Ngaio	Leaves very poisonous
Narcissus spp.	Narcissus, daffodil	Bulbs cause nausea, vomiting; may be fatal

D = *those plants likely to cause dermatitis (susceptibility varies from person to person)*

Scientific Name	Common Name	Poisonous Part of Plant
Nerium oleander	Oleander	All parts extremely poisonous, affect the heart
Nicotiana spp.	Tobacco	Foliage
Ornithogalum umbellatum	Star-of-Bethlehem	All parts cause vomiting and nervous excitement
Oxalis pes-caprae	Bermuda buttercup	Leaves
Papaver somniferum	Opium poppy	Unripe seedpod very poisonous
Pastinaca sativa	Parsnip	Hairs on leaves and stems; **D**
Philodendron spp.	Philodendron	Stems and leaves
Phoradendron spp.	Common mistletoe	Berries fatal
Pittosporum spp.	Pittosporum	Leaves, stem, and fruit very poisonous
Podophyllum	Mayapple	Apple, foliage, and roots contain at least 16 active toxic principals, primarily in the roots; children often eat the apple with no ill effects, but several may cause diarrhea
Primula spp.	Primrose	Leaves and stems; **D**
Prunus spp.	Cherry, peach, plum	Seeds and leaves; seeds contain cyanic acid
Quercus	Oak	Foliage and acorns; takes a large amount to poison
Ranunculus spp.	Buttercup	Leaves; **D**; if eaten, irritant juices may severely injure the digestive system
Rhamnus spp.	Coffeeberry, buckthorn	Sap and fruit; **D**
Rheum rhabarbarum	Rhubarb	Leaves; **D**; large amounts of raw or cooked leaves (contain oxalic acid) can cause convulsions, coma, followed by death

Mistletoe usually won't be touched by animals, but it is poisonous. However, birds eat the berries without harm.

All parts of the lovely rhododendron are very toxic if eaten.

Scientific Name	Common Name	Poisonous Part of Plant
Rhododendron spp.	Rhododendron, azalea	Leaves and all parts may be fatal
Toxicodendron pubescens	Poison oak	Leaves; **D**
Toxicodendron radicans	Poison ivy	Leaves; **D**
Ricinus communis	Castor bean	Seeds fatal
Robinia pseudoacacia	Black locust	Young shoots, bark, and seeds
Rumex acetosa	Sour dock	Leaves
Sambucus canadensis	Elderberry	Shoots, leaves, and bark; children have been poisoned by using the pithy stems for blowguns
Saponaria vaccaria	Cow cockle	Seeds
Scilla spp.	Siberian squill, autumn scilla	All parts
Senecio mikanioides	German ivy	Leaves and stems
Solanum dulcamara	European bittersweet	Leaves and berries
Solanum nigrum	Black or common night shade, poison berry	Berries
Solanum pseudocapsicum	Jerusalem cherry	Fruit
Solanum tuberosum	Irish potato	Sprouts, tubers with green skin
Tanacetum vulgare	Common tansy	Leaves
Taxus species and cultivars	Yew	Foliage, bark, and seeds fatal; foliage more toxic than berries
Thevetia peruviana	Yellow oleander	All parts
Urtica spp.	Nettles	Leaves; **D**
Veronicastrum virginicum	Culver's root	Roots
Wisteria	Wisteria	Mild to severe digestive upset; children sometimes poisoned by this plant
Zephyranthes spp.	Zephyr lily	Leaves and bulbs

D = *those plants likely to cause dermatitis (susceptibility varies from person to person)*

GARDEN PLANS

This chapter contains an assortment of garden plans to suit the different needs of different gardeners.

A Model Companion Garden

Planning your first organic garden may be a little tricky, but in time and with experience truly good gardeners get to know plants' likes and dislikes (which may sometimes vary in different localities), and they can spot just the place in their garden for any plant where it will do best.

Where it is at all possible, try to use double-purpose plants—those that are both decorative and sources of food, or those that are protective of other plants—in your garden.

Planning your garden on paper is one of the best ways to achieve success. Use large sheets ruled off into squares, using a scale of 2 feet to 1 inch. Draw large plants the size they will be at maturity. Locate existing features, such as trees, shrubs, hedges, buildings, and paths. This is important, as they may shade the garden during certain hours of the day. Mark off north and south on your plan. Then add each item you want in your garden, considering them in their order of importance to you and your family and the space and time you can give to them.

For your vegetable garden, plan on using and replacing early vegetables with those that do well in spring and summer. In turn these will be replaced with cool-season vegetables for fall.

Keep records, for you will probably want to change your plan from season to season, including some new vegetables or different varieties and discarding others that may not have done well. Note how well certain plants did on certain sites, the effect of their form and color, and the rotation of your crops. Rotation is very important in garden plots that are planted year after year.

The heavy feeders (those that need generous fertilizing, including corn, tomatoes, cabbage, and other brassicas) should be followed by legumes to help the soil recover from the demands of the heavy feeders, especially in

160

A MODEL COMPANION GARDEN

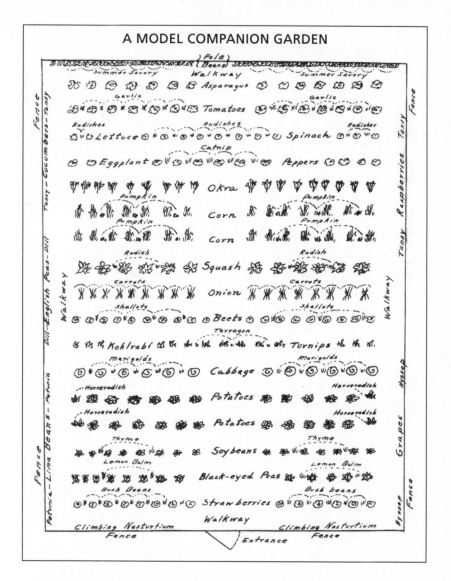

poorer soils. In a live humus soil legumes may go in third place if desired. The last group are the light feeders, which include root vegetables, bulbs, herbs, and protective flowers such as marigolds and nasturtiums.

For the serious gardener a complete record of space used and treated in a vegetable garden is an invaluable aid in planning the next year's garden.

It is not enough to place the high vegetables to the north, the corn in blocks, and the asparagus where the tractor will not run it down. A map showing the pH factor in various garden areas should be made and kept up-to-date. Application dates of slowly available materials, such as rock phosphates, should be recorded. With such, you can make a long-range plan of your garden that will serve you well in future years.

Few garden projects are more fun to experiment with than companion planting. Good plant neighbors are those that occupy different soil levels, like beets and kohlrabi, or that find in each other's company the light require-ments that best suit them. An example of this is the compatibility of celery and leek: The upright leek finds room near the bushy celery plant. Other compatible combinations are cabbage and beans, beets and onions, cucum-bers and sweet corn, carrots and peas, cucumbers and beans, early potatoes and corn, early potatoes and horseradish, tomatoes and parsley. Asparagus beetles are repelled by tomatoes growing nearby. To repel the cabbage maggot, plant mint, tomato, rosemary, or sage in the next row. Radish will grow extra well and especially tender near lettuce.

On the other hand, avoid the incompatible combinations such as sun-flowers and potatoes, tomatoes and fennel, tomatoes and kohlrabi, pole beans and beets, pole beans and kohlrabi. Note, however, that localities and seasons also seem to affect "companions." More than one gardener has gleefully told me she planted onions with peas or beans and had a wonderful crop of both.

We are sometimes surprised when we go out into our garden in the morning to find that the bugs have arrived overnight. Timing can discourage these unwelcome guests. I find squash planted in midsummer is entirely free of squash bugs. Onions around spring-planted squash help keep insects away. Aromatic herbs as border plants are very helpful to the garden.

Page 161 shows a sample companion garden like my own. The fence is used for climbers such as beans, cucumbers, peas, and grapes, and walk-ways around the inside of the fence make it easy to use a tiller.

Flowers add beauty to the garden but are chosen as well for their benefi-cial influences on crop vegetables and fruits. Note vegetable interplantings, too, in most of the rows. Perennials such as asparagus and horseradish are placed at the outsides to make tilling of the full garden easier.

The Weekend Garden

The primary object of the weekend garden is to get the "mostes' result with the leastes' effort." Remember that garden space may also be saved by doubling rows of narrow-growing crops. Thus, onions, which take only as much space as the diameter of the bulb, may be grown in double rows 6 inches apart. Mulch may be arranged between the two rows, when the plants are large enough, to cut down the amount of hand-weeding, if necessary. Carrots, beets, parsnips, and turnips may be handled in the same way. Cos or Romaine lettuce, which grows elegantly tall, is also a space saver and may be grown close together. Bush varieties of squash and cucumbers and beans also take less space.

Annual vegetables that like sole possession of their garden space for the entire summer include lima beans, Swiss chard, cucumbers, eggplant, okra, onions, parsley, parsnips, peppers, sweet potatoes, late white potatoes, salsify, squash—both summer and winter—New Zealand spinach (grows well in hot weather), and tomatoes.

Just because you are a weekend gardener doesn't mean that you cannot enjoy some of the perennial vegetables as well, if you are willing to plan for them and give them their own space. Asparagus, chives, horseradish, perennial onions, rhubarb, perennial herbs, and berries are possible, and most of these can stand a certain amount of benign neglect. They must, of course, be located where power tillage or cultivating tools will not be inconvenienced by them. Usually it's best to give them a plot of their own, removed from the rest of the garden.

Study your garden catalogs for small, space-saving varieties, even dwarf vegetables, as well as for early, midseason, and late varieties. If you have a gardening friend or neighbor, consult him as to the best varieties for growing in your area. Don't be shy—gardeners just love to give advice!

Cool-Season Vegetables

Plant	Companion Plants	Hindered by
Beets	Lettuce, onions, cabbage	Pole beans
Broccoli	Beets, potatoes, onions, celery	Strawberries, tomatoes
Cabbage	Onions, potatoes, celery, mint	Strawberries, tomatoes

THE WEEKEND GARDEN

COOL-SEASON VEGETABLES

Plant	COMPANION PLANT	HINDERED BY
BEETS	LETTUCE, ONIONS, CABBAGE	POLE BEANS
BROCCOLI	BEETS, POTATOES, ONIONS, CELERY	STRAWBERRIES, TOMATOES
CABBAGE	ONIONS, POTATOES, CELERY	"
CARROTS	peas, lettuce, chives, radishes, leeks, onions	DILL
CAULIFLOWER	potatoes, onions, celery	STRAWBERRIES, TOMATOES
CHARD	LETTUCE, ONIONS, CABBAGE	POLE BEANS
KALE overwinters	LATE CABBAGE, POTATOES	WILD MUSTARD
LETTUCE	RADISHES, STRAWBERRIES, CUCUMBERS	
ONIONS	SUMMER SAVORY, camomile	peas and beans
ORIENTAL GREENS	STRAWBERRIES	
PEAS	CARROTS, TURNIPS, RADISHES, CUCUMBERS, AROMATIC HERBS	
RADISH	redroot pigweed, NASTURTIUMS, MUSTARDS	all cole plants
SPINACH	STRAWBERRIES	
TURNIPS	PEAS, hairy vetch	hedge mustard, knotweed

WARM-SEASON VEGETABLES

Plant		HINDERED BY
BEANS, BUSH	cucumbers, strawberries, plant with corn	ONIONS
CHILI PEPPERS		
CORN	potatoes, peas, beans, cucumbers, pumpkin	TOMATOES
EGGPLANT	Redroot pigweed, green beans	
MELONS	MORNING GLORY	potatoes
OKRA	hot peppers, eggplant	
SQUASH	icicle radishes, nasturtiums	
SWEET PEPPER	basil, okra	
TOMATILLO	basil	
TOMATO	chives, onion, parsley, basil, marigold, carrot	

Do not plant next to corn.

Plant	Companion Plants	Hindered by
Carrots	Peas, lettuce, chives, radishes, leeks, onions	Dill
Cauliflower	Potatoes, onions, celery	Strawberries, tomatoes
Chard	Lettuce, onions, cabbage	Pole beans
Chinese cabbage	Bush beans, marigolds, onions, sage	Pole beans, corn, strawberries, tomatoes
Kale (overwinters)	Late cabbage, potatoes, sage, marigolds, nasturtiums	Wild mustard

Plant	Companion Plants	Hindered by
Kohlrabi	Beets, cucumbers, onions, sage	Pole beans, strawberries
Leeks	Carrots, celery, onions	Bush beans, pole beans, peas, soybeans
Lettuce	Radishes, strawberries, cucumbers	Pole beans, strawberries, tomatoes
Onions	Summer savory, chamomile	Peas and beans
Oriental greens	Strawberries	
Parsley	Tomatoes	
Peas	Carrots, turnips, radishes, cucumbers, aromatic herbs	
Radish	Redroot pigweed, nasturtiums, mustards	All cole plants
Spinach	Strawberries	
Turnips	Peas, hairy vetch	Hedge mustard

Warm-Season Vegetables

Plant	Companion Plants	Hindered by
Beans, bush	Cucumbers, strawberries, plant with corn	Onions
Chile peppers		
Corn	Potatoes, peas, beans, cucumbers, pumpkins	Tomatoes
Cucumber	Bush beans, pole beans, nasturtiums, corn, leeks, onions, peas, radishes, sunflowers	Potatoes
Eggplant	Redroot pigweed, green beans	
Melons	Morning glory	Potatoes
Okra	Bell peppers, eggplant	
Potatoes (early or late)	Bush beans, cabbage, corn, eggplant, marigolds, nasturtiums	Cucumbers, pumpkin, squash, sunflowers, tomatoes
Squash	Icicle radishes, nasturtiums	
Sweet pepper	Basil, okra	
Tomatillo	Basil	
Tomatoes	Asparagus, carrots, celery, chives, garlic, onions, parsley	Corn, potatoes

Perennial Vegetables

Plant	Companion Plants	Hindered by
Asparagus	Basil, parsley, tomatoes	—
Chives	Carrots	Bush beans, potatoes, peas, soybeans
Horseradish	Protective to potatoes	—
Perennial onions or shallots	Radish	—
Rhubarb	Seldom troubled by insects or disease	Cut off flowering growth as it reduces vigor

If your schedule allows you to garden only one or two days per week, here are some tips.

• Plan a small garden in the first year.

• Prepare your garden soil in the previous fall, incorporating compost.

• Grow the vegetables your family likes best, choosing varieties that produce well in your area.

• Interplant some companion plants in each row.

• Plan for a watering system with a timer for use when rainfall is insufficient. Use a soaker hose in the garden.

• Harvest often to keep plants producing (peas, beans, tomatoes, peppers, okra, for example).

• Pull spent plants and replace with warm-weather plants either started indoors or purchased.

• As the season advances, start a fall garden in the shade of cornstalks, etc., and pull when young plants are well started.

• Mulch plants to cut down on weeding. Plow in mulch in late fall as preparation for early-spring planting.

The Postage Stamp Garden

Whether the postage stamp garden is tucked away in the corner of a city backyard or blossoms in the rarefied atmosphere of a penthouse far above the traffic of the streets, it can bring the same rewards and satisfaction to its creator as a larger one of more impressive proportions. The small back-yard garden on the ground may be cared for much the same as a larger garden would be and with good soil, water, and sufficient sunlight may be expected to have much the same degree of success.

But city gardening in high places is different from that in a city back-yard. High above the street, dust and fumes may not be troublesome but there are other problems to deal with—excessive winds and beating sun. A certain amount of experimentation as to the placement of the garden will best determine the location most likely to be successful.

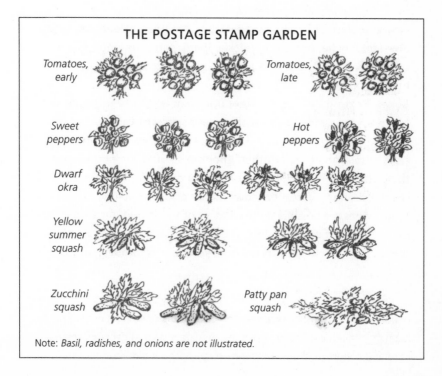

THE POSTAGE STAMP GARDEN

Tomatoes, early

Tomatoes, late

Sweet peppers

Hot peppers

Dwarf okra

Yellow summer squash

Zucchini squash

Patty pan squash

Note: *Basil, radishes, and onions are not illustrated.*

Tips for Roof Gardening

Soil. Use a good, friable loam. Fertilize with dried manure. Be sure the roof is strong enough to support a box or boxes. It's advisable to check with a structural engineer.

Mulch. A mulch will help prevent sudden drying out of the soil from wind and sun. Use peat moss, decomposed manure, leaf-mold, or black plastic.

Vegetables. With full sun, good soil, and boxes about 8 inches deep, you can grow broccoli, parsley, radishes, bush beans, endive, onions (from sets), New Zealand spinach, Swiss chard, small-size tomatoes, and short-rooted carrots and beets. Cage peppers, standard-size tomatoes, and bush squash for wind protection.

Herbs. Herbs are a good choice for shallow boxes usually used on a roof. Try thyme, chives, parsley, mint, sage, and basil.

Flowers. Marigolds, zinnias, ageratum, petunias, calendulas, sweet alyssum, lobelia, portulaca, celosia, iberis, forget-me-not, salvia, coreopsis, aster, and scabiosa are all good choices. Many now come in dwarf varieties.

Boxes. Boxes should be deep enough to hold 8 to 12 inches of soil. Make them as wide as you wish or your space will accommodate. Provide drainage holes in the bottom of each box. Paint the insides with asphaltum, or other wood-preserving compound, and the outside with several coats of quality outdoor paint. Dark green is a good color. Cypress or redwood has good durability.

Vines. Vines are possible if the wind is not too strong. An arbor over part of the roof would make a good vine support and also supply some shade and shelter. Try ivy, honeysuckle, or morning glories.

Because the roots of the plants are confined to limited quarters in their box or boxes, the soil must be prepared with care. It should be composed of sandy loam (topsoil), humus, and peat moss. Fertilizers such as decomposed manure and bonemeal should be mixed in the soil. Good drainage is essential.

Plants can be protected from high winds by windbreaks of various kinds, such as a fence of cedar saplings, a perpendicular board fence, or one of close basketweave. For a more decorative effect, use a screen of plate glass, glass brick, or other interesting material.

Plant as many started plants as possible, such as tomatoes, peppers (sweet and hot), onions, and herbs. Okra transplants well, so you might try growing a number of plants in a pot, setting them in when they are larger where you want them to grow. Yellow squash and zucchini can be grown in peat pots and handled the same way. All of these plants do well in sunshine if given wind protection. Avoid lettuce and other shade-loving plants. But you might like to tuck in here and there a few sun-loving flowers such as marigolds and nasturtiums, which can also do double duty as protection from insects. Yes, you may be surprised at the number and variety that will find their way to your garden! Some, serving as pollinators, are not all bad.

A roof garden may not be as easy as one on the ground but can be a lot of fun to plan, plant, and care for. To the confirmed gardener, having this miniature plot means not just having fresh vegetables for your table but also the joy and satisfaction of "growing your own." For the city dweller a "garden in the sky" may be just the stuff that dreams are made of.

The Kitchen Herb Garden

Why have a culinary herb garden adjacent to your kitchen door? For the simple reason of convenience and quick accessibility. Whether you are a busy house person or a busy business person, you want to give your family the most inviting and taste-tempting meals you can prepare in the shortest possible time. Those of us who cook with herbs and delight in their flavor, fragrance, and freshness find this small kitchen garden a little luxury that is pleasurable, affordable, and easily obtainable, well worth the small amount of time its maintenance requires.

Most herbs originated in rather dry, often rocky soil and do best in soil that is lightly fertilized. Most herbs like sunlight, though some will do equally well in shade. And fresh herbs add that certain something hard to achieve with dried.

Basil. You haven't really lived until you have tasted fresh basil with tomatoes. This sweet, wondrous-smelling plant is at its very best when used fresh. Pesto is that delicious Italian sauce made with fresh basil pounded with olive oil, pine nuts, parmesan cheese, parsley, garlic, salt, and pepper. Use it on spaghetti or green noodles, spread on toasted bread, or on vegetables and in salads.

Fresh bay leaves make up the ubiquitous bouquet garni that is used in many meat and vegetable dishes, noodle dishes, condiments, and soups.

Thyme. Jeanne Rose, in *Herbal Guide to Food,* describes the thyme as a little plant with a big flavor. It is available in an incredible range of scents and flavors from nutmeg, caraway, mint, pine, and pepper to lemon, citronella, and many more. Thyme is delicious in stuffings and salads, on vegetables such as beans, onions, and tomatoes, and in herb butter and breads. Try out several varieties and surprise family or guests with unusual flavors.

Mint in its variations can liven up many things besides that time-honored southern favorite, mint julep, usually flavored with spearmint or peppermint. White mint is especially useful in jellies or with lamb. Apple mint can be used to scent and flavor salads. Pineapple mint, ginger mint, and peppermint can be used in teas. Mints are good in breads and stuffings, with vegetables, in soups, or with eggs, meats, or game. Many sauces can be given a delightfully tangy flavor with the use of various mints.

Marjoram is indispensable in French cooking. It is used as a garnish, in soups, salads, eggs, cheese, vegetables, meat or poultry, stuffings, pasta, breads, and drinks.

Parsley. I need not dwell on the uses of parsley as it is so well known and universally used, so pretty and reliable in the herb garden. Just remember that it is an extremely rich source of minerals and vitamins as well, including vitamins A and C.

Chervil is another herb essential to French cooking, and one that we need to show more respect. Try seasoning peeled avocados with olive oil, pepper, lemon juice, and a really generous dose of chopped chervil. Or use it generously sprinkled over cold cream soups, fresh green salads, carrots, or potatoes.

French tarragon. You can achieve herb happiness in your culinary art with French tarragon, often called the king of culinary herbs. Its tart,

somewhat sweet licorice taste adds excitement to béarnaise or hollandaise sauce. Jeanne Rose suggests basting a roasting chicken with tarragon-mushroom butter. (See Suggested Reading.)

Sage, an old kitchen favorite, is used to counteract the greasiness of fatty meats. It is used in stuffings of goose, duck, turkey, and chicken, and it is also good with rabbit, eggs, cheeses, beans, onions, and tomatoes.

Summer savory is good chopped into salads or with baked fish or roasted pork.

Rosemary is used for everything from garnishes to desserts and jellies. Even the flowers are put into salads.

To those accustomed to using only dried herbs, the truly fresh ones from your own kitchen garden will open up a whole new world of epicurean excitement.

A Child's Garden

A little yard where children can play and have their own garden should be located where soil, sun, and drainage conditions are good—preferably in the service area where it can be watched from the house.

Water should be available nearby. Part of the playhouse may be used as a storage area.

Plant trees and shrubs that do not bear unwanted fruit for children to nibble on, fall on, or track into the house.

Plant a grass that can "take it," such as bluegrass, fescue, or Bermuda.

Use concrete or asphalt to build a "drag strip" for riding toys—making wide places for "turn-out stations."

Use chain-link fence or a similar type with clear visibility.

Floodlights, mounted on adjacent buildings or in trees or on a high pole, can add hours of evening playtime.

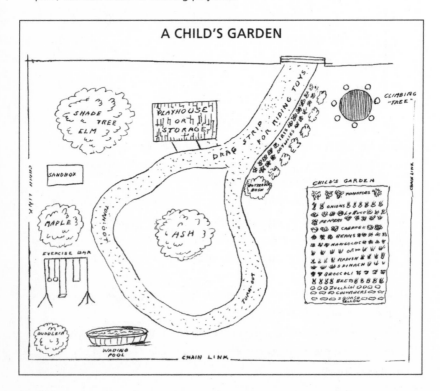

A CHILD'S GARDEN

Buy started plants for the kids' garden, varieties that are quick and easy to grow, or prolific producers such as squash. If the child is old enough to help tend the garden, be lavish with praise when something of his or hers is brought to the table.

The Able Disabled Gardener's Garden

A "no-stoop garden" is the nicest kind! Here are some suggestions to make a convenient garden for an older or disabled person.

- Box (or boxes) should be of a height easily reached by the individual.

- Boxes can be any length but should be just wide enough to be easily reached from either side.

- Box should be sturdy enough to support gardener, if using cane, walker, or wheelchair—also sturdy enough for soil when wet.

- Elevating an area improves drainage. Holes should be bored in the bottom of the box and in the pan underneath the box.

- Use a watering can with a comfortable handle and detachable sprinkler head. Try using a spray wand.

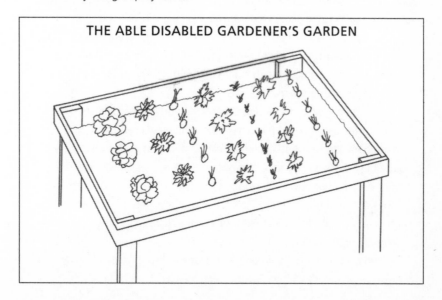

THE ABLE DISABLED GARDENER'S GARDEN

- For a larger garden you may want an overhead spray arrangement. For spot watering, use ice cubes; they'll melt slowly with little drip. Set a plant or two beneath the drip.

- Cover holes with copper screening to prevent plugging up.

- What will you grow? Just about anything you want to. Have planter boxes of different depths.

- Harvest often and keep a "floating crop game" going, quickly reseeding harvested plants.

Choose your tools carefully and buy good quality. You will not need a garden tractor but at the very least you should have a digging tool, a clipping tool, and a cultivator tool. Kathleen Yeomans, writing in *The Able Gardener,* suggests that you "test-drive" before you buy. "Lift them," she says, "and manipulate the tool from the position in which you will be using it. Whether you stand, stoop, or sit to garden will determine the handle length you require. If you expect to be sitting while you garden, sit down and go through the motions of using the tool." Buy tools that fit your hand comfortably, especially if you have unusually small or extra-large hands. A tool that seems heavy in the store will feel twice as heavy when you are working with it in your garden.

Your little garden can be caught in a late freeze just like a big one. Have some type of hot cap handy to protect tender plants. In a pinch use grocery bags, weighted down with stones or clods. Remove them during the day if the weather is warm.

Keep a sharp eye out for insects that may suddenly discover your garden, often overnight, if they are in the immediate neighborhood. Seeing them almost eye to eye in the elevated garden makes them easy to spot and you can quickly take care of the problem. Just be sure they are "enemies" and not beneficial kinds.

Give careful thought to the location of your elevated garden. Will it receive full sunlight for at least six hours of the day? Will a large tree shade it morning or afternoon? Does it have protection from the wind? And is it located convenient to your degree of disability, if you garden from a wheelchair, standing with a cane, or using a walker?

And I urge you most sincerely not to let yourself get too tired. This is supposed to be fun, but sometimes we get carried away and just want to keep going. Take it from me, for I, too, am an "able disabled" gardener with weak hands, small strength, and painful arthritis.

Vegetable Growing Guide for Minigardens

Group 1: Plant seeds of these vegetables in large containers (bushel baskets or 5-gallon buckets). They need full sunlight and warm weather. These vegetables should be planted outdoors when it is warm and the danger of frost is past. You can start tomato and pepper plants indoors 6 to 10 weeks early, then move them outdoors at the proper time for earlier harvest.

Vegetable	Days to Harvest	Planting Depth	Plants per Container
Tomato	140–150	½"	1
Green pepper	140–180	½"	1
Hot pepper	140–180	½"	1
Bush squash	50–60	¾"	1
Eggplant (use hybrid miniature)	50–60	½"	1

Group 2: These vegetables can be grown in pots 6 to 10 inches. They can withstand a little shade and do well in cool weather. Plantings should be made when it is warm and all danger of frost is past.

Vegetable	Days to Harvest	Planting Depth	Plants per Container
Mustard greens	30–60	¼"	4
Leaf lettuce	45–60	¼"	4–6
Turnips	55–65	¼"	3–4
Green onions	90	¼"	2–3
Chives	80	¼"	2–3
Radishes	25	½"	2
Beets	60–80	½"	2–3
Parsley	65–75	¼"	1

Don't plant onion seed. Use dry sets or green plants, or use an onion from the kitchen that has sprouted; usually it will divide itself into several plants.

The Tiniest Garden of All

Life can be very lonely for the elderly, housebound, or disabled. But how they can enjoy butterflies! Julia Percival and Pixie Burger give us a suggestion in their book *Household Ecology*. "Often you can persuade butterflies to settle and feed on a sunny windowsill if offered the right confection. This confection is simple: ½ teaspoon of honey and ½ teaspoon of sugar mixed in 1 cup of warm water will make an adequate butterfly nectar.

"To make a feeder use a saucer and a wad of cotton. Butterflies taste through their feet, so they need an island of cotton in the middle of the nectar sea. The cotton will absorb the honey-water so that the butterfly, when it lands on the cotton, will taste the food through its feet and then it will stand on the island and lower its rolled tongue and suck the nectar up."

The Tiniest Garden Of All

The Spirit Garden

When our ancestors first happened on the formulas for making wine and beer, it was considered a tremendous boon from the gods. Alcohol was supposed to have an in-dwelling spirit adopted from the life of the fruit and plants from which it was made. Hence the origin of our modern term "spirits" for such drinks.

Beers, wines, brandies, cordials, bitters, and whiskeys can all be made from the herbs, fruits, grains, vegetables, and trees in your garden. (See the plan on the opposite page.)

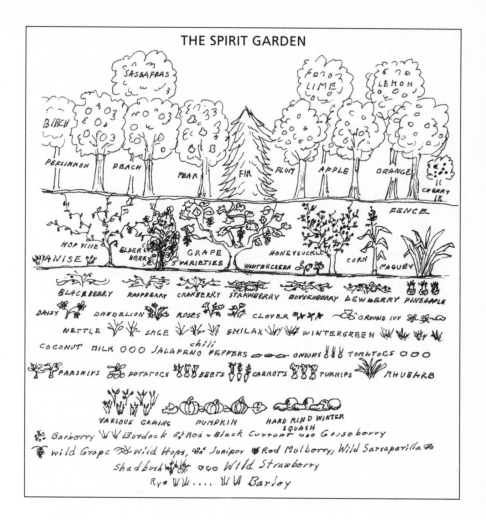

THE SPIRIT GARDEN

Trees

Persimmon, wine • Sassafras, root beer • Peach, wine, brandy; washed peelings plus a few stones can be used to make brandy • Pear, wine • Fir (Native Americans used the young treetops to make a beer said to prevent scurvy) • Plum, wine • Lime, wine additive • Lemon, wine additive • Orange, wine • Apple, wine, cider • Cherry, wine, cordial • Elderberry, wine • Birch, beer

Vines
Hop vine, beer • Honeysuckle, wine • Grape, wine

Vegetables (fully mature)
Parsnips • Potatoes • Beets • Carrots • Turnips • Rhubarb • Pumpkin • Hard-shelled winter squash

Herbs
Anise, used to flavor liqueurs • Wintergreen, wine • Ground ivy, bitters • Sage,wine • Maguey (*Agave* species), tequila • Nettle, beer

Flowers
Daisy, flower heads, wine • Dandelion, flower heads, wine • Roses, petals • Clover, flower heads, wine

Berries
Blackberry, wine, cordial • raspberry, wine • Cranberry, wine • Strawberry, wine • Boysenberry, wine • Dewberry, wine • Various hybrids

Weeds
Barberry, wine • Burdock, beer, bitters • Red currant, wine • Black currant, wine • Gooseberry, wine • Wild grape, wine • Wild hops, beer, ale • Juniper, beer • Red mulberry, wine • Wild sarsaparilla, wine • Shadbush (Juneberry, serviceberry), wine • Wild strawberry, wine

Beverage Additives
Lemon, sangria blanca • Lime, margarita • Orange, tequila sunrise • Pineapple, piña colada • Tomato, sangria • Jalapeño chile, sangria • Onion, sangria • Coconut milk, piña colada • Mint-sprigs to mint julep

Whiskey
Corn, chief ingredient of Bourbon whiskey • Rye whiskey, rye is chief ingredient • Barley malt, Irish whiskey • Barley malt, Scotch whiskey

The Aphrodisiac Herbal Window Box

Hundreds of effective aphrodisiacs have been discovered over the centuries. Many can achieve amazing results. Virility can be sustained, preserved, and even recaptured with specially selected herbs and foods.

Many herbs have proved effective as aphrodisiacs through the ages. They include basil, chervil, chives, mint, parsley, sage, tarragon, thyme, and garlic.

Soup, meat, fish, and desserts featuring the aphrodisiacs grown in your own garden can become powerful sexual stimulants. Suburbanites with a small plot of land behind their house can supply themselves with aphrodisiacs in abundance, but what of apartment dwellers who have no scope whatever for ordinary gardening? Must they be penalized because they have chosen (or been forced by circumstances) to occupy part of a building that is surrounded, not with the good earth, but with concrete, asphalt, or paving stones? Are they to be deprived of the pleasure and satisfaction of growing their own "contributory crops"?

Fortunately, the answer to these questions is an emphatic no! Experiments have shown that a great deal can be achieved in that very small garden—the window box. While such small-scale gardeners cannot expect to produce the same quantity and variety as the owner of 600 square yards, neither do they need to. They can provide themselves with a surprisingly large number of herbs and vegetables, quite often sufficient to their needs.

Assuming that the apartment has a living room, bedroom, and kitchen, and that each has an average-size window, each window box can accommodate a reasonable selection of plants. In fact, even one window box can grow a succession of herbal or vegetable plants, if you change them from time to time.

Chives can be grown from seeds. The chopped leaves give a distinctive flavor to salads and a pleasing, tangy taste to soups and egg dishes. A bit milder than onions, they improve chicken and shrimp dishes. Chicken broth with chives, yolk of eggs, and powdered almonds makes an invigorating nightcap.

Garlic as an aphrodisiac is second to none. The trouble with garlic is its powerful odor. If a garlic-flavored dish is prepared, it should be shared.

Mint, on the other hand, has a delightful fragrance. It is a powerful amatory aid and can be said to stimulate the appetite. A meal of lamb, new potatoes, and peas can be deliciously enlivened with a mint sauce, doing something more than satisfying hunger. Even boiled cabbage sprinkled with this herb may leave one in "mint" condition.

Parsley. The aphrodisiac properties of parsley were appreciated by the ancient Greeks and Romans. Parsley is just chock-full of vitamin E. Parsley is more than just a pretty face; use it lavishly as a stimulating addition to boiled chicken and omelets.

Thyme. That oldest and most attractive of kitchen herbs, thyme, can be grown either from seed or by root division. As an aphrodisiac it crops up with almost monotonous regularity in world literature. Far, far back in time it was found invigorating by the Egyptians. John Gerard in his *Herball,* written in 1633, praises it; and Benedictine monks still use it as an ingredient in their liqueurs. Use discreetly, for a little goes a long way, and it tends to kill other and more subtle flavors.

Vegetables. While some apartment dwellers may choose to concentrate on herbs and fill their window boxes with only basil, tarragon, fennel, rosemary, lavender, and so on, others might prefer to add a few stimulating vegetables such as radishes, dwarf carrots, spinach, peas, and beans, all of which have been shown to be "contributory crops."

Cress. Another possibility is cress, the aphrodisiac properties of which have been recognized for thousands of years.

Oysters. If you like raw oysters, fine; if not, try them in a creamy stew with parsley, celery, and mushrooms.

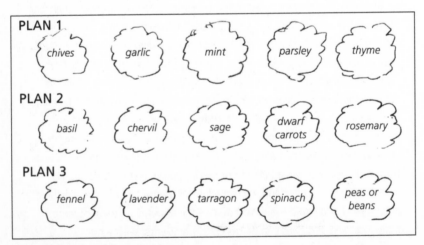

Three garden plans for window boxes

PART 2: ROSES LOVE GARLIC

THE WIDE WORLD OF FLOWERS

Flowers, from wildflowers to the finest cultivars, are beloved in every part of the world. Most are used as decorations, but many are also used for medicinal purposes or to help other flowers grow or resist disease. What is a flower? The word may mean either the blossom or the whole plant. Botanists use the word flower to mean only the blossom of a plant. They call the whole plant—blossom, stem, leaves, and roots—a flowering plant. Any plant that produces some sort of flowers, even a tiny, colorless one, is a flowering plant. Thus, grasses, roots, lilies, apple trees, and oaks are all flowering plants.

Herbaceous (nonwoody) flowering plants are generally classified by the length of time that the plant normally lives. Annuals live only one year. Biennials live for two years, blooming well only the second year. Perennials live for more than two years; they bloom the year after they are planted. Woody flowering plants—trees and shrubs—usually live for a number of years.

By any definition, flowers and flowering plants are essential to life on earth. We depend on them for our food. Flowering plants include almost all of our grains, fruits, and vegetables. We eat the roots of beets and carrots, the leaves of lettuce, and seeds of beans and peas, the fruits of apples and peaches, and the young stems of asparagus, all of which are parts of blooming plants. And artichokes, broccoli, and cauliflower are undeveloped flower clusters. Even the animals that we eat—cattle, sheep, and hogs—live on flowering plants.

We have learned many different ways in which to use plants. Dandelion and elderberry blossoms are used to make wine. Cloves, the flower buds of the tropical clove tree, are used to flavor many foods. Pickled flower buds of the caper bush are used as a relish.

Many flowers are known by their scent, which may range from pleasant to unpleasant, from the delightful fragrance of the rose to the reek of the pelican flower of South America, which smells like carrion. Scent, whether appealing to humans or not, is a key way that flowers attract pollinators.

Most flowers need soil to grow, but some can grow on tree branches, taking their sustenance from the air, while others float on lakes and streams.

Even hot, dry deserts have many lovely blossoms. During and after the rainy season, they spring up as if by magic to bloom quickly and set seed so another generation will be there to rise again when the season is right. Just about the only places flowers do not grow are in the ice-covered parts of the Arctic and Antarctic.

Some water blossoms are so small they can be seen only under a microscope; others, like the giant rafflesia, the largest flower in the world, which grows wild in Malaysa and Indonesia, may measure 3 feet across!

A fascinating aspect of flowering plants, and one with practical value to the gardener, is how the different kinds interact when planted together. This is a subject that we will be looking at closely in this book.

Companion Flowers

Companion planting is not a form of magic. It is simple and practical, making use of known factors in planning a flower or vegetable garden. Companion plantings of some kind have been practices throughout most of agricultural history.

Early settlers from Europe found the Native Americans planting corn and pumpkins together. In Holland in the 1800s a border of hemp (cannabis) was often planted around a cabbage field to keep away the white cabbage butterflies. Nature herself grows many different kinds of plants successfully as "companions." Furthermore, her plants, in most instances, grow very closely together. Instead of isolating particular kinds or varieties, she often places them shoulder to shoulder. Thus they become a source of needed shade, a climbing support, or a provider of mulch and soil-conditioning food. They may even repel other plants, preventing a sturdy, too-aggressive species from completely taking over.

Legumes such as clover and alfalfa have long been used as companion crops by farmers who have grown them to add nitrogen to the soil. The nitrogen "fixed" by legumes and other plants is not immediately available to neighboring plants but is released when the legume, or a portion of it, dies and becomes incorporated in the soil. In the flower garden lupines and other plants belonging to the Bean family perform the same function.

Some companion plants offer mechanical benefits; roots of large plants may break up the soil for smaller ones and make root penetration easier, especially in tight soil. Deep taproots of dandelions and other plants bring up minerals and make them available to plants growing nearer the surface.

Large plants (hollyhocks, sunflowers) may provide shade, wind protection, or higher humidity for small plants near them. In nature, shrubs growing around the trunk of a tree may protect it from animals. At the same time the tree, because of the shade it provides, protects the shrubs from being overwhelmed by weeds.

Another key to companion planting is controlled competition. A gardener growing a perennial flower border is engaging in one of the most complex forms of companion planting. The garden is designed not only for color, texture, height, and bloom sequence but also for controlled competition through proper spacing and varying heights. Smaller plants are protected by larger ones, but thought must also be given to the aggressive plants that will crowd out slower-growing ones if they are not kept within bounds.

We have also learned that it is unwise to plant together those plants that are susceptible to the same insects and diseases. Columbines, which are very attractive to red spiders, should not be planted near other flowers, or tomatoes, that the spider mites also find tasty.

We know too, that certain trees exude toxic substances through their roots to inhibit germination of their own seedlings beneath them. This is their natural way to reduce or eliminate competition. On the other hand, the root exudates of dahlias are helpful against certain kinds of nematodes, and they are protective to other flowers against nearby.

Many rock-garden plants also could be considered companions because they all do well in somewhat dry, sunny sites. Environmental factors make these plants companions.

Pumpkins and corn, as the Indians knew, grow well together because they are suited to the same conditions and their growing rates let them compete favorably for light, water, and nutrients. Plants that like the same growing conditions but occupy different soil strata make good companions—African marigolds and narcissus, for example—and the marigolds also repel certain nematodes that attack the bulbs.

We find many other unusual examples of "togetherness." Hawkweed or Indian paintbrush, a beautiful flower and a great attractant for hummingbirds, will not grow from seed (in cultivation) unless another plant is sown in the same pot with its seeds. The usual practice is to use blue grama (*Bouteloua gracilis*).

So now let's look at some of the flowers and plants, both cultivated and wild, familiar and unusual, that I hope you will find beautiful and useful.

FLOWER LORE

ADONIS, FLOWER OF ADONIS (*Adonis*)
This flower is named for Adonis, the beloved of Venus. According to the legend, the flower sprang from the blood of Adonis after he was killed by a wild boar.

Adonis plants belong to the Buttercup family, Ranunculaceae. The flowers are yellow or red and have 5 to 16 petals. Use annual and perennial varieties for the front border and rock garden.

AFRICAN MARIGOLD (*Tagetes erecta*)
African is a misnomer because these plants hail from Mexico. To defeat nematodes that attack narcissus, nurseries often plant African marigolds as a cover crop before planting the bulbs. To achieve satisfactory control, they plant the marigolds at least three months before planting the bulb crop.

African marigolds are also planted around apple trees or nursery stock used in grafting and budding to discourage pests. Planted near roses damaged by certain nematodes, they restore vigorous growth.

AFRICAN VIOLET (*Saintpaulia*)
This is a great favorite of indoor gardeners for its beauty. To propagate, plant the leaves in slightly moistened potting soil in a margarine tub. Slip the tub in a plastic bag and close. New plants will grow quickly and form roots.

AJUGA (*Ajuga*)
Ajuga is a delightful groundcover. *Ajuga pyramidalis* var. 'Metallica crispa' is especially lovely planted in small patches between the green varieties. It has deep purple foliage and deep blue flowers. Although it can grow in shade, it does best in full sun. *A. reptans* 'Pink Beauty' has whorls of delicate pink flowers in May and June. *A. pyramidalis,* which is larger than the others, has deep green foliage and blue flowers with purple bracts.

ALKANET, BUGLOSS (*Anchusa*)

This genus name is derived from the Greek *anchousa,* a "cosmetic plant or stain"; it may possibly have been a coloring from the blue flowers used by the ancient Greek woman for eye shadow. However, a red infusion may be prepared from the roots and, as John Gerard says in his *Herball,* "The Gentlewoman of France do paint their faces with these roots as it is said." The genus also provides showy biennials and perennials for borders.

ALOE (*Aloe vera*)

The flower of this, nature's own medicine plant, is very undistinguished, having an extremely long stem and very small blossoms. The plant, with more than 200 species, is a vegetable belonging to the Lily family.

Cut leaves exude a juice useful as a wound dressing on a tree limb after it has been cut. The healing qualities of aloe are now widely recognized, and the extracts are used in various cosmetics. It is best known for its use on burns. The juice taken internally is also healing.

ALPINE FLOWERS

These flowers know so precisely when spring is coming that they bore their way up through lingering snowbanks, developing their own heat with which to melt the snow. One, *Stellaria decumbens,* is found at 20,130 feet in the Himalayas.

ALYSSUM (*Lobularia maritima*)

The white, honey-smelling alyssums are charming with 'Martha Washington' geraniums. Or try the 'Violet Queen' variety with a 'Cecile Brunner' rose. Sow outdoors in early spring. Do not cover; the seed needs light to germinate. Pot up alyssum in August for indoor bloom in November.

AMANTILLA, SETWELL, VALERIAN, GARDEN HELIOTROPE (*Valeriana officinalis*)

Common valerian, also known as garden heliotrope, is a mainstay in the herb garden and grows from 3 to 5 feet tall. It bears tiny white or pink, highly fragrant flowers in summer, with the flowering stems rising above lacy, fernlike leaves.

The value of valerian lies in its roots, which are dug in spring before the plant has begun its growth. Dry, then pulverize the roots; store the powder in an airtight container. Valerian tea is useful for many nervous disorders such as cramps, headaches, and stomach gases. The flavor is not particularly pleasant, but it is sleep inducing and tranquilizing. Use 1 teaspoon of root per cup and steep all day in warm water. As an herbal sedative, it is very calming.

Dried valerian added to bathwater helps with skin troubles and has a soothing effect on the nervous system. Because of its sleep-inducing quality, a small amount is beneficial when added to herbal cushions or pillows. Pillows may also contain a mixture of dried peppermint, sage, lemon balm, and lavender with small additions of dill, marjoram, thyme, tarragon, woodruff, angelica, rosemary, lemon verbena, and red bergamot. (See the chapter on Cosmetics and Fragrances for directions.)

AMARANTH (*Amaranthus*)

This is the common name of a family that includes both weeds and garden plants. The family is mostly herbs. The name comes from a Greek word meaning "unfading," and is appropriate because the amaranth flowers remain colored even when dried.

A member of the Amaranth family, cockscomb (*Celosia*), is very often grown as a garden flower. *C. cristata* bears flattish, dense heads of crimson, yellow, orange, or pink flowers and is an excellent pot plant. Another type, *C. plumosa,* grows in the form of a feather plume and comes in scarlet, crimson, and gold. These plants add brilliant color to the garden.

AMARYLLIS (*Amaryllis*)

This is a genus of beautiful, lilylike plants that are usually grown indoors. Pot an amaryllis in a container only slightly larger than the bulb. Cover about one-third of the bulb with soil. For best bloom, the amaryllis should be potbound.

Potting up amaryllis

AMSONIA, WILLOW AMSONIA
(*Amsonia tabernaemontana* var. *salicifolia*)

This unusual and little-known perennial may be used as a specimen or toward the front of the herbaceous border. Its arching, willowy stems display
narrow, glossy leaves and, during May and June, clusters of small star-
shaped flowers of a strange steel blue color. The plant grows in sun but
prefers part shade, particularly in warm climates. Because it is highly resistant to wind, it grows well in the Southwest and in coastal areas. Amso-
nias grow slowly, are never troubled by insects or diseases, and rarely need
division or staking.

ANGELICA (*Angelica archangelica*)

This decorative, broad-spreading plant is the largest garden herb. Although a biennial, it will live many years if you keep the flowers cut, but
once seed develops, the plant will die. The roots and leaves have medicinal
properties. The candied stems are used in confectionery, the fruits have flavoring properties, and an oil of medicinal value is derived from the roots
and seeds. Dry seeds do not germinate well.

ANISE (*Pimpinella anisum*)

This is a white-flowered annual in the Carrot family. When thoroughly dry,
the seed germinates with difficulty. Therefore, you will get better plants
from your own fresh seed, and it will add more potent flavoring to bread,
cakes, and cookies. Use the green leaves in salads as a garnish.

    Aniseed germinates better, grows more vigorously, and forms better
heads when sown with coriander. Anise oil attracts fish.

ANTHEMIS (*Chamaemelum nobile*, syn. *Anthemis nobilis*)

The name comes from the Greek *anthemon*, "a flower," and refers to the
plant's profuse blooming. Use these aromatic perennials for the border or
rock garden. Chamomile tea is made from *Chamaemelum nobile*, and a
nonflowering variety of this species is sometimes used for lawns, particularly in very dry areas. It is said also to improve the health of other plants
when grown close to them.

ANTHURIUM (*Anthurium*)

These greenhouse plants, chiefly from tropical America, belong to the Arum family. They are grown for their brilliantly colored flower spathes, which appear in spring and summer, or for their ornamental leaves. The name refers to the tail-like flower in the center of the spathe and is derived from *anthos,* "a flower," and *oura,* "a tail." However, the tail always reminds me of Pinocchio's nose!

One of the most magnificent anthuriums is *Anthurium veitchii,* which has metallic green leaves 2 to 4 feet long.

ASPARAGUS FERN (*Asparagus setaceus*)

The plant, a member of the Lily family, is slender, with fernlike foliage on climbing stems. The fronds are very popular for floral arrangements. *Asparagus densiflorus,* an ideal plant for pots, has long branched stems clothed in narrow leaves and bears small white flowers followed by small red berries. *A. asparagoides,* the smilax of the florist, has dense minute foliage.

ASPIDISTRA, PARLOR PALM, CAST-IRON PLANT
(*Aspidistra elatior*)

Gracie Fields made this plant famous in the song "The Biggest Aspidistra in the World." During Victorian times, it was probably the most popular houseplant, gradually giving way to the philodendron, dracaena, and ivy. However, it is becoming popular again, perhaps because it is the most easily managed of all houseplants and may be kept healthy and vigorous for years with a minimum of attention.

Aspidistras are shade plants with low respiration level. Even with little sunlight, the leaves can support a steady growth of all parts of the plant. Flowers come in winter, December to March, and arise at soil level. With their magenta and gold colors, they are reminiscent of sea anemones or tiny exotic lilies, to which they are related. In their native forests of the Himalayan or Japanese foothills, the flowers are pollinated by a tiny snail crawling over them. As "potted captives" the plants seldom produce seeds but may be increased by root division.

The types with variegated leaves of cream and green are especially attractive.

ASTER (*Aster* × *frikartii*)

The plant sends up an abundance of flowers from June to November, even after a frost or two, and deserves to be seen more often in gardens. Asters are an immense group with about 160 species native to North America. On moist, low soil or by roadsides we find bushy aster (*Boltonia aster-oides*); New England aster; *Aster lateriflorus;* and willow-leaved aster; and, on banks of streams and in swamps, purple-stemmed aster (*A. puniceus*). If asters invade pastures or fields, it indicates a need for drainage.

ASTILBE (*Astilbe*)

The name is thought to be derived from the Greek word for "not shining," a reference to the leaflets. Perennials are useful for border and rock gardens; the many modern cultivars are generally the most handsome and are known as *Spiraea*.

AURICULA (*Primula auricula*)

The name comes from the Latin *auricula,* "an ear," and is a reference to the shape of the leaves, which resemble the ear of an animal. Auricula itself is one of the 30 or so classes into which botanists now divide the genus *Primula.* So-called Alpine auriculas are probably derived from *P.* × *pubescens* and what are known as florist auriculas from *P. auricula.*

BABY BLUE-EYES (*Nemophila menziesii*)

This plant shares honors with catnip as a feline attractant. In her book *The Fragrant Garden,* Louise Beebe Wilder says cats "will even dig the plants out of the ground." Baby blue-eyes, however, deserves to be more widely planted, as it makes a colorful groundcover from June to frost.

BABY'S BREATH, CHALK PLANT (*Gypsophila*)

Baby's breath is a must for dainty bouquets. In early summer these plants bear a profusion of feathery panicles of small, starry white or pink flowers on threadlike stems, creating a delicate and beautiful veil-like effect. The plant withstands cutting well and succeeds in any well-drained, not-too-heavy soil, but mix some lime into the soil before planting. *Gypsophila paniculata* 'Bristol Fairy' has large panicles of pure white, double flowers. 'Pink Fairy' produces double flowers on strong, wiry stems from June to September and adds an airy touch when placed with larger cut flowers.

BACHELOR'S BUTTON, CORNFLOWER, BLUE BONNET, BLUEBOTTLE (*Centaurea cyanus*)

Actually a beautiful weed, the cornflower is of value in supplying bees with honey, even in the driest weather. On limestone soils, the cornflowers are blue; on acid soil, they frequently develop rose and pink flowers, sometimes both colors on the same plant. The more inclined toward red, the more acidic the soil.

BALM, LEMON
(*Melissa officinalis* 'Aurea')

The flowers, which are salvia-shaped, are white, small, and inconspicuous; the heart-shaped leaves are sometimes variegated green and cream. When crushed in the hand, the leaf emits a delicious odor, suggestive of lemon-scented verbena. *Melissa* is Greek for "bee," and bees obtain large quantities of honey from the flowers. The plant will flourish in ordinary garden soil but needs a sunny, well-drained location. Balm (*Melissa*) was used by ancient Arabs as an ingredient in a cordial. Many home remedies call for it to treat vertigo, migraine, lack of appetite, and indigestion.

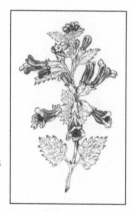

Lemon balm

BAT FLOWER (*Tacca chantrieri*)

The bat plant is said to have "the blackest flower in the world." It hails from Malaya and Burma. Some call it the devil's flower, and the many strange stories told about it probably originate from the malevolent way in

which the eyes in the bloom seem to follow your every move. Sometimes
its curious influorescence looks batlike. To some it resembles an aerial jelly-
fish. It is indeed an awesome flower and a prize for those who want to
grow something different.

BALSAMROOT *(Balsamorhiza sagittata)*

A powder of stems and leaves is somewhat toxic to pea aphids. The seeds
are edible and may be roasted, ground, and mixed with flour to make a
bread, according to author Nelson Coon.

BEGONIA *(Begonia)*

There are many varieties of begonias—tuberous begonias, wax or fibrous-
rooted types, and those grown for their ornamental leaves such as *Begonia
masoniana* 'Iron Cross' as well as less well-known types.

All begonias grow well in pots, porch boxes, or hanging baskets. The
best potting compost consists of 2 parts fibrous loam, 1 part leaf mold or
peat moss, ½ part well-decomposed manure, and a sprinkling of sand.
Add ¼ ounce of bonemeal to each quart of compost. Keep the atmosphere moist and shade the plants from hot sunshine. Begonias do well
planted with *Achimenes* (*a gesneriad*) in pots or boxes, as both take the
same culture and will bloom in shade.

BELLADONNA LILY *(Amaryllis belladonna)*

The common garden amaryllis may be grown permanently outdoors in
California and Florida, but in most places the large tuberous bulbs are
taken up and stored during the winter. Store them with caution because
the alkaloids present in the bitter-tasting bulb cause trembling and vom-
iting if inadvertently eaten. The showy, sweet-scented flowers are typically
rosy pink and trumpet-shaped, which makes for a beautiful pot plant.
Some members of the Amaryllis family, such as the century plant and the
Cuban and Mauritian hemp, are sources of useful fibers.

BERGENIA *(Saxifraga)*

These handsome plants, about 1 foot tall, have masses of decorative broad,
deep green foliage and clusters of pink flowers that appear in early spring

from March to May. They are fine for the front of the border, to "face down" shrubs, as an informal groundcover, and for the rock garden.

BIBLE LEAF, COSTMARY, ALECOST
(*Tanacetum balsamita*, syn. *Chrysanthemum balsamita*)
Used as a bookmark, the bible leaf provided some distraction for children to smell during long church services in colonial days. The plant will grow in some shade but will not bloom there. The flower heads are golden yellow, small, buttonlike, and in loose clusters.

BLEEDING HEART (*Dicentra*)
This old-time favorite is still very popular. It may have red, pink, or white flowers. *Dicentra spectabilis* is the old-fashioned showy bleeding heart with long, graceful, pendulous racemas covered with heart-shaped pink flowers on plants about 2 feet in height. Of easy culture, these plants increase in size but do not need transplanting or dividing very often. However, since they do go dormant early in fall, it is wise to set another plant close by as a filler; *Anemone vitifolia* is recommended for this purpose.

BLUE FALSE INDIGO (*Baptisia australis*)
This perennial of unique appeal makes an outstanding cornerstone in the perennial border. Its blue-green leaves stay handsome all season, and its 9- or 12-inch spikes of intense blue, pealike flowers bloom in late spring and summer. It is splendid as a companion for Oriental poppies and grows best in a lime-free soil in a sunny location.

BORAGE (*Borago officinalis*)
Borage is the common name of a familiar herb whose leaves and flowers have traditionally been used in claret cup and other beverages, to which it imparts a cucumber-like fragrance and refreshing flavor. The blue flowers are also dried for use in potpourri. It is an annual and easily raised from seed sown in spring in ordinary garden soil. The plants require a sunny position, but the blue coloring of the flowers is finest when the plants are grown in poor soil.

For many centuries, borage has been used medicinally; in the preparation of various cordials and cups, it is believed to have an exhilarating effect. Pliny had a high opinion of its virtues "because it maketh a man merry and joyful." However, present-day herbalists advise against using borage for long periods.

BOUNCING BET, SOAPWORT (*Saponaria officinalis*)

This showy flowering plant grows almost too readily. Its great virtue lies in its sudsing quality, the bruised leaves acting as a soap when agitated in

water. The lather may be used as a shampoo. When decomposed, the lathering substance (saponin) helps retain soil moisture. (See also *Cuckoo Flower,* on page 205.)

Bouncing bet

Bouncing Bet was brought to the New World more than 300 years ago for its valuable lathering qualities. It was once used extensively for washing fine silks and woolens. The carnation-like pink and white flowers cover the plant during its long blooming season.

BROMELIADS

Did you know that you can force a bromeliad to bloom by covering it for five days with a plastic bag that has an apple inside? A bromeliad is any plant belonging to the Pineapple family. Typical bromeliads are aechmea, billbergia, cryptanthus, nidularium, tillandsia, and vriesia.

If you're cramped for space indoors, try miniatures. Among my favorites are the little bromeliads, specifically those known as *Cryptanthus,* or earth-stars. The species *C. bivittatus minor,* for example, forms a rosette 3 or 4 inches across. It hugs the ground and is composed of leathery-stiff leaves, the edges of which have tiny spines. If you run your index finger lightly along one, it will remind you of a cat's tongue.

The leaves are striped lengthwise in color that varies depending on age and growing conditions. Usually they are a combination of green with red or pink suffusion. The flowers, which are white and typical of cryptanthus, grown from the center of a mature plant; hence the name of the plant.

BUCKBEAN, BOGBEAN, MARSH TREFOIL
(*Menyanthes trifoliata L.*)

This perennial plant found in marshes and bogs has white flowers tinged with rose borne in a terminal cluster on a stalk 4 to 12 inches long. Rootstalks used as an emergency food must be dried, ground, then washed several time to leach out the bitter principle, and then dried again. Fernald describes the bread made from such flour as "thoroughly un-palatable but nutritious." In Europe the roots have been used to replace hops in making beer.

Buckbean

BULBS

You can have bulbs flower in succession, from snowdrops in early spring to lilies in late summer. There are bulbs for every purpose:

Bed and borders. Try hyacinth with English daisies in pink or white, pansies in selected colors, or forget-me-nots for color combinations. Tulips give a medley of color in a good-size tulip bed edged with pansies.

Rock gardens. Small bulbs bring color and early-season interest. Snow-drops lead off; then come the crocus species in white, yellow, and lilac shades; followed closely by little *Iris reticulata* in very dark purple with orange veinings. Other suggestions are grape hyacinth (*Muscari*), glory-of-the-snow (*Chionodoxa*), Siberian squill (*Scilla*), small narcissus (*Narcissus minimus*), hoop-petticoat daffodil (*N. bulbocodium*), and angel's-tears (*N. triandrus*), to name but a few possibilities that do well together.

Forcing bulbs for winter bloom. Narcissus, such as paper-whites and Chinese sacred lily, are the easiest and earliest to bloom. Set them in bowls of pebbles and water in September and they'll bloom for Thanksgiving. Later starts will prolong the season. Hyacinths, white Roman and the miniatures in yellow, pink, and blue, grow well in bowls of vermiculite or the special bulb fiber obtainable from dealers. Daffodils and tulips are mostly grown in standard pots or bulb pans in a good soil mixture. Use precooled daffodil bulbs or cover pots outside until after a heavy freeze before trying to force them.

After blooming, spring bulbs should have the dying flowers cut off, though not the whole stem, so that the plant's strength is not wasted in seed production. Work the soil lightly between the bulbs with a hand fork or hoe before planting with annual bedding plants for summer display.

Naturalizing bulbs. Some bulbs grow and flower for many years under natural or "wild" conditions; their only need is good soil. Snowdrops often grown and flower on the north side of a slope for generations. Grape hyacinths grow and spread. So will Siberian squill and glory-of-the-snow. Crocuses bring color and interest to bare ground. Plant poet's narcissus (*Narcissus poeticus*) near water. Spanish bluebells (*Hyacinthoides hispanica* syn. *Scilla hispanica*), in delicate shades of white, pink, and blue, enliven the somber green of ferns in a moist, shady spot. Daffodils naturalize readily although they will not increase as rapidly in grass as under cultivation. Trumpet or large-cupped daffodils look enchanting on a grassy slope.

CACTUS

The origin of the bizarre Cactus family is shrouded in mystery. Botanists theorize that the first cacti evolved from roses because their lavish, showy flowers closely resemble roses in shape and structure. Cactus blossoms are unbelievably beautiful, especially the fragrant flowers that bloom at night.

Native Americans and other southwesterners find a variety of uses for cacti. Many species have delicious edible parts: The pads of opuntia taste like green beans, and the flowers may be used in the same way as squash blossoms. The fruits of some species taste like strawberries. The pulp of the barrel cactus is used to make cactus candy.

Some of the giant cacti are used as lumber, to roof houses and to make cradles. Peyote cactus buttons have been used for centuries for medicinal and religious purposes. Many cacti are beautiful and interesting even when not in bloom and are widely used in the Southwest for landscaping. Many cactus fanciers in more northern climates enjoy potted cactus or indoor cactus gardens.

Cacti make up a perennial family of herbs and shrubs equipped with areoles. No other plants have these unique organs of growth. From them spring branches, flowers, glochids, spines, and, when present, leaves. If areoles are removed, the cactus will die.

CACTI: THE "OTHER ROSE" AND SOME POLLINATORS

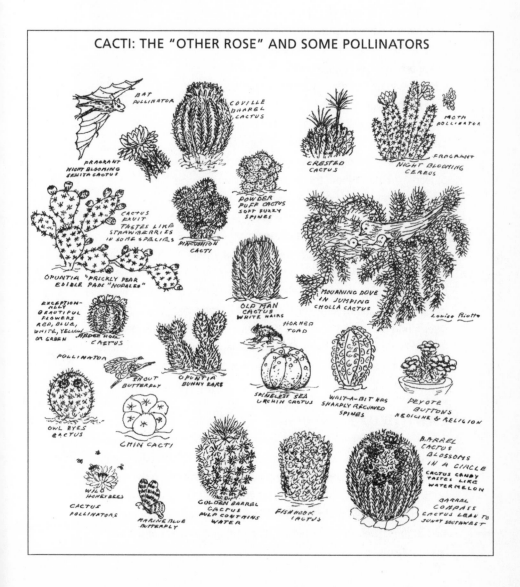

CALLA LILY, ARUM LILY
(Zantedeschia)

This plant is easily grown in mild climates such as in California, where it sometimes "escapes" and grows wild. The most popular kind is the common calla, which has handsome green leaves and bears large white flower spathes. It is cultivated in large quantities by market growers for the spathes, which are in great demand for decorative purposes, particularly at Easter.

Calla lily

The common calla and its varieties are moisture-loving plants and must be grown in rich, loamy soil that does not dry out quickly. In parts of North America where the winters are mild, the common calla and its varieties may be grown outdoors. They thrive in moist or wet soil on the edge of a pool or pond, or may even be planted in shallow water.

The lovely golden calla (*Zantedeschia elliottiana*) with its white spotted leaves is the most popular of the yellow zantedeschias and ranges from pale to golden yellow. *Zantedeschia albomaculata* includes ivory, pale yellow, or pink flowers, some with black throats. Depending on variety, they may have either spotted or unspotted leaves. There are now many cultivars available, with colors including pure white, pale yellow, golden, pale pink, rose, carmine, and even a few bicolors.

CANDYTUFT *(Iberis)*

This plant is practically foolproof, even for "purple thumbers," and does well in any soil in sun or light shade. To extend the blooming period, shear off the spent blossoms. Use this evergreen subshrub for edging borders and garden walks, in the rockery, or for mass plantings. Most candytufts have white flowers: *Aethionema cordifolium* (sometimes sold as *Iberis jucunda*) is noted for its dense pink flowers and blue-green foliage.

CANNA, INDIAN SHOT (*Canna*)

These tender herbaceous perennial plants from South America and the West Indies have unbranched, stately stems springing from a fleshy root-stock. During the summer they have large, ornamental foliage and brilliant, showy, gladiolus-like flowers in dense terminal clusters.

The leaves and stems of cannas have insecticidal properties. They are particularly useful as a greenhouse fumigant to control aphids, ants, and mites. Place newspapers in the bottom of a pail with a metal grating on top of them. Above the grating put a layer of straw to act as a buffer; then place leaves and stems on top. Light the newspapers; the leaves will smolder rather than burn, producing thick smoke. Close the greenhouse door tightly and smoke the leaves for about 30 minutes.

CARNATION (*Dianthus*)

These charming plants of cheerful colors and spicy fragrance have attractive, often bluish green foliage. They are fine for borders, edging, and rock gardens. The hardy border carnation, *D. caryophyllus,* blooms almost perpetually. Grow it in well-drained soil with lime added and protect it in winter with a dry mulch. Cut back the plants after blooming.

Carnations are very popular for Mother's Day.

CARPETWEED (*Mollugo verticillata*)

This is a low-growing weed that forms mats in gardens and on paths. It thrives well on lighter, sandy soils but will not resist hoeing and cultivation.

Charles Harris, in his book *Eat the Weeds,* states that carpetweed is good steamed or boiled, or used as a potherb.

CARRION FLOWER (*Stapelia*)

This cactus looks and smells like carrion and attracts carrion flies from a long distance. Bluebottle flies also seek it out and have been known to lay their eggs on the leathery petals, as they do on another foul-smelling plant, *Amorphophallus rivieri,* or devil's tongue.

CARROT, WILD

See *Queen Anne's Lace* in this chapter.

CATCHFLY (*Silene*)
This group of plants often infests dry meadows, clover, and alfalfa fields. It may become a pest if not controlled by early cutting. The sleepy catchfly, forked or hairy catchfly, and night-flowering catchfly are all annuals and propagate by seed. The sleepy catchfly opens if flowers only to bright sunshine. The name catchfly is derived from a gluelike substance on the stem, which does catch flies.

CELANDINE, GREAT (*Chelidonium majus*)
This plant grows here and there in barnyards, pastures, and roadsides. It contains a yellow, slightly poisonous juice, which was once used against warts (hence the popular names wartweed, killwart, and devil's milk). It has also been recommended that the freshly oozing juice be applied daily to corns until cured, or used on sores of horses. There is a caustic quality in the juice (even more in the roots than the upper parts) that could be of medicinal value for skin conditions. Cultivation and early cutting before it goes to seed will keep celandine from spreading.

CERATOSTIGMA (*Ceratostigma plumbaginoides*)
This plant is truly one of the most rewarding of all perennials. Growing only 6 to 8 inches high, it forms mats 12 to 18 inches across, solidly covered with clusters of intense peacock blue flowers in the late summer and fall. Simultaneously the interesting leathery foliage turns deep mahogany red. Use the plant for groundcover or for underplanting shrubs. It will grow virtually anywhere, good soil or bad, well drained or poorly drained, in heat or cold, sun or shade!

CHEROKEE ROSE (*Rosa laevigata*)
This charming decoration of our southern states originally came from China, but is now widely and permanently naturalized. The foliage is evergreen and shining and the immense white single blooms have the rich fragrance of the gardenia.

CHICKWEED MOUSE-EAR (*Cerastium arvense*)
The field mouse-ear chickweed is a beautiful flowering native plant used in gardens. It derives its name mouse-ear from the shape of its leaves. The

blossom is large, white, and star-shaped. Because of its creeping rootstock (every joint can produce a new plant), it can become troublesome in pastures. Plow and cultivate to control it. In *Gardening Without Poisons,* Beatrice Trum Hunter says that "rye overpowers chickweed." Chickweed likes roadsides, sunny hills, and even grows way up on high mountains.

Common mouse-ear chickweed (*Cerastium fontanum*), which propagates only by seed, has smaller leaves and blossoms than the field mouse-ear. It likes to grow in fields and on roadsides. Early cultivation in grainfields will lift up the shallow roots and so eradicate this perennial.

CHINESE LANTERN (*Physalis alkekengi*)
Chinese lantern is often grown for its large, showy calyxes, which are attractive as winter decorations. The white flowers are followed by dense clusters of bright orange-scarlet, lanternlike husks enclosing scarlet berries. Cut the stems in autumn and dry to preserve. Chinese lantern is an aggressive plant that will rapidly take over, so grow it in a waste spot to prevent crowding other flowers.

CHOCOLATE FLOWER (*Berlandiera lyrata*)
This branched perennial has pale yellow, daisylike flowers that smell like chocolate. The underside of the "petals" have brown veins. The large green bracts below the flower are attractive for dried arrangements.

CHRISTMAS ROSE (*Helleborus niger*)
This perennial herb is often cultivated in gardens for its midwinter bloom. Because it blooms in early spring in some regions, it is often incorrectly called the Lenten rose. The white or pink-white flowers, about 2 inches across, become purplish with age. The thick but fibrous rootstalk, which is blackish brown, yields drugs for commercial use. However, the rootstalk is violently poisonous if eaten and, as a warning, emits an unpleasant odor when cut or broken. It has a bitter, slightly acrid taste. The poisonous leaves may cause dermatitis on contact.

CHRYSANTHEMUM
See the chapter on Companion Planting with Flowers and Herbs.

CLOVE (*Syzygium aromaticum*)

This is the name given to the flower buds of a tropical tree that grows wild in the Moluccas (or Spice Islands), and in Sumatra, Jamaica, the West Indies, and Brazil. The tree's purplish flowers grow on jointed stalks and are picked before they open. Reddish when first picked, they turn dark brown when dried. The dried buds are used as a spice. They have a fragrant odor and a warm, sharp taste. Maybe you know them best as decorations for a fine ham.

CLOVE PINK (*Dianthus caryophyllus*)

This is the gillyflower of medieval Europe, popular for its variety of color and sweet, spicy fragrance. Before oriental spices become available to everyone, the flowers of clove pinks were used with foods and for flavoring wine and vinegar. These lovely old-fashioned pinks bloom from early summer until late fall and keep the air delightfully perfumed—and the plants are winter hardy.

CLOVER (*Trifolium*)

Like other legumes, clover fixes the nitrogen from the air by means of bacteria growing on its roots. When the clover is plowed under, the nitrogen enriches the soil. Many of the clovers are used for cover crops, including the common red or sweet clover.

There are also bur, crimson, Egyptian, and Persian or Wood's clover, but red clover is the most important member of the family. The flowers of red clover will not be fertilized unless a bee pollinates them. When red clover was first planted in Australia, there were no bumblebees to carry the pollen; not until they were introduced did the red clover produce seed. Crimson clover is much used for soil improvement. Its flowers are often red but may be white or yellow.

Red clover herb tea is especially good for canaries. To make: Steep 2 teaspoons in ½ cup hot water and allow to stand 15 minutes. Put a few drops a day in the canary's drinking water.

Clover honey is very delicious and one of the best-known flavors.

COLEUS (*Coleus*)

These are superb, colorful foliage plants for shady spots in the garden—and they are fine houseplants as well. The luxuriant foliage displays shades of red, green, crimson, yellow, white, pink, and combinations thereof. As exotic in color and form as they are, coleus are probably due for even bigger changes. They react strongly to radioactivity, and many new forms have been seen at the nuclear installation outside Knoxville.

COLUMBINE (*Aquilegia*)

These hardy perennial plants bear spurred, beautifully colored flowers from May to July. They grow wild in North America, Siberia, and other north temperate countries and belong to the Buttercup family, Ranunculaceae. The word *Aquilegia* is derived from *aquila,* "an eagle," an indication of the spurlike petals.

Columbines will thrive in ordinary garden soil and are easily raised from seed. A sunny or partially shaded location suits them. But keep them to themselves—they do not companion well with other plants and are also very attractive to the red spider. However, the hummingbird finds their red and yellow bells irresistible.

Columbine

COMFREY, KNITBONE
(*Symphytum officinale*)

This plant was believed to aid the knitting of fractured and broken bones and has been used for this purpose for centuries. The leaves make a poultice for bruises, swellings, and sprains. It has also been used for lung disorders, internal ulcers, external ruptures, burns, and splinters. However, it has been found to contain compounds that damage the liver and is no longer recommended for internal use.

Grown in the garden, comfrey is beneficial to other plants—a sort of "plant doctor." Because it is deep rooting, comfrey does not rob minerals in the surface soil from other nearby plants. It keeps the surrounding soil rich and moist and gives protective shade and shelter with its large, rough leaves. Flowers are pale blue pink, bell-form, and borne in drooping clusters.

COMPASS PLANT, ROSINWEED, PILOTWEED
(*Silphium laciniatum*)

The scientific name refers to the resinous juice of these plants. Although the genus includes 15 species of perennials in North America, only 2 are likely to be seen in cultivation.

The plant bears yellow flowers that look much like sunflowers. It is a coarse plant and sometimes grows to be 10 feet. The leaves are about 1½ feet long and are cut into several lobes. The petioles, or leaf stalks, bend so that the leaves, by pointing in a north-south direction, escape the strong midday sun, but get the full early-morning and late-afternoon light. Frontiersmen and hunters in the prairies of the Mississippi Valley noticed that this plant's leaves, pointing north and south, accurately indicate the points of the compass.

CORIANDER (*Coriandrum sativum*)

Grown for its savory seed, coriander is not suitable for the flower garden because the foliage and fresh seed have an evil smell. However, the foliage is delicate and lacy, and the rosy white flower umbels are beautiful. The ripe seeds are fragrant and the odor increases as they dry. Use them in cooking.

COTTON, ORNAMENTAL (*Gossypium*)

These 2-foot plants have pink buds and creamy blossoms, followed by big white bolls of cotton.

CREEPING CHARLIE, GROUND IVY, GILL-OVER-THE-GROUND, CAT'S-FOOT
(*Glechoma hederacea*)

This member of the Mint family is a ground-hugging vine that returns year after year to produce pretty purple flowers and an aroma that protects nearby plants from insects. It spreads rapidly, sending down roots wherever it touches the ground, and likes a partially shaded location. According to some sources, sniffing the crushed foliage relieves headaches, and the roots contain a substance that stops bleeding.

CROWN VETCH (*Coronilla*)
Crown vetch is wonderful for preventing soil erosion along steep banks and for choking out unwanted weeds. It increases quickly by sending out roots both above and below ground. The neat, billowy foliage grows no higher than 2 feet and, in season, is a mass of lovely pink flowers.

CUCKOO FLOWER (*Lychnis flos-cuculi*)
Originally from Asia Minor and Siberia, this plant prefers moist meadows and has some value for feeding livestock. Its blossoms are bright red. (The familiar garden varieties, pink, white, or blue, are known as phlox.)

The roots of all *Lychnis* species contain saponin, which produces a soapy foam if stirred in water. Before the discovery of soap, it was used along with the true *Saponaria* for washing. To this same group also belong red campion, found on grainfields and in pastures, and white cockle or evening lychnis, so called because its white blossoms open in the evening and close at sunrise.

CUPID'S DART, LOVE PLANT
(*Catananche caerulea*)
The flower spikes, springing from silver-green foliage, are an exquisite cornflower blue and stay in bloom from early June right up to late September. These are superb as cut flowers, either fresh or dried to preserve the glorious color in long-lived arrangements. Plant in full sun in well-drained soil; they will seldom need dividing.

CUP PLANT (*Silphium perfoliatum*)
This plant has yellow flowers, and its leaves join together to make a cup around the stem.

CYCLAMEN (*Cyclamen*)

This is a genus of dwarf tuberous plants from the Mediterranean regions. They are exotically handsome and best grown in a cool temperature. *Cyclamen* is from the Greek *kyklos,* "a circle," referring to the coiling of the flower stems in some species after flowering. This is the plant's method of bringing the seed capsules down to soil level.

Do not keep these plants near any orchid plant at any time, as they give off ethylene gases that kill the orchid and its blossoms. However, cyclamen has long been esteemed for killing parasites on fruit trees. The principle, saponin, found in the bulbs, is effective fresh or dried.

DAFFODILS (*Narcissus*)

Daffodils announce the advent of spring. Their cheerful yellow combines well with the violet-purple of the grape hyacinth, or you might finish the

Daffodils

bed off with a ring of crocuses. Daffodils discourage moles. In Europe and England the interest in daffodils has sometimes assumed the proportions of a craze. Rival enthusiasts grow and cross daffodils, getting extravagant prices ($500 to $2,000) for no more than five or six bulbs. In America, we have never taken these flowers so seriously, but we love them nevertheless and plant them widely in our spring gardens.

DAHLIA (*Dahlia*)

If you want an ideal flower, try the vigorously growing dahlia for an abundance of beautiful blooms and, best of all, relative freedom from diseases and pests. They come in many sizes, many colors, and single or double flowered. Cactus varieties are particularly attractive.

Dahlias grow best in deep, fertile, well-drained soil in a sunny location; the plants are easily damaged by cold. Separate root clusters and plants about the time of the last killing frost. Space them 3 or 4 feet apart if you want large exhibition flowers. For large blooms, allow only one stalk per root to develop. Remove all small, weak sprouts. When the shoot is about 6 inches tall, pinch it back to the third set of leaves to promote branching.

Dahlias protect nearby flowers against nematodes.

DAISIES, WHITE OR OX-EYE
(*Chrysanthemum leucanthemum*)

These are the well-known white flowers with long stems and yellow centers that often infest pastures, hayfields, and lawns. They are frequently planted with grass seed but should be avoided in the lawn. They increase with increasing acidity of the soil, standing surface moisture, and loss of lime. Good neutral compost with lime in it, bonemeal, surface harrowing, or frequent raking of the lawn to break the upper crust and root felt will take care of this problem.

Cultivated daisies are something else again, and there are many beautiful varieties for beds and borders.

DAMIANA (*Turnera aphrodisiaca*)

This Mexican and African plant has been widely recommended for treating impotency and for its tonic effect on the nervous system. Long ago, the Aztecs used the leaves of damiana as an aphrodisiac. A commercial liqueur made from damiana, called Liqueur for Lovers, can be purchased in the United States.

DATURA, THORN APPLE, JIMSONWEED
(*Datura stramonium*)

Both seeds and leaves yield narcotic drugs used in medicines. The strong odor of the plant causes drowsiness and, if used sparingly, smoke from dried datura leaves is calming to honeybees when opening a hive. Sucking nectar from the flowers has poisoned children, and some have died from swallowing the seeds. Even eating the boiled plants produces irrational behavior. Early California Indians knew this and gave their children potions made from the plant to obtain visions, especially on reaching puberty, but expert shamans regulated the dosage.

Datura is a beautiful plant and because of the large white blossoms is also called angel's-trumpet. Datura helps the growth of pumpkins when planted nearby.

DAYLILY (*Hemerocallis*)

The daylily has long been a mainstay of perennial plantings. The usual method of propagation is by clump division, easily done in autumn, but for best results a clump should not be divided more often than every two or three years.

Another reproduction method is by proliferation. The sturdy little rootless plants that develop along the flower scapes (stalks) in the axil between the rudimentary stem leaves and the stalk eventually die if they do not touch the earth; nature probably intended them to spread the plant when the flower stalk finally breaks and falls to the ground. Give nature an assist. When the flowering period is past and the stalk starts to dry, cut off the proliferation by severing the stalk about 2 inches below. Insert the proliferation in a pot full of good growing mix with the base just below the soil surface. Keep soil moist. Roots will soon develop from the base. After good root development, set the plant out in the garden.

Daylily

If you have a slope or terrace too steep to mow, try planting daylilies and iris together. Their roots will hold the soil, and their blossoms, which arrive successively (the iris blooming first), will delight you all summer. Daylilies are unaffected by the juglone washed from the black walnut's leaves. The flower buds are delicious dipped in batter and fried.

Not generally known is the fact that daylilies are a tasty food. Buds and blossoms can be sautéed in butter with a little salt and used alone or added to a zucchini and tomato dish. They may also be dipped in batter and fried. They may even be dried for later use in soups and stews.

DEATH CAMAS (*Zigadenus venenosus*)

Sometimes cultivated in gardens, this bulbous perennial plant is found below 8,200 feet in meadows, pastures, open slopes, and along roadsides

from Canada to Florida, Texas, New Mexico, Arizona, and California.
All parts of the plant are toxic; poisoning has followed even the eating of
the flowers. The onionlike bulb has a dark-colored outer coat but lacks
the onion odor.

Another variety of camas, the Indian quamash, Queen of the Bulbs,
played a vital role in the history of some Native Americans. With breadroot
and cous, camas constituted their basic starch food.

Camas of this type are found as high as the subalpine zones, growing
along streams or in moist meadows. They cover vast areas so closely that
in springtime the effect is that of lakelets of brilliant blue. Their flowers,
grouped on spiny racemes, are spectacular. The ovate bulbs also look like
onions. The big blue flowers clearly differentiated this plant, the "bringer
of life," from the death camas, the "bringer of death," with its small
yellow blossoms.

DELPHINIUM, LARKSPUR
(*Delphinium, Consolida*)

Widely cultivated for their beauty, many of these plants have escaped to
roadsides and fields. One gardening book calls the delphinium "a confused hybrid of uncertain origin." No matter; it's a hardy flower that's
easy to grow. The prevailing color is blue, but cultivated forms come in
many colors—some even have beautiful
double blooms. All species of the plant
contain alkaloids of varying quantities.
Ingesting young leaves before the flowers
appear causes poisoning, but the plants'
toxicity decreases as they age. Leaves and
seeds may cause dermatitis on contact.

Common field larkspur (*Consolida
ambigua*) yields the alkaloids delcosine and
delsoline, found effective against aphids and
thrips. Powdered roots are toxic to bean leaf
rooler, cross-striped cabbageworms, cabbage
looper, and melonworms.

Delphinium

DENTARIA (*Dentaria*)

The name is from the Latin *dens,* "a tooth," and refers to the toothlike scales of the roots. The plants are uncommon but useful little perennials for shade. Try them in porch boxes located on the north.

DITTANY, BURNING BUSH (*Dictamnus albus*)

The Greek name, *dictamnos,* refers to Mount Dicte, Greece, where, according to legend, Zeus was born and where this plant once grew. One species, a curious perennial, gives off a volatile oil from the upper part of its stem that may be ignited in hot weather and will burn without harming the plant.

DOG FENNEL, STINKING CHAMOMILE, MAYWEED (*Anthemis cotula*)

This plant is sometimes mistaken for chamomile, but once you know the true chamomile fragrance you will never confuse the two; dog fennel has a rank, weedy odor. To tell them apart, cut one of the well-developed little daisylike blossoms vertically through the middle. True chamomile has a hollow center; dog fennel is solid.

DRAGON'S HEAD (*Dracocephalum*)

From the Greek *drakon,* "a dragon," and *kepthale,* "head," the name refers to the gaping flower mouth. Both annuals and perennials are useful for the front border.

DUMBCANE (*Dieffenbachia*)

These evergreen foliage plants are widely grown in greenhouses, homes, restaurants, and lobbies as potted ornamentals. The two most commonly cultivated species are *Dieffenbachia maculata* and *D. seguine.* They thrive planted outdoors in the southern part of the United States. The flowers are tiny; the fruit is fleshy.

All species contain calcium oxalate as needlelike crystals in the stems and leaves. This toxic plant is called dumbcane because chewing on it causes temporary speechlessness. The irritation of the mouth can be fatal if the base of the tongue swells enough to block the throat.

DUTCHMAN'S-BREECHES (*Dicentra cucullaria*)

This plant is a native to the eastern woods and has a surprisingly sweet scent. But beware: Eating the leaves and roots produces poisoning and such nervous symptoms as trembling, loss of balance, staggering, weakness, difficulty in breathing, and convulsions.

EDELWEISS (*Leontopodium alpinum*)

This hardy perennial likes sun and a dry location. Do not cover the seed because light promotes germination, which takes place in 15 to 20 days. The gathering of edelweiss has long been symbolic of daring achievement, since it is native to the high rocky ledges of the Swiss Alps. The plant grows 12 inches high and is a soft gray-green.

EGLANTINE, SWEETBRIER (*Rosa rubiginosa*)

Eglantine is lovely but can become too much of a good thing. It immigrates from hedgerows to pastures where it shoots up quickly. This indicates that the pasture has not been grazed sufficiently and should be mown and harrowed. The prickly canes can be troublesome to cattle and sheep. If they become established, they will protect the growing and seeding of other weeds. Cut the plants while they are still young and the canes are soft.

ELEPHANT'S HEAD (*Pedicularis groenlandica*)

This large clumping perennial grows to 3 feet with delicate fernlike leaves. The rose-pink flower resembles an elephant's head with its large floppy ears and long trunk. Flowers in tall spikes emerge from the ferny foliage in August. The plant is nice for naturalizing, as it grows well in moist meadows, bogs, streams, and lakeshores from Greenland to Alaska and south to the mountains of New Mexico. Seed should be cold stratified for 30 days before sowing in spring.

EVENING PRIMROSE (*Oenothera*)

Oil of evening primrose is said to be the world's richest source of natural, unsaturated fatty acids. It is helpful in cases of obesity, mental illness, heart disease, and arthritis and for relief from postdrinking depression.

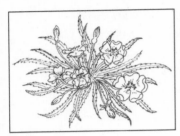

Evening primrose

The cultivated *Oenothera tetragona* 'Fireworks' is a cheerful plant with dark green dwarf foliage, tinted purple in the spring. The red buds open to profuse silky bright yellow flowers borne during June and July.

O. macrocarpa (formerly *O. missouriensis*), evening or Missouri primrose, is classed as a wildflower. It has immense 4- to 5-inch, cup-shaped yellow flowers. A low-growing species, it is ideal for border or rock garden. The winged seedpods are tan, often streaked with crimson, and are excellent for dried arrangements. The plant likes fertile, well-drained soil and a sunny location.

EVERLASTING, FRAGRANT (*Gnaphalium polycephalum*)

This is a fragrant immortelle of the autumn fields, spicy and sweet, often growing in old fields and woods. *G. ramosissimum* is a fragrant pink-flowered everlasting.

EVERLASTING PEA (*Lathyrus grandiflorus*)

Use this plant as a control against field mice.

EYEBRIGHT (*Euphrasia officinalis*)

Eyebright is characterized by lovely white flowers, striped violet, that open in summer. The plant was once thought to be effective in treating ophthalmia. Culpeper writes: "If the herb was but as much used as it is neglected, it would half spoil the spectacle makers' trade. The juice or distilled water of eyebright, taken inwardly in white wine or broth or dropped into the eyes for diverse days together helps all infirmities of the eyes that cause dimness of sight. It also helps a weak brain or memory."

Some people smoke dried eyebright as they would tobacco. A kind of wine can be made by cooking it with unfermented wine at harvest time; however, eyebright should not be taken internally because there have been cases of stomach disorders. The juice is acrid and has an unpleasant taste.

FALSE SAFFRON, SAFFLOWER (*Carthamus tinctorius*)

False saffron is a thistlelike plant with large, attractive flower heads varying from white to brilliant red. It has been grown in dry areas of Asia, Africa, and Europe for hundreds of years. The chief value is in the oil and meal made from the seeds. The oil is used in the diets of persons suffering from heart disease and hypertension, and is a valuable source of linoleic acid. The meal is fed to livestock.

FENNEL, SWEET (*Foeniculum officinalis*)

In the garden, fennel is valued for its masses of fringed foliage. In times past the fragrant seeds were made into a tea for babies' colic. Mixed with peppermint leaves they also make a delicious tea, calming to the nerves. The Italians consider fennel valuable as a key to longevity and virility; the Egyptians, Greeks, and Romans all vouched for fennel, using it in salads, fish sauces, and fennel soups. Fennel oil, which contains estrogens, has a reputation as an antiwrinkle agent.

The Florence fennel, prized for its enlarged leaf bases and used like celery, is a close relative and is grown the same way. These fennels are poor companions for bush beans, caraway, kohlrabi, and tomatoes. The black swallowtail butterfly lays its eggs on fennel, dill, and parsley, and its beautiful black and gold caterpillars are ravenous.

FEVERFEW (*Chrysanthemum parthenium,*
syn. *Tanacetum parthenium*)

Feverfew, an old favorite for edging, has little white buttons ¾ inch across, and yellow foliage with a characteristic strong, bitter odor.

This little daisylike flower is sometimes incorrectly called pyrethrum. However, like the real thing, it works as a bug chaser and can be planted as a border around roses or scattered throughout the garden. Feverfew grows in tufts, becoming bushlike and occasionally attaining a spread of as much as 3 feet. The single-flowered form was the feverfew cultivated in old physic gardens. The modern forms, largely used for cutting, are double-flowered. Feverfew is an accommodating flower that will grow in any soil, gives generous bloom all summer, and self-sows readily.

FEVERWEED
See *Joe-pye Weed* in this chapter.

FIG (*Ficus*)
Figs have their flowers inside the fruit. The fig wasp, blastophaga, lays its eggs at the base of the flowers, thus ensuring pollination. Figs have long been considered love food by primitive peoples.

FINGERLEAF WEED
This weed lives on acid soil and indicates increasing acidity. Weeds are specialists and close observation tells us a great deal about the soil they grow on.

FIREWEED, GREAT WILLOW HERB (*Epilobium angustifolium*)
The large, showy flowers are borne in terminal spikelike clusters. They have four purplish or rose-colored petals that may be occasionally pink or rarely white. The plants are common along roadsides and on open hillsides from southeast Alaska north to the Arctic and west to the Alaska Peninsula and the Aleutian Islands. They also are found in the mountains, along streams, and in clearings throughout North America and Eurasia. Fireweed was one of the first living plants found after the Mount St. Helens volcanic eruption.

Native Americans collect the young shoots in spring and mix them with other greens; they peel and eat the young stems raw. Like other tender young greens, these are a good source of vitamin C and vitamin A. Fireweed honey is one of the finest.

FLAX (*Linum*)
Narbonne flax is an elegant, free-flowering perennial with feathery, blue-green, evergreen foliage, just right for a sunny spot or a rock wall. The flowers, 1¾ inches across, are sky blue with dainty white eyes. The 1½ foot plants with slender, erect stems bloom generously.

Dwarf golden flax (*Linum flavum* 'Compactum') has myriad bright yellow translucent flowers, 1 inch across, and makes a fine border plant. Give these full sun and a moist but well-drained location.

FLEABANE (*Erigeron*)

This weed invades relatively good land and is one of the few weed "presents" the American continent has given Europe. It was inadvertently introduced about 1655 in a stuffed bird. Its acrid oil is used against mosquitoes, hence the name fleabane. Sensitive people may be allergic to this weed; however, it is still collected for medicinal purposes. Canadian fleabane (*Conyza canadensis,* formerly *Erigeron canadensis*) can be ground up to make melonworm repellant.

FORGET-ME-NOT (*Myosotis sylvatica*)

These lovely, dainty plants are most effective in the garden when planted in large drifts. The popular variety 'Blue Bird' is a compact, attractive plant with the bluest of all blue flowers. The flowers are too fragile for indoor arrangements and do not last well when cut. If cut as they fade, the flowers will bloom over a long period of time. The variety *alba* has white flowers.

Botanists say the hairy stems of many forget-me-nots are intended to keep ants and similar insects from stealing the nectar reserved for flying insects that pollinate.

FUNKIA, PLANTAIN LILY (*Hosta crispula*)

This plant is characterized by its bold, colorful, fragrant flowers that are excellent for cutting and its leaves that are dramatically splendid in cut arrangements. *Hosta crispula,* of Japanese origin, has 2-inch flowers of funnel form and lavender color, and each leaf has a serrated look. Protect from strong winds.

GENTIAN (*Gentiana*)

Ordinarily the gentians are without fragrance, but the rare perennial fringed gentian has a delightful scent suggestive of strawberries.

Septemfida var. *lagodechiana* is an enchanting summer-flowering gentian with deep, true blue, 1-inch blooms with a white throat. The dense, heart-shaped foliage, which likes summer heat, forms an attractive groundcover.

To sprout gentian seed, get a box about 6 inches deep, put in 2 inches of coarse gravel or broken crock pieces, then 3 inches of fine, porous soil. Plant the seeds in the soil and cover with 1 inch of peat moss. Sink the box in the garden in filtered shade and keep it moist until the seeds germinate. Treat other hard-to-sprout seeds like cyclamen, smilax, violet, canna, Christmas rose, and nasturtium this way as well.

GERANIUM (*Pelargonium*)

The hybrid geraniums in scarlet, cherry, salmon, coral, and white are simply gorgeous. In addition to these introductions, geraniums that could only be propagated by cuttings can now be grown from seed.

Geraniums make good companions for roses because they repel Japanese beetles. Geraniums also repel cabbageworms when planted among the brassica, and the white geranium is helpful when planted near corn.

GERANIUM, SCENTED-LEAVED (*Pelargonium*)

In competition with the gorgeous Zonals, the sweet-leaved geraniums fell out of fashion, but the wheel is turning again. The sweet-leaved geraniums offer the best of scents and immense variety, including the filbert, nutmeg, cinnamon, almond, lemon, orange, apple, anise, rose, pink, musk, violet, lavender, balm-scented, and many more. All of these also bloom but the flowers are inconspicuous.

A leaf of the rose-scented geranium in the bottom of the glass imparts a delicious flavor when making apple jelly. Use the oil of geranium as an insecticide against red spiders and cotton aphids.

GLADIOLUS (*Gladiolus*)

Thanks to Dr. Forman McLean of the New York Botanical Gardens, we now have sweet-scented gladiolus. Gladiolus are gorgeous flowers—tall and dramatically beautiful, coming in many brilliant colors. But keep them away from peas and beans, which dislike their presence.

GLOBE ARTICHOKE

This elegant perennial dates back to the 16th century and is a form of cardoon, a favorite of the ancients. It prefers cool, often foggy summers and is grown mainly in California. If allowed to blossom, the great thistlelike flowers are gorgeous. The edible buds are a great delicacy and one of our most expensive "vegetables."

GLOXINIA (*Gloxinia*)

One of the most popular gift plants, this native of tropical America is remarkable for its richly colored, velvety leaves and large, bell-shaped flowers. It is widely cultivated as a houseplant and in greenhouses.

GOLDENROD, SWEET (*Solidago odora*)

There are more than 100 species of goldenrod. It's used for dyeing, giving color from mustard to brown olive, depending on the strength used. Anise-scented goldenrod is a delightful beverage plant. Collect the leaves when the plant is in bloom and use them fresh or dried, with peppermint leaves, as an after-dinner summer tea.

Goldenrods are also medicinal, for not only did Native Americans employ this common plant as a cure for sore throat and for pain in general but herbalists also recommend it today as a diaphoretic in colds and coughs and as an aid in rheumatism. In fact, the generic name, *Solidago,* means "I make whole," that is, heal.

Goldenrod is incorrectly blamed for hay fever and similar allergies; ragweed, which blooms at the same time, is the real culprit.

Goldenrod

GOLDENSEAL, INDIAN TURMERIC, ORANGEROOT, YELLOW PUCCOON
(*Hydrastis canadensis*)

A perennial flowering forest plant of the eastern United States, goldenseal has showy, 8-inch-wide leaves and greenish white, ¼-inch flowers, followed by large clusters of red fruit. Native Americans dried and ground the thick yellow root for medicinal purposes.

GROUNDSEL (*Senecio vulgaris*)

This common plant of waste places and pastures is of erect growth with grayish green leaves, branching with jagged lobes. Flower heads are in close terminal clusters, the individual flowers being of tubular shape, solid, and yellow, like minute candles, and possessing no ray petals. This herb is particularly rich in minerals, especially iron. Animals seek it out as

Caged birds seem to appreciate the value of mineral-rich groundsel, or ragwort.

a tonic, especially poultry, and it is relished by caged birds kept as pets. The herb has powerful drawing and antiseptic properties. Mixed with ground ivy, it makes an important poultice. It strengthens the eyes and reduces inflammation.

GUNNERA (*Gunnera insignia*)

These ornamental perennials are mostly from the Southern Hemisphere. They prefer to live on the inaccessible mountainside of Guatemala and Panama. The huge leaves measure from 4 to 5 feet across and resemble our common pot geranium. Gunnera is of such massive proportions that usually it is planted in public grounds for landscape effect.

One leaf is capable of serving several people as an umbrella. Gunnera bears a large stalk covered with thousands of small, brownish, wind-pollinated flowers. A similar species of gunnera is found in the mountains of Hawaii.

HENS AND CHICKENS, HOUSELEEKS
(*Sempervivum tectorum*)

These plants grow without care in a sunny spot. There are 60 or more varieties, from tiny green cobwebby ones to sturdy copper-colored ones like 'Heuffell'. They need little soil and are fine for covering walls, rocky hillsides, or small pockets in rocks or paving. They are hardy and colorful in late winter. The Romans thought they would protect a house from lightning if they grew on the roof.

HEPATICA (*Hepatica acutiloba*)

These perennials bloom in early spring. Their soft pink flowers resemble anemones and are surrounded by sharply pointed leaves that often persist through winter. Just 6 inches tall, these wildflowers are fine for naturalizing.

HOLLYHOCK (*Althaea rosea*)

This native of China is a tall perennial herb that is usually treated as a biennial in our gardens. The roots are demulcent and emollient, thus of good use as cold and diuretic remedies. The generic name, *Althaea,* is derived from the Greek *althainein,* meaning "to heal." Hollyhock is good to rub on bee stings if the leaves are bruised in oil and made into an ointment.

HOLY HERB, YERBA SANTA (*Eriodictyon californicum*)

The small white and lilac flowers are fragrant. The leaves are strongly aromatic when crushed and were used medicinally by Native Americans for colds.

HOPS (*Humulus lupulus*)

Prepare the young sprout like asparagus or use as a potherb for soup. At one time, ground hops were a substitute for baking soda. The greenish yellow flowers are used commercially in the preparation of beer and ales.

A pillow of hops will induce sound, refreshing sleep. Powdered hop leaves are toxic to southern armyworms and melonworms.

HOREHOUND (*Marrubium vulgare*)

The wild and woolly horehound is not a very decorative herb; the flowers are small, mintlike, and white, while the leaves are covered with a dense, felted wool that gives the whole plant a downy, whitish appearance. Unlike most mints, the flavor and medicinal properties of horehound are not volatile or easily lost, so the plant can be used fresh or dried or boiled without driving off the flavor. Horehound tea is good for colds and horehound candy is good for coughs.

To make horehound candy, prepare a small decoction by boiling 2 ounces of the dried herb in 1½ pints of water for about 30 minutes;

strain this and add 3½ pounds of brown sugar. Boil over a hot fire until it reaches the requisite degree of hardness (testing from time to time in a cup of cold water). Then pour into well-greased, flat tin trays. Mark into sticks or squares with a knife as it becomes cool enough to retain its shape.

HYACINTH (*Hyacinthus orientalis*)

These fragrant flowers are natives of eastern Mediterranean lands and western Asia. The plant has a large bulb with a purple or white scaly covering and from four to six linear-lanceolate, hooded, bright green leaves. The flowers are produced in spring on a terminal cylindrical cluster on a central stem from 6 to 12 inches high. In cultivated varieties the flowers are white or various shades of red and blue. The white and blue companion well with tulips by offsetting their flashing, brilliant colors.

HYSSOP (*Hyssopus officinalis*)

This plant, with its dark blue blossoms, neat linear leaves, and bushy growth, is attractive in the flower garden. It may be clipped like boxwood to make a low hedge, although this will be at an expense of the blossoms. There are also white and pink varieties. Hyssop is good to plant with grapevines. Planted with cabbage, it will deter white cabbage butterflies.

ICE PLANT (*Mesembryanthemum crystallinum*)

This only is a little trailer and creeper grown for its thick, succulent foliage and tiny, white blossoms. The fat, fleshy leaves are covered with glistening dots that have the appearance of ice, thus the popular name. The plant may be started from cuttings and does well in the driest and thinnest of soils. Try it in hanging baskets, window boxes, and rock gardens.

IDENTIFYING SPECIMENS

To identify poisonous plants, contact your local Poison Control Center or your local Cooperative Extension office.

INDIAN BLANKET (*Gaillardia pulchella*)

Our native gaillardias have been developed into a variety of horticultural forms. The rich yellow and red daisies thrive in hot, dry places and bloom all summer long. They are fine as cut flowers and are easy to grow. Plant Indian blanket with calendula, Iceland poppy, and Maltese cross.

INDIAN LICORICE (*Abrus precatorius*)

Also known as rosary pea, this plant is a native of India. Generally the seeds are brought into the United States by tourists who have visited tropical areas. The hard, attractive, bright scarlet or orange seeds hold abrin, one of the most deadly poisons known, but they have been used for rosaries, necklaces, bracelets, and toys.

Indian paintbrush

INDIAN PAINTBRUSH (*Castilleja*)

These perennials grown to 1$\frac{1}{2}$ feet tall. Their large, showy, brilliant red to orange, leaflike bracts surround the tiny, inconspicuous flowers.

Indian paintbrush is thought to be parasitic on other plant roots. It is difficult to transplant, but it can be grown from seed if seed of another plant is grown in the same pot. Blue grama (*Bouteloua gracilis*), a grass, will encourage the seed to sprout.

Seed may be sown outdoors on bare ground around sage or clumps of native grasses. The plant has a long period of bloom and is attractive to hummingbirds.

INDIAN PIPE (*Monotropa uniflora*)

These strange parasitic ghost flowers never fail to shock when we unexpectedly come upon them, clustered in seemingly sinister closeness, deep in some somber woodland. At certain periods of their development the waxen white or flushed flowers give forth a "delicate and wholly sweet scent." They are often found in dry woods, usually under pine or beech trees, and get their food from decaying plants in the soil.

IRIS, FLEUR-DE-LIS (*Iris*)

The name comes from the Greek word for "rainbow." These perennial plants in many attractive colors grow throughout the temperate region, blooming in spring, summer, and a few in autumn. Many of the tall bearded varieties are gorgeous, earning for themselves the name poor man's orchid.

Iris

Their worst enemy is the iris borer. If you find these, dig up the plants, clean out the borers, and dip the rhizomes (roots) in diluted chlorine bleach (1 part bleach to 10 parts water) before replanting.

Iris is one of the earliest plants to bloom in the garden and teams well with grape hyacinth and daylilies. The dried root of the Florentine iris is called orrisroot. It is used in perfumes, powders, and medicines.

ISMENE (*Hymenocallis*)

Ismene is a real treasure for southern gardens, where the bulbs are hardy. In June four or five giant funnel-shaped flowers are borne on each stem. The flowers are richly perfumed and uniquely shaped, having a delicately fringed cup framed by five long sepals, and come in shades of white and yellow. This exquisitely different flower may be grown as a pot plant in the North.

IXIA (*Ixia hybrida*)

Ixia, boasting the greatest range of color of any bulb, is cool-loving, which makes it very desirable for indoor cultivation. The stems are slender and graceful; the flowers offer white, yellow, purple, ruby, blue, and green, in many shades and variations, usually with a black eye. The flower spikes contain 6 to 12 flowers, each 1 or 2 inches in diameter. Pot the bulbs in late autumn in a mixture of loam, leaf compost, and sand, placing eight or nine in a 6-inch pot. Keep cool and dark until growth starts, then bring into light and warmth.

JADE PLANT (*Crassula*)

Jade plants have been popular for a long time, and are favorites with plant lovers in North America. Some advanced growers are even treating it as a tropical bonsai, yet the jade plant is easy for beginners to grow as well. Common jade, *Crassula ovata,* has dark green leaves that become red-edged in sufficient sun. Mature plants bear clusters of star-shaped white or pale pink flowers at the branch tips in winter or spring. Dwarf cultivars are simply smaller in all their parts.

Jade is virtually trouble-free. Mealybugs are the chief insect pest; eradicate them by using a cottom swab dipped in alcohol. Malathion should not be used on jade or other crassulas.

JEWELS-OF-OPAR (*Talinum paniculatum*)

This garden treasure has bright, waxy green foliage with a myriad of cameo pink flowers that open every afternoon. It is 1^1/$_2$ feet tall and good to use for borders and rock gardens. It is heat resistant and naturalizes well.

JEWELWEED
(*Impatiens capensis or Impatiens pallida*)

This tender, succulent, tall-growing annual is often found in extensive patches in damp woods. The expressed juice is a light orange color and is an antidote for poison ivy, which it often grows near.

The jewelweed has pretty flowers, butter yellow in color, followed by slipper-shaped seedpods about ¾ inch long. When ripe they will suddenly split, the two sides curling back into tight spirals with an audible snap, and throw their seeds in all directions.

Jewelweed is so called because its leaves are unwettable; rain will stand on the leaves in round drops, shining like jewels, without ever wetting the leaf surface.

JIMSONWEED (*Datura stramonium*)

This weed spreads usually by having its seeds carried by birds. It is very poisonous, causing a kind of intoxicated state, but has some medicinal value. It is helpful when grown with pumpkins. See *Datura* in this chapter.

JOB'S TEARS (*Coix lacryma-jobi*)

This 3-foot ornamental grass bears hard, pearly white seeds that make distinctive necklaces.

JOE-PYE WEED (*Eupatorium purpureum*)

The Joe-pye weed grows mostly in moist damp thickets, ditches, and streams, invading only badly drained meadows. The plant is named for a Native American herb doctor, but is also called feverweed. The juice is said to heal open sores and bruises; hunters have observed that wounded deer will search for it and eat it. (Early peoples leaned much about the medicinal qualities of plants by watching animals when they were sick or injured.) Feverweed is closely related to thoroughwort (*E. perfoliatum*).

JUNGLE CACTI, ORCHID CACTI (*Epiphyllum*)

These strange flowering plants are almost unbelievably beautiful. Somewhat resembling orchids in their delicate loveliness, they include both day- and night-blooming varieties. They are every color of the rainbow except blue. If you are a cactus lover, these will delight you.

To grow these, begin with the right potting mix: 4 parts leaf mold, 1 part fine redwood bark, 1 part well-aged steer manure, 1 part perlite, and 1 part horticultural charcoal. (If you cannot obtain leaf mold, commercially packaged camellia-azalea mix is a good substitute). For each cubic foot of mix, add ½ cup bonemeal. If you live in an area that has alkaline water, add ½ cup garden sulfur per cubic foot of mix.

Keep the plant in a pot a bit small for the size of the plant, and do not make the soil too firm. During November and December, keep the soil nearly dry (but do not let the skins shrivel); at other times keep it moderately moist. Grow in a temperature of 60° to 70°F except when resting the plant (when it should be kept at 50° to 55°). Give all light possible in winter, but from March through September shade it lightly from strong sun. It may be put outside in summer.

KALANCHOE (*Kalanchoe*)

These showy, winter-blooming plants make long-lived pot plants. Tender succulents with attractive flowers, they are chiefly natives of the tropics and South Africa and belong to the Crassula family. Sow in spring for winter and spring bloom. The flowers are orange-scarlet, pink, orange-red, and white.

KISS-ME-AT-THE-GARDEN-GATE (*Viola tricolor*)

This is the wild pansy sometimes called Johnny-jump-up. It is a hardy perennial but rather short-lived. Sow the seeds of this charming little flower outdoors; it will need no special care.

KITAIBELIA (*Kitaibelia vitifolia*)

This hardy perennial flowering plant with ornamental foliage is a native of eastern Europe and belongs to the hollyhock family, Malvaceae. The stems, which grow about 8 feet in height, have large, vinelike leaves; the large pink flowers open in summer.

This plant thrives in a sunny position and is fine for planting at the back of the herbaceous border. It prefers light, well-drained soil, and may be planted in autumn or spring. Propagation is by division of the roots at planting time or by seeds outdoors in summer.

KNIPHOFIA (*Kniphofia*)

This member of the Lily family is sometimes called torch lily or red-hot poker. Hailing from Africa and Madagascar, these perennials are also called tritoma. They are very showy plants for borders, mostly in the red and yellow color range. In cold areas they will need winter protection.

LADY'S SLIPPER (*Cypripedium*)

These beautiful orchids are found in swamps and wet woodlands, most numerous in the eastern and southeastern states. The flowers that grow on straight stems have a pouchlike lip, hence the name lady's slipper. They are beautiful, but be careful in handling; all species contain a poisonous substance in the stalks and leaves that frequently causes dermatitis.

LAVENDER (*Lavandula*)

Lavender has a long and creditable history as a stimulating or medicinal plant. Among its many virtues are its soothing effect on the stomach, its use as a disinfectant, and its power to relieve sprains, headaches, and toothaches. Flies steer clear of it and, according to old demonologists, the fragrant odor of lavender is guaranteed to ward off evil spirits. The essential oil of lavender is produced from the leaves and has been an ingredient of love philtres from earliest times. The oil also stimulates the generation of new cells and in so doing helps to preserve the health and youth of the skin.

Three varieties, used in industry, medicine, and household preparations, are beautiful additions to the garden. These are spike (*Lavandula latifolia*), true or English (*L. angustifolia*), and French lavender (*L. stoechas*). Both flowers and leaves are fragrant. Plants must be grown in poor soil to produce the most fragrance; in good soil they grow more luxuriantly but fragrant essential oils are lacking.

LICORICE (*Glycyrrhiza glabra*)

The wild licorice of North America is *Glycyrrhiza lepidota*. From licorice roots we get a valuable flavoring material that is 50 times as sweet as sugar. But oddly, although sweeter than sugar, licorice has the power to quench thirst. In 1951 it was discovered that licorice root contains the female sex hormone estrogen, used in the treatment of menopausal problems.

Chewing a licorice stick helps those who wish to stop smoking.

Even the fibers that remain after the licorice is extracted from the roots are valuable. They are used in making fire-fighting foam, boxboard, insulation board, and other products.

LIGULARIA (*Ligularia dentata*)

This bold perennial has large, heart-shaped leaves, bronzy green on the upper surface and a rich mahogany red below. In July and August, orange daisy flowers brighten the plant. It does best in a constantly moist soil in partially shaded location.

LILY OF THE VALLEY (*Convallaria majalis*)

Lily of the valley, a single species, is a lovely, fragrant spring-blooming herb. It is commonly cultivated in the partial shade of gardens everywhere and is a great favorite. It looks best in mass plantings but makes a fine groundcover for narcissus. However, if narcissus and lily of the valley flowers are put together in the same vase, they will soon wither.

To get flowers by Christmas, professional Dutch growers of the 1870s planted moss-wrapped pips (rooted buds arising from the rootstalk) in sand about a month before the holiday, giving them bottom heat and liberal watering until sprouts appeared.

The leaves, flowers, berries, and rootstalk are well known for their toxicity. They contain dangerous amounts of cardiac glycosides (convallarin and convallamarin). Children have lost their lives just by drinking the water from a vase containing a bouquet of lily of the valley.

Lily

LILY (*Lilium*)

This perennial comes in many colors and is lovely used with delphinium, aster, and marigold. It grows well in raised beds and is ideal for mass plantings along a wall. *Lilium longiflorum,* the Easter lily, is pure white with a powerful fragrance, and is a very popular gift at Easter time. Easter lilies should be watered regularly through their flowering period. They've been cultivated for more than 3,000 years.

LIVE FOREVER (*Sedum telephium*)

This hardy succulent lives up to its name. It displays heads of pinkish bloom above gray-green foliage in late summer. The plant will tolerate poor growing conditions, is transplantable in almost any season, and propagates readily by leaf cuttings. Recently developed varieties have flower colors ranging from ivory to rosy red.

LOBELIA, INDIAN TOBACCO (*Lobelia inflata*)

This field plant of North America is commonly called Indian tobacco. The leaves are pointed and yellowish green with hooded flowers of a brilliant blue. This was one of the most important herbs of Native Americans but can be harmful in internal use.

Since the lobelia blossom is such a lovely blue, it combines well with white alyssum and red rosebuds for bouquets or tiny place markers. And lobelia, like alyssum, is an excellent plant for "clothing" a pot, planter, or window box planted with larger plants. Keep well pinched to promote shapeliness and persistent bloom.

LOVE-IN-A-MIST (*Nigella*)

Whether you think of it as love-in-a-mist or devil-in-a-bush, its other name, this charming blue, purple-blue, or white annual makes a hardy plant for the border, where it is very beautiful for several weeks. Well-grown plants bear finer blooms and remain in flower longer than those that are crowded.

LOVE-LIES-BLEEDING, TASSEL FLOWER
(*Amaranthus caudatus*)

This native of India, the Philippines, and other warm countries has drooping stems bearing dark, reddish purple blooms.

LUPINE (*Lupinus*)

There are about 100 species of lupines found in the United States. They are showy, hardy, and grow profusely in fields, on ranges, and on mountainsides. The pealike blossoms, in loose clusters at the ends of the branches, may be blue, purple, yellow, pink, or white. Some but not all species of lupine hold toxic alkaloids throughout the entire plant and are a common cause of stock poisoning.

Lupine, which is a legume, is helpful to the growth of corn as well as most cultivated crops. Plant in full sun or light shade.

Cultivated lupines come in a gorgeous assortment of colors—blues, pinks, reds, purples, maroons, and many striking bicolor combinations. The individual flowers are large, with some standards up to 1 inch across.

MARIGOLD (*Tagetes*)

The marigold, "herb of the sun," is as helpful in the flower garden as in the vegetable garden, serving the same purpose of driving away nema-

todes. They are particularly useful with chrysanthemums, calendulas, and dahlias. Brown areas
on lower leaves signal the underground
feeding of the foliar nematodes.

Marigold

MARTYNIA, DEVIL'S CLAW, UNICORN PLANT (*Martynia*)

Sonoran tribes domesticated this plant for the
long black fibers of the "claws" on the fruit.
They were used in basket weaving. The seeds
may be eaten like sunflower seeds or pressed
for oil. The fuzzy green seedpods are picked when half-grown if wanted
for pickling. The strange mature pods are often used for decorations.

MEADOW SAFFRON (*Colchicum autumnale*)

This plant blooms in fall, and will do so even without soil around its poisonous bulbs. Colchicum, derived from the bulbs, is the source of a present-
day drug used to treat gout and also has been used to produce doubling
of chromosomes in plants.

MEADOWSWEET (*Filipendula ulmaria* or *Spiraea*)

Meadowsweet is perhaps the most important of all nature's remedies.
Chemists discovered acetylsalicylic acid in this plant, synthesized it, and
called it aspirin. But the dried leaf in its natural form can be ground with
mortar and pestle and used wherever aspirin is recommended without
aspirin's side effects. Meadowsweet is particularly recommended by
herbalists as an antidote for rheumatism, arthritis, gout, and all kidney
and bladder complaints, particularly gravel and cystitis.

MESCAL BEAN (*Sophora secundiflora*)

The beautiful green glossy leaves and the display of fragrant sweet-
pea–like lilac flowers, borne in a drooping inflorescence, are very impressive. This excellent landscape plant is particularly good for poor alkaline
soils and tolerates heat and drought well.

The beautiful seeds are large, brilliant, and coral, and even though they are very poisonous, the Mexicans use them for necklaces. Powdered seeds of this flowering shrub are toxic to armyworms.

MEXICAN HAT (*Ratibida columnifera*)

This flower, dark mahogany red or yellow with a red blotch, is a form of the prairie coneflower, the dramatic yellow daisy with cone-shaped brown center. The perennial plant is about 1 foot tall, and given full sun it will bloom continuously from summer to fall. Grow it with the purple cone-flower (*Echinacea purpurea*), which is taller (3 to 5 feet) and also flowers from summer to fall. The lavender-purple daisies make excellent cut flowers.

MIGNONETTE, LITTLE DARLING (*Reseda odorata*)

The mignonette has a low bushy mass of smooth, soft green leaves. The tiny flowers, growing on tall spikes, are yellowish white with reddish pollen stalks inside. The larger-flowered varieties are prettier but less fragrant. The mignonette is a good border plant that grows best in cool temperatures and light soil. If the seed fails to sprout, check for ants; they have been known to carry off the seed.

MILKWEED (*Asclepias*)

These rather coarse erect plants grow in dry fields, on hillsides, in woods, and along roadsides. Their profuse milky juice accounts for their name. Milkweed juice is said to remove warts.

There are about 60 species distributed throughout the United States but only a few are cultivated. The white-, pink-, or rose-colored flowers develop in round clusters. The large, rough-surfaced, flat seedpods are filled with many seeds, each with a tuft of long, silky hairs. Divested of the seeds, the silky hairs are sometimes used as background for pictures made with butterflies and dried flowers.

An old-time method of trapping cutworms consisted of placing compact handfuls of milkweed in every fifth row or hill of cultivation and tamping them down. Cutworms gathered in this trap material where they could be easily collected. Clover and mullein were also used for this purpose.

Some species of milkweed have medicinal value, but all are known for their content of resins and most of them are exceedingly poisonous to humans and livestock. The poison is concentrated in the stout stem and in the leaves.

MIMOSA, HUMBLE PLANT, TOUCH-ME-NOT, SENSITIVE PLANT (*Mimosa pudica*)

The sensitive plant is so called because its leaves will fold together with sufficient irritation or cloudy weather, representing one of the most re- markable cases of physiological response in the plant kingdom. The timo- rous plant has a mechanism that reacts whenever a beetle, ant, or worm crawls up its stem toward its delicate leaves; as the intruder touches a spur, the stem raises, the leaves fold up, and the assailant is either rolled off the branch by the unexpected movement or is obliged to draw back in fright.

MINT (*Mentha*)

Many delightful mints are used medicinally, for cooking, and for fragrance. These seven are among the most frequently grown:

Apple mint (*Mentha suaveolens*) has stiff stems growing 20 to 30 inches tall. The rounded leaves are slightly hairy and gray green, about ¼-inch long. This mint has purplish white flowers and is not good for culi- nary use, but American apple mint (*M. gentilis* 'Variegata') has a fruity, re- freshing odor and taste.

Corsican mint (*M. requienii*) is a creeping sort that grows only about 1 inch high. The tiny, round leaves form a mosslike mat. Small, light purple flowers appear in summer. When bruised or crushed underfoot, the foliage has a delightful minty or sagelike fragrance. Plant it between stepping-stones.

Golden apple mint (*M. gentilis*) has smooth, deep green leaves, varie- gated with yellow. It grows about 2 feet high and makes an attractive groundcover where taller, spring-flowering bulbs are planted.

Orange mint or bergamot mint (*M. piperata* var. *citrata*) has lavender blossoms in dense flower spikes and a characteristic minty odor.

Pennyroyal (*M. pulegium*) is an attractive plant with small, rosy lilac flowers, blooming late in the summer and early autumn. Pennyroyal is believed to repel insects in the garden and is good to rub on a cat's collar to repel fleas.

Peppermint (*M.* × *piperita*) or its flavor is familiar to many people. It has strongly scented small purple flowers and 3-inch leaves with toothed edges. Peppermint grows to about 3 feet.

The plant produces many creeping stolons (runner) that spread quickly in favorable environments. It is propagated by pieces of the runners and, like many of the mints, can be increased from slips (clones) planted in moist, sandy places.

Mint

Leaves are darker than those of the spearmint, larger, and not so crinkly. The plant has a reddish tone. Even the small leaves near the rosy lavender flower spikes have red on their margins.

A medicinal oil is expressed from the dried leaves of the flowering plants and is considered a powerful analgesic. The oil is also used to flavor tooth powders, pastes, and washes. The leaves of peppermint make a wonderful tea, reputed to be good for the nerves. One author has called this tea one of nature's best tranquilizers. Peppermint is used in cooking for flavoring candy.

Spearmint (*M. spicata*) is another familiar species. It is usually used in mint jelly. The leaves are dark green; the mature plant is 1^1/$_2$ to 2 feet tall.

Mints have been used medicinally since ancient times and modern commerce still makes use of them. Spearmint and peppermint are two of the most common flavorings for everything from chewing gum to toothpaste.

Most mints do best in light, moist, moderately rich soil, and in shade or partial shade. They spread underground by means of stems and runners. Use the leaves fresh or dried; add them to potpourris, lamb, and jelly. Spearmint is the best for garnishing iced drinks (the mint juleps of the South). Add fresh leaves of peppermint, pineapple, apple, and orange mints to fruit cocktails or sprinkle over ice cream.

MONEY PLANT, HONESTY (*Lunaria annua*)

This purple or white biennial is best as a filler plant until fall, when the silver pods begin to show at their best. Use dried in a winter bouquet or mixed with other dried material.

MONKEY FLOWER (*Mimulus*)

These perennials with large, showy, snapdragon-like flowers like to grow beside streams and ponds, even on swampy land. The bush monkey flower (*Mimulus aurantiacus*) is a flowering shrub and grows to 4 or 5 feet. It blooms over a long period with flowers in shades of orange, yellow, and red. Grow in full sun in well-drained soil; the plants are drought tolerant.

MOTHER-OF-THOUSANDS, STRAWBERRY GERANIUM (*Saxifraga stolonifera*)

Many admirers have grown mother-of-thousands in hanging baskets or window boxes. Prettily colored, the leaves are light green, variegated, with silver above and reddish on the undersides. The flower stalk rises about a foot and produces white flowers in loose panicles.

From the rosette of leaves come runners that, as they touch moist soil, root and produce new plants. When a young plant acquires six leaves, it may be broken off from the parent and started on its own. This plant grows best in rich, sandy soil with a little filtered sunlight. Mother-of-thousands may be planted outdoors and will survive mild winters, even in the vicinity of New York.

MULLEIN (flannel leaf, beggar's blanket, Adam's flannel, velvet plant, feltwort, bullock's lungwort, clown's lungwort, cuddy's lungs, tinder plant, rag paper, candlewick plant, witch's candle, hag's taper, torches, Aaron's rod, Jacob's staff, shepherd's club, quaker rouge) (*Verbascum thapsus*)

For sheer diversity of names, this one wins the prize.

Mullein is valuable in alleviating human ills and has long been known to many people of varying cultures, all of whom agree on its value as a healing herb. The fresh and dried leaves and the fresh flowers are the parts used in home remedies. Officially, mullein has been recognized as a valuable demulcent, emollient, expectorant, and mild astringent.

NARCISSUS, DAFFODIL, JONQUIL, PAPER-WHITE (*Narcissus*)

This lovely perennial in solid borders and masses is one of the first flowers to welcome spring. The flowers may be yellow, yellow with white, orange, pink, apricot, white, or cream. But beware, the bulbs are poisonous and must never be eaten.

Sow African marigolds (*Tagetes erecta*) before planting narcissus bulbs to defeat certain nematodes that often attack the bulbs. Sulfur-containing substances called thiophenes are present in root exudates of African marigolds, as well as in many other plants from the Compositae and Umbelliferae families; these repel nematodes.

NASTURTIUM (*Tropaeolum majus*)

Forms of this South American tendril climber brighten a fence or trellis with brilliant shades of yellow, orange, and red. It flowers best in full sun and clean soil. Potash added to the soil aids bloom development. *Nasturtium,* an old Latin word used by Pliny, was derived by him from *nasus,* "the nose," and *tortus,* "twisted," in reference to the supposed contortions of the nose caused by the hot, pungent odor and taste of these flowers. Nasturtium and rose geranium in a nosegay complement each other.

Nasturtium is an amiable flower that gets along in the worst sort of soil. It was an inhabitant of many an early American garden, being ideal for bouquets, and the flowers, leaves, and seeds were eaten. Flowers and leaves went into salads, and the seeds were pickled after being picked green; so treated, they were thought to be a nice substitute for capers.

Sow nasturtium seed around apple trees in spring to combat the woolly aphis. Sow a few in each hill to repel cucumber beetles. Sow with broccoli against aphids. Though nasturtium often have aphids of their own, they seem to keep them away from their companion plants.

NETTLE, STINGING NETTLE (*Urtica dioica*)

The name *diocia* indicated that the staminate and pistillate flowers, instead of being together on the same plant, are on separate plants. Therefore, look for seed only on some, not all, nettle plants. Both flowers and seed are greenish and inconspicuous.

This weed causes much discomfort when touched. The stinging hairs cover practically the whole plant and prick the skin with their tiny silica tips, letting in enough formic acid to be felt. Relieve the itching with yellow dock leaves (usually found growing nearby). Crush and apply as a poultice.

Even so, the young leaves are very good to eat, though you must wear gloves when picking them. Wash and cook quickly. Boiling renders the stinging hairs harmless. Nettle, which is slightly laxative, is a healthy and easily digested vegetable, very high in iron. In England, a pleasant drink called nettle beer is relished by ailing aged folk.

This is a valuable plant for gardeners, especially good when used in the compost pile. Nettles help plants withstand lice, slugs, and snails in wet weather. They also strengthen the growth of mint and tomatoes and increase the aromatic qualities of many herbs such as sage, peppermint, marjoram, angelica, and valerian. Fruit packed in nettle hay will be free of mold and keep longer.

NIGHT-BLOOMING CEREUS (*Cereus grandiflorus*)

No mention of night-blooming flowers would be complete without this one, a strange cactus of the West Indies. Its bristling, tortured stems give birth in the darkness to the most spectacular of blossoms.

Truly a night bloomer, often at midnight, this lovely flower with the saxophone stem also emits a delightful fragrance. The waxy white blossoms with delicate pink sepals are breathtakingly beautiful. Once it has bloomed, however, the flower's sense of humor seems to assert itself; the spent blossoms hang limply, looking a bit like the legs of a freshly plucked chicken, and the rose sepals descend like bedraggled feathers from the yellow tubular stems.

NIGHT-BLOOMING JASMINE (*Cestrum nocturnum*)

This member of the Nightshade family is cultivated for its fragrant, night-blooming, trumpet-shaped flowers. The leaves are somewhat oval; the fruit, following the flowers, is small and berrylike. All cestrums are extremely poisonous if eaten.

ORCHID (JUNGLE) CACTI
(*Epiphyllum*)

These incredibly beautiful plants are the so-called
jungle cacti or leaf-flowering cacti. The name
Epiphyllum is derived from *epi,* "upon," and
phyllon, "a leaf," and refers to the location of
the flowers. Epiphyllums may be grown in baskets suspended from the greenhouse roof. They
require:

Orchid cactus

Light. Filtered light preferred, never direct
noonday sun.

Humidity. Approximately 50 percent. Mist during summer months.

Temperature. 45° to 70°F preferred greenhouse temperature. Protect
from frost.

Watering. Water when the soil surface has dried to a depth of
1½ inches. Water less frequently during winter.

Potting mix. Must be coarse and fast draining. Packaged indoor
planter mix, if coarse enough, may be used or a mix of the following proportions: 4 parts leaf mold, 1 part perlite, 1 part medium bark, 1 part horticultural charcoal.

Fertilizer. Mild fertilizer (with no higher than 10 percent nitrogen),
once a month starting in April, ending in fall. Once in February and again
in November, apply low-nitrogen fertilizer to promote blooms.

The orchid cactus comes in almost all of the colors of the rainbow. It
should be kept potbound to induce flowering. Blooms normally occur in
two- to three-year-old root-bound plants.

ORCHIDS

All members of the family look a bit like the gorgeous flowers found in
greenhouses and florists' shops. The family relations number more than
6,000 species, and many are dainty wildflowers that grow in cool, damp
American woods and swamps. These include the white, yellow, and purple
lady's slippers; the calopogon; the violet-pink arethusa; the calypso; and
the fragrant, pale pink moccasin flower.

In the tropical countries many orchids are air plants, attaching themselves to the bark of trees and sending roots into the air from which they take their food.

The orchid takes many unusual forms; the blossoms of one species look like butterflies. Another species furnished vanilla. The tubers of still another are dried for their nourishing starch. They are sold on the market as "salep," which is used in medicine as a lubricant.

Orchids have the longest blooming period of any flowers. The blossoms of certain kinds may remain open for five weeks or even more.

This pink lady's slipper is a North American woodland orchid, and is as lovely as its many tropical relatives.

The largest orchid is *Grammatophyllum speciosum,* native to Malaysia. A specimen recorded in Penang, West Malaysia, in the 19th century had 30 spikes. The plant was 8 feet tall with a diameter of more than 40 feet. The largest orchid flower is that of *Phragmipedium caudatum,* found in tropical areas of America. Its petals are up to 18 inches long with a maximum outstretched diameter of 3 feet.

OSWEGO TEA, MONARDA (*Monarda didyma*)

This perennial, which grows to 4 feet under ideal conditions, has very aromatic foliage. Large, brilliant scarlet flowers from summer to early fall entice hummingbirds.

Oswego tea (so named by the Shakers, who found the herb growing in profusion near Oswego, New York) was later known as bee balm. The healthful refreshment and delicious fragrance of this plant have made it a favorite among Native American teas, and its down-to-earth beauty is valued in the flower garden. Sage or basil, freshly ground, or dried peel of orange or lemon, gives variety to the tisane.

PANSY (*Viola × wittrockiana*)

The "flower with a face" is really a cultivated variety of violet. The lovely blossoms may be purple, violet, blue, yellow, white, brown, or a mixture of

all these colors. In some climates pansies are best
planted in fall. The more you pick, the more pansies
will bloom. If allowed to go to seed, they will stop
blooming. You can increase pansies by taking cuttings from the center of the plant, but side shoots
and branches will also grow. Place in a mixture of
sand and loam, shade them from the sun, keep
moist, and they will soon strike root.

Pansy

Wild pansy (*Viola tricolor*) germinates well if
grown near rye, the growth of rye is improved by a
few pansies. The same is not true if pansies are
grown with wheat.

PAPER FLOWER, WOOLLY (*Psilostrophe tagetina*)

This very showy perennial covers itself with bright yellow flowers, forming
a cushionlike mound. It is excellent as a border plant. Flowers dry and stay
on the plant, making them valuable for use in dried arrangements.

Pasque flower

PASQUE FLOWER, EASTERN WILD CROCUS (*Pulsatilla patens*, formerly *Anemone patens*)

This wildflower, one of the first of spring, has lovely,
large, pale blue or violet, bell-shaped flowers. They are
2 to 3 inches in size and appear before the foliage.

Sow outdoors in fall for germination the following
spring, or cold stratify 30 days and sow in spring. The
plants are nice in rock gardens and also for dried arrangements; the decorative seedheads look like fuzzy pompoms.
The name of the pasque flower means Easter flower, and
the plant is sometimes used to dye Easter eggs green.

PASSIONFLOWER, MAYPOP (*Passiflora incarnata*)

According to legend, early Roman Catholic missionaries named these
plants. They thought the 10 colored petals represented the 10 apostles

present at the Crucifixion. Inside the flower, colored filaments form a showy crown, which was thought to represent the crown of thorns. The five pollen-bearing anthers suggested Christ's wounds. The division of the pistil represented the nails of the Cross. The bladelike leaf was symbolic of the spear that pierced his side. The coiling tendrils suggested whips and cords. Be this as it may, the flower is truly beautiful. The giant granadilla is red, violet, and white and is grown extensively in certain tropical countries for its fruits, which are believed to be aphrodisiac.

PATCHOULI (*Pogostemon cablin,* syn. *P. patchouli*)
The fragrant oil of this shrubbery East Indian mint is favored for perfumes.

PEPPERMINT
See *Mint* in this chapter.

PERUVIAN GROUND CHERRY (*Physalis peruviana*)
This is a pretty little plant with small, pale blue flowers. It is also said to mean death to any bug partaking of its foliage. It grows in shade but produces more flowers in full sun. Set the seeds in fairly rich garden soil. Its blooming period (and its effectiveness) can be extended by keeping fading blossoms picked.

PETUNIA (*Petunia*)
This, one of the world's greatest summer plants, falls into four types: grandiflora doubles, grandiflora singles, multiflora doubles, multiflora singles. Petunias can "take it" but will bloom better and longer if fed liberally once a month—or if diluted fertilizer is added each time you water. Petunias live over the winter in mild climates. They attract beautiful moths at night, and the fragrance of some kinds is very pleasing.

Never underestimate petunia power; they perform well everywhere in the garden. They thrive in pots, flowerbeds, greenhouses, or wherever they have a sunny location, and they can be grown from cuttings or from seeds. Petunias help to protect beans from the Mexican bean beetle.

PHLOX, ANNUAL (*Phlox drummondii*)

The word *phlox* comes from the Greek word for flame. Its brilliant flower, however, never becomes flame-colored. Phlox are a true North American species and are favorite garden flowers because they are hardy and grow well in fertile soil. All annual phlox are derived from Drummond phlox, a species that grows wild in Texas. The familiar sweet William, whose bluish or pale lilac flowers are among the early summer blossoms, also belongs to the phlox group. Phlox are delicately fragrant and for a long season in summer they dominate the garden.

Phlox are also attractive in hanging baskets with browallia or lobelia. They make a good groundcover with nicotiana or zinnia borders and are nice for beds and edgings. Sow where they are to grow; they dislike transplanting.

PIGGYBACK PLANT (*Tolmiea menziesii*)

The piggyback has the fascinating habit of growing baby plants on top of the mature leaves. The other surprising fact is that it grows wild along the Pacific coast from northern California to Alaska. If you live where winter temperatures seldom fall below 10°F, plant piggyback outdoors. In a shady, moist rock garden, a single plant will soon multiply into a colony. Scattered about the floor of a woodland, piggyback looks lovely in the company of hardy ferns.

Indoors it is definitely a winner. Give it good light and keep it a bit on the cool side, with a good soil mix consisting of equal parts all-purpose potting soil, sphagnum peat moss, and vermiculite. Don't let it stand in water but keep it evenly moist at all times.

PINE

Pine tar oil improves standard codling moth baits.

PLANTAIN

This weed is often troublesome to gardeners. Yet it has value as an emergency measure to stop bleeding. Crush or bite the leaves to let out the juice and apply directly to the wound. Bleeding will stop, even from a deep cut.

Plantain has been used for hundreds of years for healing broken bones. Keep a few plants in the garden in case of need.

Add the tender heart leaves to a green salad in early spring.

POCKETBOOK PLANT (*Calceolaria*)

This multicolor perennial is spotted orange, yellow, or red with blossoms that mimic miniature pocketbooks. The plant is very effective in mass beds or along a shaded lawn area.

POINSETTIA (*Euphorbia pulcherrima*)

This plant of the Spurge family has tiny flowers surrounded by large, colored bracts, or special leaves. The bracts are usually bright red but may be yellow or white. The brilliant red bracts contrast with the green leaves and make the poinsettia popular during the Christmas season. In tropical and subtropical regions, the poinsettia thrives outdoors. It may grow 2 to 10 feet tall. It is a popular garden shrub in the southern states and California. In cold climates it must be grown indoors. As a potted plant it grows from 1 to 4 feet tall.

The mealybug is sometimes a problem. Alcohol is lethal to the mealybug. Dip a small stick wrapped with cotton in alcohol and touch it to a pest for just an instant.

Root aphids may cause your plant to become weak and stunted, and plants may die in severe cases. To make it difficult for these pests to get to the roots, pack the soil firmly around the plant.

Poinsettia scab is sometimes prevalent in summer. Prune and burn scab-infested branches as soon as noticed.

POOR-MAN'S-WEATHERGLASS, PIMPERNEL (*Anagallis arvensis*)

The small starlike, bright blue flowers of this low-spreading annual close with the coming of bad weather—believed by many to be a sure sign of rain. Plant seed in spring in full sun and poor, sandy soil. The plant will flower from May to August.

POPPY (*Papaver, Eschscholzia, Romneya*)

The poppy has a bad name because opium is made from one species. Many varieties look good in groups in the perennial border or in rock gardens. Many different flowers are popularly called poppies, some natives, some introduced. In this flower family are the California poppy

(*Eschscholzia californica*), corn or Flanders poppy (*Papaver rhoeas*), Iceland poppy (*Papaver nudicaule*), matilija poppy (*Romneya coulteri*), Mexican gold poppy (*Eschscholzia mexicana*), and the yellow poppy (*Papaver radicatum*).

Opium comes from the young seed capsule of *Papaver somniferum*. To obtain it, workers slit the capsules late in the day. The milky juice that seeps out solidifies overnight and is collected by hand the next day. It takes about 120,000 seed capsules to yield 25 to 40 pounds of opium.

Poppies are actually robbers of the soil and inhibit the growth of winter wheat. They dislike barley but will lie dormant until winter wheat is sown in the field.

PORTULACA, MOSS ROSE
(*Portulaca grandiflora*)

This most gaudy of coverings for very dry spots is first cousin to the weed pussley (or purslane). It grows and does well in hot, dry, shallow soil where no other flower will; for seaside gardens it is indispensable. Portulaca grows 6 to 8 inches high and is of trailing habit. The blossoms are red, magenta, orange, and white, appearing from July to October. Culture is simple. Just scatter the seeds over the surface of raked ground when the weather is warm.

Portulaca

POT MARIGOLD (*Calendula officinalis*)

This annual with yellow or orange blossoms is a nice background flower

for pansies and candytuft. Plant it in fall for color through winter and spring. Pot marigold is good against asparagus beetles, tomato worms, and many other insects.

The pot marigold, or calendula, has been a popular annual for centuries. In the Tudor period it was known as the Sunne's hearbe or the Sunne's bride, and the name marigold is linked with the Virgin Mary.

Pot marigold

PRIMROSE (*Primula parryi*)

In summer this flower has intense cerise flowers with a yellow eye. The plant likes full sun or partial shade and wet feet! It grows well with candytuft, pansies, calendulas, and violas.

PURSLANE (*Portulaca oleracea*)

Purslane is a persistent weed. Do not put it in your compost heap; it will survive and again be planted in your garden.

PYRETHRUM (*Chrysanthemum coccineum*, syn. *Tanacetum coccineum*)

This interesting perennial grows to 2 feet with red, pink, or white daisies of 3 inches across. The crumbled flowers are made into a spray to control aphids and other soft-bodied insects. It is of low toxicity to man, animals, and plants and is also useful as a powder to control insect pests on pets or farm animals.

Pyrethrum

QUEEN ANNE'S LACE, BIRD'S NEST (*Daucus carota*)

The tiny white flowers are exquisite used in arrangements with larger, coarser flowers. Allowed to go to seed it will become a pest, but the seeds serve as seasoners for soups, stews, and baked fish; fresh or dried, they make a fair substitute for anise or caraway seeds.

The herb is the wild prototype of our table carrot. It is found in rich soil, the root is sweet and palatable; located in sandy, hard soil, it is small and hard. It is edible if steamed or cooked in a little water or cut into 1-inch lengths and added to a soup or stew. Use a reliable plant guide for positive identification if you try this, however, because Queen Anne's lace closely resembles other Umbelliferae that are deadly poisons.

Queen Anne's lace

A far better substitute for garden use is bishop's weed (*Ammi majus*). This lovely 2½-foot annual has lacy, white flowers like Queen Anne's lace and is widely cultivated for cut flowers. It will grow just about anywhere.

RAFFLESIA
This small genus of plants has huge flowers but no leaves or stems. The flowers grow as parasites on the stems and roots of several cissus shrubs in Malaya. One species of rafflesia produces flowers more than 3 feet wide and weighing up to 15 pounds. The stamens and pistils grow on separate flowers, and need some agent to pollinate them. The flowers have five wide, fleshy lobes and an unpleasant odor.

RANUNCULUS (*Ranunculus*)
This lovely perennial in yellow, orange, red, white, and pink is sensational in masses, or use with snapdragons, pansies, and daffodils.

RATTLEBOX (*Crotalaria*)
See the chapter on Companion Planting with Flowers and Herbs.

RESURRECTION PLANT, ROSE OF JERICHO
(*Anastatica hierochuntica, Selaginella lepidophylla*)
Two different plants bear these common names and they come from two different parts of the world. *Anastatica hierochuntica* is from the Middle East. At maturity, the leaves fall from the plant and the branches curl around the seedpods, forming a ball. The ball is blown about on the desert wind until the rainy season. If placed in a saucer of water, the seemingly lifeless ball opens into a beautiful rosette of fernlike leaves, appearing to be "resurrected." *Selaginella lepidophylla* grows from Texas and Arizona south into El Salvador and Peru. It, too, curls into a ball during the dry season and can remain dormant for months. When moistened, it uncurls, thus also appearing to have been "resurrected."

ROSE (*Rosa*)
See the chapter "The Queen of Flowers;" see also *Eglantine* in this chapter.

RUBBER PLANT (*Ficus*)

This common houseplant is related to the fig. It does well in indoor heat and lack of humidity, growing tall rapidly and living a long time. It grows even better if the pot is rich in minerals and the plant is given enough sunlight, water, and room. Place it outdoors in summer so it will get enough sunlight to last during the winter months. Should it be attacked by scale insects, spray the plant with a control approved for indoor use.

Commercial rubber does not come from these rubber plants, but from a tropical tree that belongs to the Castor Bean family.

SAGE (*Salvia*)

These aromatic plants of the Mint family have large, showy flowers and require only the simplest of care. The salvias are a large group with both herbaceous and woody members. Among the shrub salvias is the "purple sage" so often referred to in songs and stories of the Old West.

Salvia officinalis is the source of the spice. Sage is a delicious flavoring for sausage, pork, duck, and poultry dressing. A favorite herb of Marcus Aurelius, it has figured frequently in love potions, recipes, brews, and stews throughout the ages.

SEA ONION, SQUILL (*Urginea maritima*)

Urginea belongs to the Lily family. Its name is derived from the name of an Arabic tribe in Algeria, who were probably the first to use the bulbs medicinally. These are tender bulb plants from the Mediterranean region, the tropics, and South Africa. Tiny new plants grow on the "mother" under a thin skin. From time to time they pop out and take root as they fall off. They are easily propagated by the gardener if the small bulbs are picked off or allowed to drop and set in moist soil.

Squills have a history of medicinal usage dating back about 4,000 years to the Egyptian Ebers Papyrus, which lists as a remedy for heart troubles a prescription including the bulbs of squills. Hippocrates, Theophrastus, and Dioscorides, as well as Pliny, all had much to say about squills, most of it wrong. From their day to the present, the squill has been used in the preparation of medicines and also as ornamental in our gardens.

A preparation made from squill is useful for colic in cattle, but should be used only under the direction of a veterinarian, for it can be poisonous. Indeed, red squill powder is manufactured as a poison for rodents.

SEDUM (*Sedum*)
These succulent plants have attractive foliage and showy flower heads borne in late summer and fall. Plant in pottery containers, old shoes, boots, driftwood, or other unusual "pots" for a conversation piece, as well as in the rock garden. They are absolutely maintenance- and trouble-free.

SERPENTARIA, VIRGINIA SNAKEROOT
(*Aristolochia serpentaria*)
This small, aromatic perennial herb grows to a height of 8 to 15 inches. The flowers are usually hidden beneath the dry leaves and loose top-mold. Serpentaria is a traditional remedy for snakebite.

SESAME (*Sesamum indicum*)
The tiny but exquisite flower may be pink or white. This plant is grown mainly for its delicious seeds used to flavor bread, cakes, candy, and biscuits. The oil obtained from them is a flavorful addition in cooking. Sesame will grow in the southern states and should be planted at about the same time as cotton. It increases the effectiveness of pyrethrins.

SHAMROCK (*Oxalis* or *Trifolium*)
There has been much argument over which plant is the true shamrock. Some say it is a small clover plant with green leaves consisting of three leaflets; others insist it is the wood sorrel. The flowering shamrock sold by florists blooms best in warm weather and with generous sunlight. Water when necessary and do not repot often. Pot binding tends to encourage blooming.

The shamrock is the national flower of Ireland and appears with the thistle of Scotland and the rose of England on the British coat of arms.

Woodsorrel (*Oxalis acetosella*), sometimes called shamrock, is a delicate little wild plant that grows in shady places, often in backyards or along fences. It also has three leaves, each of which is heart-shaped.

SNAPDRAGON (*Antirrhinum*)

The Greeks named this one *anti,* "like," and *rhinos,* "snout," to describe the curious shape of the flowers. However, nurseries are now producing penstemon- and azalea-flowered type of very different shapes. A little-known trailing variety, *A. hispanicum* spp. *hispanicum* (cream and yellow), is a delightful addition to the rock garden. Old-fashioned snapdragons are compatible with nicotiana, baby blue-eyes, and alyssum.

SNOWDROPS (*Galanthus*)

Winter jewels, undaunted by snow, snowdrops last a long time in bloom. They are lovely in patches of woodland under deciduous trees. Plant in fall so they have a long growing period. Increase your stock right after they bloom; replanting them pays off in a much bigger crop.

SOAP PLANT, SOAP ROOT (*Chlorogalum pomeridianum*)

The powdered bulbs of this native from California and Oregon are toxic to armyworms and melonworms.

SOWBREAD

See *Cyclamen* in this chapter.

SPIDER PLANT (*Chlorophytum comosum* 'Variegatum')

This interesting houseplant from South Africa develops young plants at the end of its flowerstalks. The spider plant needs good light with or without direct sun, a cool room, and moderate humidity. It grows well in ordinary soil if kept moist. Propagate by removing young plantlets and potting them separately. The plants grow 1 foot tall with wider spread. The leaves are green with broad white center stripes.

SPURGE (*Euphorbia*)

This plant has been rated by one of the world's outstanding nursery experts as among the 10 best perennials for its long-lived reliability, ease of cultivation, neat impressive form, and outstanding color. It looks a bit like cactus but is totally unrelated. Shapes run from clean geometrics through organ pipes, fat balls, and cylinders. Most strange of all are those with convoluted crests and snakelike, Medusa-head forms.

All euphorbias have small flowers without petals, enclosed in a cup-shaped, leaflike structure with five lobes and a honey-secreting gland. The single female flower is normally surrounded by numerous male flowers.

STAPELIA (*Stapelia*)

The enormous, hairy, star-shaped blossoms of this plant have a real stench. Actually "stink bomb" seems more appropriate than "blossoms," for when *Stapelia gigantea* opens one of its blooms, from 11 to 16 inches across, there is no doubt why it is called the carrion flower. But if you can overlook the odor, the bloom is remarkable—the sort of thing you'd expect to find in a science-fiction garden.

Giant stapelia isn't a good choice for a houseplant, but one *Stapelia* relative, *Orbea variegata,* is fun to grow and the odor of the strange blossoms isn't nearly as potent. These plants are, however, very attractive to flies.

There are about 90 species of stapelia, mostly from South Africa. They belong to the Milkweed family. Other, more refined members include hoya and stephanotis, which have sweet-smelling flowers, and the ceropegia or rosary vine.

STAR-OF-BETHLEHEM (*Ornithogalum*)

This early-spring bloomer with white blossoms lightly striped with green, is excellent alone, in masses, or as an edging for a bed of daffodils.

STATICE, SEA LAVENDER (*Limonium*)

The stiff flower stems bear literally hundreds of dainty flowers in many branched panicles. Statice is particularly good for drying as well as being graceful in the garden combined with larger flowers. The plants are superb for seaside gardens, as they are unaffected by salt wind or salty soil. They grow well in garden soil; the richer the soil, the larger the flower heads.

STINGING NETTLE

See *Nettle* in this chapter.

STINKING WILLIE, TANSY RAGWORT (*Senecio jacobaea*)

This weed, poisonous to cattle, causes a hardening of the liver.

SUNFLOWER (*Helianthus annuus*)

This is one of our most valuable flowers. Its bright and cheerful blossoms dutifully follow the sun. Their blooms are visited by bees for pollen and nectar. The seeds, loved by birds, are rich in vitamins B_1, A, D, and F and make a fine vegetable oil for cooking and salad dressings.

Also try sprouting the black sunflower seeds and using the sprouts with other greens as a salad. The black seeds are more delicious and nutritious than even the striped ones.

Sunflowers can provide not only a windbreak but also a quick-growing screen for any portion of the garden where visibility is undesirable—a compost heap, for instance. (See also the chapter "Companion Planting with Flowers and Herbs" for more information on garden use.)

SWAN RIVER DAISY (*Brachycome iberidifolia*)

This native of Australia is pretty in rock gardens with poppies and sedum, or used as an edging plant. Plant in masses. The name comes from the Greek, *brachys,* "short," and *comus,* "hair."

SWEET FLAG, CALAMUS (*Acorus calamus*)

The alkaloid root works as a contact poison to insects, even though it is edible to humans. It commonly grows in swamps and along brooks.

TELEGRAPH PLANT
(*Codariocalyx motorius*, syn. *Desmodium gyrans*)

This native of India belongs to the Pea family, Leguminosae. It has trifoliate leaves; the center leaf is elliptical and about 2 inches long, the two side leaves are about ½ inch in length. The leaves are usually in constant motion, rising and falling alternatively, but not in regular time. The rise and fall of the leaves has been compared to railway telegraph signals. The plants are most active in the early morning, especially the young ones. They make a wonderful conversation piece.

Use sandy soil when sowing the seeds of annuals in pots or when rooting cuttings.

TEXAS BLUEBONNET (*Lupinus texensis*)

The state flower of Texas is beautiful grown in masses. It belongs to the Lupine family, and the spikes of sweet-pea flowers are bright, rich blue. Nitrogen-fixing legumes, they thrive on well-drained, poor, sandy soils in full sun. These plants are easy to raise from seed but difficult to transplant.

THISTLE (*Onopordum*)

Thistles, though beloved of butterflies, have never been popular with people. In spite of their beautiful flowers, the prickly leaves are unap-

pealing. Thistles are rich in potassium (good in the compost heap) and would have high feeding value if it were not for their thorns. In grainfields they take away food and moisture, and in pastures they protect and thereby increase the spread of other weeds.

To get rid of thistles, timing is important. If cut before the blossoms are open, the thistles will spread from the rootstocks. If cut after the blossoms are pollinated, the situation is a little better. But if the blossom heads only are cut off shortly after pollination, the plant will bleed to death and wilt.

Thistle

THYME (*Thymus serpyllum* and *T. vulgaris*)

This very valuable plant has been used in medicine since the very earliest days of herbal treatment. It is a powerful antiseptic and general tonic. As an aphrodisiac, thyme crops up with almost monotonous regularity in literature throughout the ages.

Thyme yields an essential oil that accounts for its antiseptic properties and is a good vermifuge. The oil, called thymol, is found in may orthodox preparations such as disinfectants, dentifrices, and hair lotions.

TILLANDSIA (*Tillandsia*)

These tender, evergreen plants have attractive flowers and large, beautifully colored bracts. They belong to the Bromelia or Pineapple family. *Tillandsia usneoides*, the Spanish moss, is another family member.

TOADFLAX (*Linaria vulgaris*)

Toadflax, found in waste places and often growing among corn, has powerful dissolvent properties and has traditionally been used to treat obstructions in all parts of the body, particularly the intestines, kidneys, and bladder. The leaves are small and flat; the flowers are in racemes of yellow and orange, marked white, and of the familiar snapdragon form. The plant is also considered one of the best jaundice remedies known to the herbalist.

TULIP (*Tulipa*)

The name comes from a Turkish word for "turban." Between 1634 and 1637, tulips become so fashionable in Holland that the craze was called tulipomania, and the bulbs brought fantastic prices. The Tulip Festival, which takes place in May when the flowers bloom, is a renowned event in the Netherlands.

Though they are sun lovers, tulips grow better in the North than in the South, for the bulbs need a period of cold. Since mice like to eat the bulbs, it sometimes is advisable to plant them in a small wire cage sunk in the earth and covered with soil. Scilla bulbs may also be planted with them as a protection against mice. Do not plant tulips near wheat, as they discourage its growth.

TURKEY MULLEIN, DOVEWEED
(*Eremocarpus setigerus*)

Greenish flowers; dark gray, shining seeds; and stinging hairs characterize the turkey mullein. The leaves contain a narcotic poison and were used by Native Americans to stupefy fish and poison their arrow points.

Other plants used to stun fish were blue curls (*Trichostema* spp.), vinegarweed or camphor weed, wild cucumber (*Marah* spp.), and members of the Gourd family. Turkey mullein is also toxic to cross-striped cabbageworms.

UNICORN PLANT, DEVIL'S-CLAW, ELEPHANT-TUSK (*Proboscidea*)

The showy, reddish purple to coppery yellow flowers are large and attractive but few in number. More spectacular are the large, black, woody pods ending in two curved, pronglike appendages that hook about the fetlocks of burros and the fleece of sheep. In this way, the pod is carried away from the mother plant and the seed is scattered. The attractive pods are used for many decorative purposes; some even are painted to resemble birds. Young pods are eaten by desert Indians as a vegetable. The mature fruits are gathered by the Pima and Papago Indians, who strip off the black outer covering and use it for weaving designs into basketry.

VANILLA (*Vanilla*)

The extract from this group of climbing orchids is used to flavor chocolate, ice cream, pastry, and candy. The vanilla vine, cultivated in Mexico for hundreds of years, has been introduced into other tropical areas, mainly Madagascar and the Comoro and Reunion Islands. However, it is said to set seed naturally only in Mexico. Elsewhere it must be hand-pollinated, adding greatly to the cost of its production. The cultivated plant lives for about 10 years, producing its first crop at the end of 3 years.

The flowers, though dull in color, are very fragrant. Vanilla is obtained from the prepared seed capsules of *Vanilla fragrans* (*planifolia*), which are 6 inches long and beanlike in shape. To grow vanilla in a greenhouse, a tropical atmosphere is required; in winter, a temperature of 66°F is suitable.

VERBENA (*Verbena*)

Vervain or verbena was the holy herb used in ancient secret rites; it was also supposed to cure scrofula and the bite of rabid animals, to arrest the diffusion of poison, to avert antipathies, and to be a pledge of mutual good faith—hence it was worn as a badge by heralds and ambassadors in ancient times.

Most of our perennials come from South American and are hardy only in favorable climates. They come in many colors and types, and are nice for edging or hanging baskets. Verbena is attractive grown with yarrow and dusty miller.

VICTORIA WATER LILY, ROYAL WATER LILY
(*Victoria amazonica*)

The plant, a member of the Water Lily family, was named *Victoria regia* in honor of Queen Victoria. The round leaves with upturned edges measure up to 7 feet across and are strengthened by a marvelous network of veins capable of sustaining weight up to 150 pounds. A child can easily sit on the floating leaves.

The huge flowers are nocturnal; it is a breathtaking sight to watch them open in early evening, rapidly moving from a bud to a creamy white, wide-open, deliciously scented flower. Closing the next day at about noon, they open again three or four hours before dusk, with the color turning to a definite pink. They fade the next morning and sink below the surface of the water.

VIOLET (*Viola*)

Violets are among the "artillery flowers"—the seedpods, when ripe, split apart and the seeds are flung hither and yon to begin new plants.

Violets have a delightful, fresh, springlike fragrance, and the edible leaves and blossoms are so rich in vitamins C and A that Euell Gibbons

(*Stalking the Healthful Herbs*) calls them "nature's vitamin pill." The violet blossoms are three times as rich in vitamin C, weight for weight, as oranges.

Violets are used in many delicious recipes, which include violet syrup, candied violets, and even a violet bombe made with candied violets, ice cream, and whipped cream.

Violets are a favorite flower of almost everyone. They're beautiful, no matter which species is grown.

VIPER'S BUGLOSS (*Echium vulgare*)

These plants are biennials, usually blue with gray-green foliage. They are fine for rock gardens and especially so for seacoast gardens. For a pretty combination, plant with columbine or armeria.

WALLFLOWER (*Erysimum cheiri*, formerly *Cheiranthu cheiri*)

This perennial is usually orange or golden. Before sowing seed, water the drills with a spray of rhubarb leaves boiled in water to protect against clubroot.

Combine wallflowers with daffodils and tulips; in cool climates combine with snapdragons and dusty miller. New strains exist in lovely pastel shades of cream, lemon, apricot, gold, salmon, light pink, rose, ruby, purple, copper, and rust. Wallflowers are good grown with apple trees.

WANDERING JEW (*Tradescantia*)

The endearing habit of the wandering Jew is that they are luxuriant in their growth habits. They are many-branched with a compact leafing pattern that is enhanced by a variety of color forms—from shades of green, to variegated white and green, to green and red.

Wandering Jews do best under filtered light because they originate in the rain forests of the tropics and semitropics. Use potting mediums of equal parts loam, peat moss, and perlite. Pinch them back occasionally to keep them looking tidy.

WANDFLOWER (*Dierama*)

From the Greek *dierama,* "a funnel," the name describes the shape of the individual flowers hanging from long, slender stems. These perennials from South Africa cannot withstand wet, cold winters; use as houseplants in northern areas.

WATER LILY (*Nymphaea*)

Water lilies, as with many other plants, have been hybridized and now come in an almost endless variety of magnificent creations, beautiful in form and color. The lilies mentioned here are less showy but interesting nonetheless in their own way.

European white lily (*Nymphaea alba*) grows wild on ponds, lakes, and other still waters. The name is from the Greek for water nymph. The flowers are large, solitary, rounded of form, and sweetly scented with prominent yellow stamens. The root is soothing and astringent with antiseptic properties. The leaves are sometimes used for binding over wounds or inflamed areas of the skin.

The yellow pond lily (*Nuphar lutea*) is also medicinal. Its common name is brandy bottle, from the brandylike scent of its flowers and the shape of its seed vessels, which are like the traditional brandy flagons. These lilies grow wild in the shallows of lakes.

WHITE HELLEBORE, FALSE HELLEBORE (*Veratrum*)

This was a safe, popular insecticide against slugs, caterpillars, and other leaf-eating pests in early American gardens. It was used as a dust or dissolved for a spray: 1 ounce to 3 gallons of water.

WILD CUCUMBER, MANROOT (*Marah* spp.)

The powdered root is toxic to European corn borer larvae.

WILD MUSTARD (*Sinapis avensis,* syn. *Brassica arvensis, B. kaber*)

Wild mustard is fairly common in cultivated areas and waste places. Gather the young leaves in spring and they may be steamed, or use them in a cold salad or soup. Wild mustard growing among fruit trees or grapevines is beneficial, according the Beatrice Trum Hunter in her book *Gardening without Poisons.*

WINDFLOWER (*Anemone*)

The name is from the Greek *anemos,* "wind," and *mone,* "habitation." The plant is so called because some species are found in windy places. The windflowers are suitable for border, for the rock garden, and for cutting. Their blue, pink, and white coloring combines well with narcissus.

WORMSEED, JERUSALEM TEA
(*Chenopodium ambrosioides*)

Some parts are toxic as extracts or dusts on several species of leaf-eating larvae.

WORMWOOD (*Artemisia absinthium*)

This name is loosely applied to many artemisias, but it properly belongs only to *Artemisia absinthium,* a hardy perennial with woolly gray leaves and a strongly bitter odor. As a tea, spray it on the ground in fall and spring to discourage slugs and on fruit trees and other plants to repel aphids.

Southernwood is a close cousin, growing about 3 feet tall with gray-green divided foliage. Silky wormwood (*A. frigida*) is excellent against snails. Plant it among flowers in full sun and a dry location. Other family members include tarragon, mugwort, silver mound, fringed wormwood, and dusty miller. These are somewhat more moderate in odor than wormwood and southernwood.

YUCCA (*Yucca*)

Yucca has many names, the loveliest being the Spanish *candelabra de Dios* (candles of the Lord). Yucca's great clusters of creamy, bell-shaped florets appear on the stiff, woody stalks ranging from 3 to 9 feet tall. In some species the immense symmetrical flower heads from the distinct shape of a cross. The blooms, which last from four to six weeks, are very fragrant, particularly in the evening hours.

Yucca makes an excellent specimen plant as an accent for rock gardens, having the same cultural requirements as many others used for this purpose. It grows well with pine tree moss, an upright groundcover that grows 6 to 15 inches high with little tufts of needlelike foliage. Tall tulip varieties and daffodils interplanted around yucca in clumps add color while the moss is greening up.

The fibers of yucca leaves have long been used by Native Americans for making rope, matting, sandals, basketry, and coarse cloth.

In the Southwest, many varieties of succulents and cactus grow well with yucca plants, which are valued for their ability to bind a sandy soil, particularly in areas of high winds.

Yuccas are both beautiful and edible; Native Americans even eat the flowers. The stalks are rich in sugar. The leaves produce a fiber used in making baskets and mats.

There are many species of yucca. Soap tree yucca (*Yucca elata*) is treelike, often branched, and has the tallest flowerstalk of any of the yuccas. It is very ornamental with creamy white, lilylike flowers. Native Americans beat its roots in water, using the milky liquid produced for washing their hair. This shampoo is thought to help the hair retain its natural color well into old age.

The large, pulpy fruits of *Y. baccata* can be eaten raw or roasted, or cooked and dried for future use. Cattle eat the flowers.

The powdered leaves of Spanish dagger (*Y. schidigera*) are toxic to melonworms, bean leaf rollers, and celery leaf tiers.

The hardiest type, Adam's needle (*Y. filamentosa*), will grow in the North if given some winter protection.

The yucca plant and the yucca moth are symbiotic. The word *symbiosis* means "living together," and any organisms that do so are referred to as symbiotic, whether they benefit one another, harm one another, or have no effect at all. In the instance of the yucca, the moth feeds but also pollinates, leaving enough seeds to start new plants.

ZANTEDESCHIA

See *Calla Lily* in this chapter.

ZINNIA (*Zinnia*)

This is the easiest and most satisfactory annual to grow, and the hybridizers have made them so elegant that they can surely find a place in every garden.

Zinnias are bright and cheerful. The older varieties were beautiful in their day, but zinnias are now available in a rainbow of colors from white to purple. They even come in many striking bicolors.

The prairie zinnia (*Zinnia grandiflora*) is a spectacular bedding and border perennial shrublet. It forms ground-hugging cushions less than 6 inches tall and is completely covered with deep yellow flowers from midsummer through fall. Plant this close relative of the annual garden zinnias in full sun. Prairie zinnia is slightly toxic to celery leaf tiers.

Lamb's-quarters gives added vigor to zinnias, as well as marigolds, peonies, and pansies.

THE QUEEN
OF FLOWERS

The Rose family, one of the most important in the plant kingdom, includes about 2,000 species of trees, shrubs, and herbs. Some of the loveliest flowers and most valuable fruits belong to it. A few members of the family are apple, apricot, blackberry, cherry, cinquefoil, eglantine, peach, pear, plum, quince, raspberry, spiraea, and strawberry. Its many ornamental plants include the meadowsweet, mountain ash, and hawthorn. Plants of this family give us many useful products such as attar, an oil from rose petals, used to make toilet water and perfume. Several fine woods are used in cabinetmaking.

Plants of the Rose family have regular flowers. Each has five petals, a calyx with five lobes, many stamens, and one or more carpels. They bear seeds, so they are classed as angiosperms. The sprouts have two seed leaves, therefore they belong to the dicotyledonous plants.

Different species of wild rose are native to every state of the union except Hawaii. Often they represent the so-called transition shrubs between forest and meadow or prairie. Many sucker freely and can be invasive, but they are much hardier and more disease resistant than their hybrid relatives.

Old Roses of Romance and Legend

Once upon a time, roses were different from what they are today—less dramatically beautiful in form and color but far, far more fragrant. Most of our modern roses are descended from these older types.

Fragrance is the rightful heritage of the rose. In the minds of most of us, the ideas are inseparable. Even long ago when the rose was a simple flower, it was known as the Queen of Flowers. Surely it must have been the unsurpassed quality of its fragrance that gave it this prestige.

Recently flower lovers have been uneasy because of the scentless, or nearly so, roses appearing on the market. This trend toward mere beauty in roses is greatly deplored.

What is meant by the pure odor of roses, sometimes referred to as the "true old rose scent"? This is the property of that famous trinity: *Rosa × centifolia,* the cabbage rose; *Rosa × damascena,* the damask rose; and *Rosa gallica,* the French rose. This lovely scent has been inherited by many modern Hybrid Perpetuals (H.P.) and also by some Hybrid Teas, though in a lesser degree. The old H.P. 'General Jacquemont', which first saw the light in 1852, is the parent of a long time of deliciously scented roses, and it is still popular today.

Rose

The fragrance of a rose flower is in its petals. Red roses, perhaps because they are closer to the grand varieties of early times, are generally the most richly endowed with fragrance. Next come the pink varieties. Yellow roses are the least scented, and white almost scentless.

Old-Fashioned Roses

Old-fashioned roses have been hard to find but are enjoying a revival and are now offered by many nurseries. Aside from the fragrance, they have another advantage: Many are hardy not only where winters are severe but also where summers are hot. They are excellent for different landscaping effects.

Roses will grow, and grow well, practically anywhere if you are careful about a few things. Buy only first-quality bushes, plant them with care in a sunny, well-prepared bed, maintain a regular diet or spray schedule, water and feed at correct intervals, and remove spent blossoms.

The American Rose Society suggests these planting pointers for roses: Plant them where they'll get sun at least half the day. Plant during winter where ground isn't frozen. A raised bed works well for roses in many areas.

PRUNING

It is not easy to give definitive rules for the pruning and care of old-fashioned roses. Each old, rare, or unusual rose is an individual, with its own type and habit of growth.

Old, shrub, and species roses should not be pruned in spring as with the Hybrid Teas, for if you do, you remove the canes that would have produced their great spring flowering.

However, roses that bloom repeatedly should have weak growth removed and be trimmed to shape the plant. Pruning them is more a matter of shaping and thinning than of cutting back. Removing spent flowers encourages the growth of new flowering stems.

Treat varieties with but one annual flowering like flowering shrubs. Leave them alone, and put away your pruning shears until after they bloom.

Some of the loveliest and most intriguing of the old-fashioned beauties have long canes that arch over naturally from their own weight. Others have canes that grow straight up, which will bloom only at the top unless pegged or pruned.

To achieve a bushy, many-branched plant, shorten the long canes by one-third after the plant blooms and shorten lateral canes by a few inches. If you desire, keep this up until late summer, then leave the plant alone until after it blooms in spring.

Climbers

The various types of climbers behave quite differently, but all must have a support to look their dramatic best. Support for climbers can be of conventional patterns, or design and build a support that is suited to your own needs.

Species Roses

Botanists have discovered species of rose in various parts of the world and brought them into cultivation. These roses are fasci-

Miniature roses may be used as edgings or in beds, or when potted as points of emphasis on patios. Give them a rich loam, and mulch plants in the winter with soil or straw.

nating individualists, for all developed distinct characteristics enabling them to survive in their native habitats. Some are extremely hardy and have one annual flowering; others, native to the subtropics, are tender and bloom repeatedly.

Rosa glauca is a beautiful shrub (originally named *R. rubrifolia* because everything about it is red) from the soft pink 1½-inch single flowers with their reddish brown calyxes, to reddish brown canes, dark greenish red foliage, and bright hips the color of Queen Anne cherries. Blooming off and on through the season, it grows to 6 feet, is quite hardy, and is native to the mountains of China and southern Europe.

Rosa roxburghii is more commonly known as chestnut rose, for the unopened buds look like little chestnut burrs. It is one of the most beautiful and unusual roses in existence. Its light green foliage has new tips shaded with copper and gold. The silvery gray branches shed their bark as many trees do, and the very double 2½- to 3-inch flowers of many small petals glow pink at the center with silvery pink on the outside. It blooms repeatedly and makes an excellent bank cover. It grows quite large in mild climates.

Rosa soulieana was discovered in western China. Its relaxed canes lie on the ground and grow from 12 to 20 feet long. In June and July they are covered with corymbs of 1½-inch, single white flowers with a distinct fragrance that perfumes the air. It makes an excellent bank cover, for over the years its canes will take root and hold the soil. Grown on a retaining wall, it will cascade to the ground like a waterfall. Growing up a tree such as a weeping willow, the canes will fountain down very dramatically. When the petals fall, small orange hips form. Occasional flowers appear after its mass summer blooming.

Japanese Roses

These are the pure rugosa roses that have been developed from species originally found in Japan; sometimes they are known as Japanese roses. They are hardly anywhere and very disease resistant. They should not be confused with hybrid rugosas, which have been developed by cross-pollination with other types of roses.

Grow Your Own Vitamin C

Vitamin C, found in greater concentration in rose hips than in oranges, is essential to good health and may have benefits that are not yet fully understood. Most animals have the enzymes to synthesize their own vitamin C, but humans and apes do not. Since the body does not store vitamin C, our supply must be constantly replenished; there is little danger of an overdose, as the body eliminates what it does not use.

Some rose hips, those of the *Rosa rugosa,* contain 20 times as much vitamin C as citrus fruits, and the wild Scandinavian types are even richer. The hips are rich in vitamin E as well. The rugosa blossoms with a single-petaled rose. Rugosas make a dense hedge, molding a desired contour in the garden. Planted 18 inches apart, they make a bright living fence. Bees hover around the flowers because of their intense perfume.

Some kinds include 'Will Alderman', clear lilac pink; 'Blanc Double de Coubert', pure white; *Rosa rugosa* 'Magnifica', carmine; and 'Fru Dagmar Hastrup', with five petals of clear pink and lower growing than most of this type.

For either a hedge or a specimen planting, the method is much the same. If possible, set your roses out immediately after arrival. For individual plants, dig holes; for a fence, it's better to dig a trench about 1 foot across and 1 foot deep.

Harvest your own supplies of vitamin C by collecting rose hips, and use them in jam, soup, syrup, marmalade, and rose-hip tea.

Put some well-rotted manure or compost several inches below where the roots will rest; this promotes a stronger start and quicker results, helping the roses produce usable hips much sooner.

Rugosas need little care but will establish more rapidly after transplanting if cut back. Leave three or four buds or leaf nodes on each stem. Rugosas, like other roses, respond well to mulching to retain moisture, especially during summer. As it decomposes, the mulch also feeds the plants.

USING ROSE HIPS

To receive the most benefit from this fantastic source of vitamin C, remember that the more roses you pick, the fewer the hips (these are the
fruits that mature after the flower petals fall). Gather the hips when they
are fully ripe, but not overripe. If they are orange, it is too early. If dark red,
it is too late. In the North, the hips usually ripen after they've been
touched by the first frost.

After picking, cook your rose hips immediately and quickly to retain the
greatest amount of vitamin C. If this is not convenient, pack them in tight
containers and keep refrigerated.

The hips, taken when fully ripe, can be split longitudinally and the inner
seedlike structures removed. This gets rid of the hairs that are attached to
them. The blossom end is usually removed, and the pump can be eaten
raw or stewed, or can be used to make jam or jelly. Rose juice blended
with apple juice makes a different but very tasty jelly. Be sure to cook your
rose hips (and jelly) in glass or enamel saucepans.

To make rose-hip marmalade, soak the cleaned rose hips for 2 hours in
plain cold water then let simmer for 2 hours; and strain. Measure the
purée and add 1 cup of brown sugar to each cup of rose-hip purée. Boil
down to thick consistency. Pour into sterilized glasses and seal.

ENJOYING TREES AND SHRUBS

ACACIA, WATTLE, MIMOSA (*Acacia*)

These tender trees and shrubs with ornamental foliage have attractive flowers in spring. They may be grown outdoors in mild climates. Some species seem to know which ants will steal their nectar; they close when ants are about, opening only when there is sufficient dew on their stems to keep the ants from climbing. The sophisticated acacia actually enlists the services of certain protective ants, rewarding them with nectar in return for protection against other insects and herbivorous mammals.

White-thorn acacia (*Acacia constricta*) has fragrant masses of yellow flowers, amply protected by long, straight, white thorns. It grows in Texas, Arizona, and Mexico, and makes a good barrier plant for traffic control.

AZALEA (*Rhododendron*)

Botanically, all azaleas are rhododendrons, but most gardeners call the smaller-leaved and deciduous types azaleas. Azaleas are one of springtime's delights, blooming early in May and June in a wide range of colors. About 40 species grow in North America. Azaleas are truly gorgeous, and blossoms range in color through pink, red, white, yellow, and purple. Their long pollen stalks extend beyond the petals. Some of the leaves are narrow, others egg-shaped. In some azaleas, the flower has a covering of sticky hairs that keep ants away from the sweet nectar. A long, slender pod with hairs holds the seeds. The plants live best in acid soil and partial shade.

Azalea

The Arnold Arboretum in Boston, Massachusetts, established in 1872, was the first extensively organized effort to control and introduce ornamental plant varieties from foreign countries. Of the several collectors sent out by the Arnold Arboretum, Ernest H. Wilson was the most famous. During his travels throughout Asia, Wilson gathered one of the finest collections of the azalea varieties of Japan. Since then azaleas have been hybridized into the glorious flowers we have today.

BLOOD-TWIG, SIBERIAN DOGWOOD
(*Cornus sanguinea, Cornus alba*)
These shrubs are unbelievable for winter color. Their bright red stems provide intense contrast against evergreens or winter snow. See also *Dogwood* in this chapter.

BONSAI
Bonsai are miniature trees grown in pots. The aim of bonsai culture is to develop a tiny tree that has all the elements of a large tree growing in a natural setting. Over the years, the Japanese have devised standards of shape and form that gradually became the classic bonsai style. Many ordinary shrubs and trees take to bonsai—quince, forsythia, even a scrubby little American elm shoot.

Begin by cutting back the roots; if your plant has a taproot, cut it off to the end. Trim other roots if numerous, but not by too much. Shape the top and put the plant in a clay pot with a hole in the bottom for drainage. Rocks on the bottom are helpful as well as a screen to keep out bugs.

The soil mix is a third each of sand, compost, and soil. Screen out the mix in three sizes, through ¼-, ⅛-, and ¹⁄₁₆-inch screens, with the larger lumps at the bottom.

Set the plant in the soil, water well, and in the beginning limit sunshine to mornings. Branches may be shaped by clipping, or trained on an attached wire covered

Bonsai were developed in the Orient, but now their popularity has spread around the world. Some bonsai are hundreds of years old—and still tiny—but don't let this deter you from trying to grow your own shrub or tree.

with twig tape (a greenish brown color) to hide the wire. Twist-ties may also be used to hold branches in the unnatural position. Plants should not become root-bound. Nor should the pot be too big. The pots may be sunk in the ground to winter the plants from November 1 to March 1. If the plant is in a container that might crack, move it to a clay pot.

BROOM (*Cytisus, Genista*)

Plant these lovely spring-flowering shrubs in a hot corner on sandy or gravelly soil if you wish them to flower lavishly. If fertilized heavily, they will not bloom. Brooms come in brilliant colors and are breathtakingly fragrant. They grow very quickly, filling in just after azaleas finish. The branchlets do not lose color in winter and, of course, you can make your own brooms from them!

BOX, BOXWOOD (*Buxus*)

Because of their handsome appearance and beautiful foliage, the common box and the edging box are greatly valued as garden ornamentals. Most will thrive in any good garden soil, but are particularly useful for planting on limestone ground. Many are grown for topiary work, either to stand out individually or for hedges and dwarf borders for garden beds and paths.

CACAO (*Theobroma cacao*)

The source of chocolate is the seeds or "bean" of the cacao tree. Native to tropical America, the trees have been cultivated for more than 4,000 years. After the canes are dried, they are shipped to chocolate factories, cleaned, roasted, and ground into a pastelike substance called chocolate liquor. Pressing out the fat from this produces dry cocoa.

Researchers have discovered that cocoa contains sizable amounts of phenylethylanine, a substance produced by human brain cells during emotional episodes.

CALIFORNIA BUCKEYE (*Aesculus californica*)

Flours that are made with meat and hulls of the nuts are toxic to larvae and adults of Mexican bean beetles; also, certain parts of this plant are toxic to humans.

CERCIS
See *Redbud* in this chapter.

CHERIMOYA (*Annona cherimoya*)
This small, unusual, tropical American tree grows wild in Peru and is now cultivated in California and Florida. The tree bears fragrant yellow flowers followed by egg-shaped or heart-shaped fruit weighing a pound or more. Its white smooth pulp tastes like a mixture of pineapple, peach, and banana.

The tree grows quickly and has very ornamental foliage; the fruit is about 4 inches across. However, it normally takes 2 years to fruit and the flowers must be hand-pollinated with an artist's paintbrush for the flowers to set.

CHINESE WINGNUT (*Pterocarya stenoptera*)
The powdered leaves of this ornamental tree are slightly toxic to Mexican bean beetle larvae.

COFFEE (*Coffea*)
Coffea has shiny dark green foliage and white, starlike, fragrant flowers, followed by berries, which are harvested when they are scarlet. Then the fleshy outer pulp can be removed and the "beans" dried. Finally roasted and ground, they make real coffee.

COLD-CLIMATE TREES
Southern gardeners delight in magnolias, but farther north the autumnal foliage of such trees as silver, red, and sugar maples; red and white oaks; and white birch brings a similar pleasure.

For flowering trees there are black and honey locusts as well as bristly locust, which has large, deep rose-colored flowers from late May to mid-June. Other "blossomers" include eastern redbud, tulip poplar, wild black cherry, catalpa, white-flowering dogwood, and little-leaf linden with its inconspicuous but very fragrant flowers.

CORNELIAN CHERRY (*Cornus mas*)

This shrub will cheer you up when the delightful fluffs of yellow bloom dot every leafless branch in February. They are followed by green foliage and, in turn, by scarlet fruits attractive to birds and great in jellies and preserves. The purple-red fall foliage makes this shrub of year-round interest. Specimens can be pruned to produce alluring stem and bark patterns. Cornelian cherry is also excellent for spring forcing.

CORYLUS, HARRY LAUDER'S WALKING STICK (*Corylus avellana* 'Contorta')

Here's something very unusual for your flower arrangements. Like the crooked cane of the old-time Scottish comedian Harry Lauder, its branches are so fantastically twisted and contorted that it is almost corkscrewlike in appearance. Plant it where you can enjoy its strange silhouette against the winter snow.

COTINUS (*Cotinus*)

Unequaled for its lovely display in early spring is the cotinus, the so-called smoke tree. *Cotinus coggygria* 'Royal Purple' is regal indeed with its coppery purple-black foliage and plumed inflorescence of the same color.

CRAPE MYRTLE (*Lagerstroemia*)

Masses of spectacular flowers make this shrub a southern favorite; it is dramatically beautiful when in bloom in summer. And it comes in lovely colors of white, pink, watermelon red, and royal purple.

Crape myrtle stands heat and drought well and is not only easy to grow but also easy to root. Cut off a branch, strip the lower half of leaves, and insert the cutting (or clone) in moist soil.

DAPHNE, WINTER (*Daphne odora*)

This small evergreen shrublet has white or purplish flowers in January. It is said that the daphne "can boast of being the most powerfully fragrant plant in the world." It grows as far north as Washington, D.C., and persists over winter if given a warm wall to sun its back against.

Daphne laureola (spurge laurel) grows luxuriantly in shrubberies where it is hardy and often produces its small green flowers as early as January. The plants have a delicious scent like primroses that can be detected at a distance of 30 yards.

D. mezereum is often grown as an ornamental. However, it produces poisonous berries, and eating even a few can be fatal to a child. Its fragrant, lilac-purple flowers in stalkless clusters of three bloom before the leaves come out.

DOGWOOD (*Cornus*)

The western dogwood, *Cornus nuttallii,* a very handsome tree, has beautiful blossoms and an agreeable, honeylike fragrance. *C. amomum,* the red-stemmed dogwood, has fragrant inner bark that Native Americans use for smoking.

EUCALYPTUS, GUM TREE (*Eucalyptus*)

This tree has a remarkable capacity for storing solar energy. Experiments in South Africa have shown that a forest of such trees produces yearly approximately 20 tons of fuel per acre. The dry timber is heavier than coal and gives out as much heat when burned.

These trees thrive best in hot, moist regions, but some varieties are extremely drought-resistant. Foliage of most eucalypti is fragrant. The lance-shaped leaves are long, narrow, and leathery. The feathery flowers look like bells and are filled with nectar. In California, eucalypti are planted around orange and lemon groves as windbreaks.

The resin, called Botany Bay kino, protects wood against shipworms and other borers. The bark of some species furnishes tannin, which is used medicinally. The leaves contain a valuable oil that smells somewhat like camphor and is used as an antiseptic, deodorant, and stimulant.

Gather bark, stems, leaves, and seeds of long-leaf eucalyptus and make a decoction by boiling. Use to spray plants affected with aphids.

Young blue gum (*E. globulus*) is handsome for a houseplant or for planting in the garden for summer foliage effects.

EUONYMUS (*Euonymus*)

Euonymus alata (winged spindle tree) is a prize for flower arrangements. The twigs develop pronounced corky wings that are very well defined. Plant this shrub in sun for spectacular fall foliage as well as for the color effect of the orange fruits.

EVERGREENS

Needles are useful for soil building and make good humus for azaleas. Evergreen plantings make good windbreaks.

FORSYTHIA, GOLDEN BELLS (*Forsythia*)

The yellow blossoms of this lavishly beautiful shrub are one of the joys of February. Forsythias are outstanding as specimens and excellent for forcing. Branches cut in January and February will force in just a few days. Prune older wood immediately after blossoming to keep the shrub in good health and heavy flowering.

When food is scarce, birds may pick and tear at the unopened buds, but happily the plant has a reserve set that are rapidly brought into action if the season's normal quota is pillaged. Almond trees make good neighbors.

FOTHERGILLA (*Fothergilla gardenii*)

This early-blooming deciduous shrub is noted for its 1-inch spike of honey-scented, cream-white flowers appearing in early spring. An outstanding characteristic of this shrub is the change of its leathery, dark green summer foliage to a spectacular display of brilliant yellow and orange red in the fall. To achieve this, plant the shrub in full sun and in an acid soil with good drainage.

FRANGIPANI, PERFUME TREE (*Plumeria*)

There are more than 40 species of these warm-weather ornamentals. With their exceedingly sweet-smelling flowers, they are considered the most fragrant of ornamental plants. Their waxy blooms of deep rose and white consist of five petals overlapping in star fashion to a narrow throat supported by a thick short stem. Blooms form bouquets in clusters often 8 to 11 inches across and continue opening in the same cluster for many weeks. In some parts of the country they bloom year-round.

Frangipani cuttings root easily, and if desired, the tree can also be propagated by seeds from the occasional paired, tightly filled seedpods. Sizable trees are sold by nurseries to furnish immediate beauty for outdoor gardens and, where they are not hardy, they make lovely pot plants.

FRUIT TREES, FLOWERING

Few things are lovelier than the blossoming of fruit trees in spring. Not to be overlooked are the marvelous, easily grown flowering crab apples. Some bear fruit that makes a delicious jelly, but they are grown mostly for their beauty. Fiery crimson crab apple is grown for its gorgeous blossoms but is laden in fall with small scarlet crab apples that cling for a long time after the leaves have fallen.

Another valued ornamental is the Bradford pear, which is one of the earliest trees to bloom in the spring. The abundant blossoms appear in clusters of 10 to 12. They are off-white, nonfragrant, and borne on short spurs. Collectively, they appear as a solid mass of white in vivid contrast to other spring foliage and flowers. The glossy green, thick, and broadly oval leaves appear just as the flowers start falling. Their wavy margins cause them to flutter in the wind and, like the flowers, they are abundant. The Bradford pear is gorgeous again in fall. Early frosts bring about changes in the color of the leaves to deep hues of purplish red, then crimson. The tree rarely fruits, and when it does so, the fruit is inedible.

Other fruit trees often grown as ornamentals include almond, cherry, peach, plum, and quince.

Dwarf fruit trees, grown espaliered against a wall, provide an interesting and attractive way to use a narrow space.

FUCHSIA (*Fuchsia corallina*)

This vigorous, nearly prostrate shrub spreads to 3 feet or more. Its branches, covered with large, dark green leaves, arch gracefully. All summer long there is a sparkling display of long flowers with brilliant red calyxes and a rich purple skirt. The shrub is beautiful for banks or to overhang a wall, and grows in sun or shade.

GARDENIA, CAPE (*Gardenia jasminoides*)

This is a beautiful broad-leaved evergreen shrub, 2 to 6 feet tall, with dark lustrous leaves and exquisite large, white, waxen flowers of enchanting fragrance. The double-flowered form is famous as a buttonhole flower. The shrub blooms from May to September in the South, where it is often used for hedges. Use as a background for lower-growing flowers as the white blossoms blend well with all colors.

GUELDER ROSE, HIGHBUSH CRANBERRY (*Viburnum opulus*)

The highbush cranberry is a valuable wild plant, yielding food, drink, medicine, and beauty, but it is not a cranberry, nor is the almost identical guelder rose of England a rose.

These tall shrubs, reaching from 6 to 10 feet, are related to the honeysuckle, the elderberry, and the blackhaw. The attractive white flowers appear in showy cluster 3 to 4 inches across, with large sterile blossoms about the edge of the cluster and much smaller fertile ones near its center. The flowers are followed by bountiful clusters of bright red berries that become better-tasting and soft when touched by frost.

The berries hang on the bushes all winter. Birds eat them, but not until early spring, when other food is scarce. This is an excellent plant for bird lovers to place in a wild garden.

HAWTHORN (*Crataegus laevigata*)

The hawthorn is a beautiful May-blooming shrub with sweet-scented blossoms and lovely pink, rose, or white double flowers. Not all varieties are fragrant, however, and the blossoms of some of the American hawthorns have a disagreeable odor. Be sure of the kind you plant.

Hawthorns make excellent hedges around the flower garden and for windbreaks, shade in hot weather, and protection against intruders.

HEATHERS AND HEATHS (*Calluna* and *Erica*)

These dwarf evergreen shrubs are excellent for edging or in front of taller evergreens in a foundation planting. Their foliage, which persists in winter, takes on attractive shades of green, bronze, and gold.

Heathers and heaths resemble each other closely in their growth characteristics; however, the heathers (*Calluna*) are hardier than the heaths (*Erica*). Heathers flower for the most part in summer and fall, some continuing into winter, while the heaths bloom in late winter and spring. They come in shades of deep rosy red, brilliant pink, and white. Heather may be lilac mauve, silvery pink, red-purple, and pure white. The heath *Erica carnea* 'Springwood White' is one of the first shrubs to bloom in winter.

HONEYSUCKLE (*Lonicera*)

There are numerous kinds of honeysuckles, characterized by the sweet honeysuckle scent and full of nectar for the bees.

In fairly recent times a decoration of the stems was used for the gout, while an infusion of the flowers was believed helpful for asthma sufferers.

The winter honeysuckle (*Lonicera fragrantissima*) suggests the scent of roses. Blooms usually occur before leaves open, and the fragrance of the profuse, tiny blossoms carries for yards from late February to April. The plant grows well in sun or shade even in a northern location, and sometimes is evergreen in a sheltered spot.

HYDRANGEA (*Hydrangea macrophylla*)

As many as 35 species of these deciduous shrubs or vines with large bold flower clusters and leaves are found growing in the United States. *Hydrangea macrophylla,* widely called hortensia, is the pot or tub hydrangea that florists force for spring bloom.

Potted hydrangeas are often made to produce blue flowers by adding aluminum sulfate to the soil weekly at the rate of ½ pound to 5 gallons of water. After two or three applications, apply 4 ounces of ferrous sulfate per 5 gallons of water for a few weeks. This treatment is continued as long as it is necessary to keep the soil acid.

Blue-flowered hydrangeas produce flowers of various pinks if grown in nonacid soil; a neutral or slightly alkaline soil will give the results you desire. Add sufficient lime to raise the pH of the soil to a figure between 6.7 and 7.2. The addition of lime works best in fall. Lift out the hydrangea. Shake the roots free of as much soil as is safely possible. Then mix the lime thoroughly with the soil before replanting. Have your soil tested if you want to be certain.

H. macrophylla is one of the several that are extremely poisonous. Cyanide compounds are present mostly in the leaves and branches.

IDESIA (*Idesia polycarpa*)

This attractive deciduous tree is found wild in southern Japan and in central and western China. The Chinese type is hardy as far north as Boston, Massachusetts. The yellowish green flowers are followed by bunches of small fruits resembling bunches of grapes. Fruits are red when ripe. Some trees produce all female flowers, others all male, and yet others, both. The all-male-flowered trees do not fruit, and the all-female-flowered ones only do so when a male tree is nearby.

INDIGOBUSH (*Amorpha fruticosa*)

Indigobush is a deciduous shrub with narrow spikes of lovely, tiny purple flowers in late summer. Butterflies adore it. It is drought-tolerant and can become invasive in more favorable conditions. Plant in full sun. *Amorpha canescens* is similar but much smaller.

JASMINE (*Jasminum nudiflorum*)

The so-called winter jasmine is one of our most brilliant winter-flowering shrubs. The cheerful butter yellow flowers and red-tinged buds appear in profusion in midwinter. Before October is over, if you grow it on a south or west wall, the earliest flowers will brighten a garden already entering its first bare stages of winter. Then, until February or March, your winter jasmine should be a never-failing source of blossom for your house. Most generous flowers appear when this handy plant is grown in a lime-free soil in a position fully exposed to the sun.

JOJOBA (*Simmondsia chinensis*)

Jojoba is a dense, mounding, evergreen desert shrub that may grow as high as 8 feet and equally wide. It is an excellent landscape plant even for a formal garden, for hedges, background, foundations, and screens. Mature plants are hardy to 15°F, but seedlings are sensitive to frost.

JUJUBE, CHRIST'S-THORN
(*Ziziphus jujuba* and *Z. spina-christi*)

These may be either shrubs or small trees. They bloom late, small greenish white flowers giving way to elongated fruits sometimes called Chinese dates. These can be used in the kitchen or as a tonic food for people and animals.

It is an attractive plant with shiny, bright green leaves, and the small, woolly flowers in clusters are richly honey scented. The fruit is a prized delicacy of the Bedouins. However, the thorns of this shrub dig cruelly into human flesh. Legend says that Christ's crown of thorns are made of its branches. Legend also says that the *christi* name was given to the shrub because Christ loved its fruit. They are refreshing and alleviate fatigue in the heat of summer.

KAVA KAVA (*Piper methysticum*)

Kava and ava are the names of two shrubs related to the pepper plant. People have cultivated them for centuries in the Pacific Islands and Australia. The kavas are erect shrubs and may grow as tall as 5 feet. They have small yellowish cream flowers and round leaves. They may be easily raised in greenhouses and can be grown from stem cuttings.

The roots yield a juice called kavaic acid. Peoples of the South Pacific use the roots to make a fermented drink called kava, ava, or kavakava.

The kavas are in the family Piperaceae. The two kinds are *Piper methysticum* and *P. excelsum*.

KERRIA (*Kerria*)

A valuable plant that flowers well even in dense shade, this small, tough shrub grows upright with thin branches that remain bright green all winter in all but the coldest regions, and even there it grows when given protection. It blooms in mid-May with a wealth of 1½- to 1¾-inch bright yellow flowers. And it doesn't stop there—the spring-blooming period is followed by light, sporadic flowering through summer and an impressive show again in early fall.

KOCHIA, BURNING BUSH, SUMMER CYPRESS
(*Bassia scoparia,* formerly *Kochia scoparia*)
For a quick-growing ornamental, try kochia. It grows 30 inches tall and makes a nice annual hedge. The feathery foliage turns red in the fall.

KOLKWITZIA, BEAUTY BUSH (*Kolkwitzia amabilis*)
This is a handsome flowering shrub from China. Its clean foliage is untroubled by insects and diseases. In June the whole plant becomes a fountain of bell-shaped, light pink flowers. It reaches a height of 7 to 8 feet and will grow anywhere, thriving even in dry, sandy, poor soil.

MAGNOLIA (*Magnolia stellata* and *Magnolia soulangeana*)
Magnolias are of special interest because they have the largest flower of any tree in our gardens. The lustrous evergreen leaves, the big deliciously fragrant white blossoms, the conelike fruits that flush from pale green to rose—all have helped give the magnolias a preeminent place in every country where ornamental planting is valued.

Magnolias are reasonably hardy and in sheltered locations may be planted as far north as Massachusetts. They prefer a rich, moist soil. Transplanting is, however, a difficult operation and is best done when new growth starts. The flowers show up marvelously against a dark background of evergreens.

MAHONIA (*Mahonia*)
These evergreen shrubs have compound, hollylike leaves, fragrant yellow flowers, and berries that are both edible and delicious. Plant mahonia for foundations, background, screens, and groundcover.

MESQUITE, HONEY POD (*Prosopis*)
These interesting and beautiful multiple-trunked deciduous trees or large shrubs are found growing in desert areas. The clusters of creamy white

flowers attract bees, which make an excellent honey from their nectar. The fine-textured, fernlike foliage gives light shade. These plants grow slowly in nature but faster in cultivation. They do well in lawns and are nitrogen-fixing members of the Bean family.

The mesquite is almost an object of worship to desert dwellers. The long, fat pods supply a nutritious food. Cattle thrive on the young shoots when other foliage is lacking. The deep-reaching roots, 60 feet or more in length, are hauled out of the ground for fuel, posts, railroad ties, furniture, and paving blocks. The wood is also cut into building and fencing materials, two great needs of the desert.

OLEANDER (*Nerium oleander*)

Beautiful but poisonous, this houseplant may be grown outdoors in the South. It makes a shrub about 15 feet tall, with leathery lance-shaped leaves and showy, roselike flowers in red or white. Oleander is easily grown from cuttings. All parts of the plant are poisonous but effective against codling moths.

Yellow oleander (*Thevetia peruviana*) also is reputed to have insecticidal properties. All parts except leaves and fruit pulp are used to make cold-water extraction effective against a number of insect pests, especially aphids.

OSAGE ORANGE, BODARK, BOIS D'ARC, BOW WOOD (*Maclura pomifera*)

The name refers to the Osage Indians, who used the wood for bow-making. The plant grows wild in Oklahoma, Texas, and Arkansas. The yellow fruit looks much like an orange but is inedible. Cut and oven-dried, the fruit is sliced and used to make lovely flowers which may be painted for decorations.

The pioneers planted Osage orange for a living fence around their farms before barbed wire came into general use. Posts sprout easily and soon became trees. The wood is also used for making wagon wheels. A yellow dye is made by boiling chips in water.

Roots, wood, and bark repel insects, particularly crickets and roaches.

OTAHEITE ORANGE (*Citrus × limonia*)

This plant is sometimes classed with limes, but the purple flower buds and outer petal surfaces indicate a lemon relationship. However, because of the fruit's shape and color, it is commonly known as Otaheite orange.

This naturally small shrub, usually raised from cuttings and grown in pots, is very ornamental. It is fragrant when in flower and attractive in fruit. The fruit is small to medium in size, round, orange-colored, with orange, juicy, blandly sweet pulp.

PHILADELPHUS, MOCK ORANGE (*Philadelphus*)

Mostly hardy, these deciduous shrubs vary in size from small bushes 2 to 3 feet high to large ones 15 to 20 feet high and equally wide in diameter.

Breeders have produced many beautiful hybrids of mock orange. One of these, *Philadelphus × virginalis,* is among the best and most fragrant. The sweet-scented mock orange, *P. coronarius,* is the most common. Its flowers are strongly scented, and although they are delightful in the garden, their scent is too strong indoors for many people. The double-flowered varieties are less strongly scented than the common kind and last longer as cut flowers.

Mock oranges are among the oldest shrubs in cultivation, dating as far back as the 16th century. Like the lilac, they were brought to America and planted in dooryards of the early settlers. They're tolerant of a wide variety of soils but do require good drainage.

PHELLODENDRON, CORK TREE (*Phellodendron*)

These are handsome, deciduous trees with short trunks and widely spreading branches. Male and female flowers are borne by different trees in summer.

Phellodendron belongs to the Rue family, Rutaceae, and several kinds have the aromatic odor peculiar to other family members. They also share in the repellent properties of rue and a decoction made from the bark is repellent to insects.

The name is taken from the Greek *phellos,* "cork," and *dendron,* "tree," and refers to the corky bark of several kinds. The Amur cork tree

(*Phellodendron amurense*) is one of these and is prevalent in Manchuria, northern China, Korea, and Japan.

PRUNING PRINCIPLES

Early-flowering ornamental trees and shrubs form their buds in summer and fall. Therefore, do any necessary pruning only during the month after they have flowered; if you prune them in late winter or early spring, you will be cutting off the buds. Some early-flowering trees and shrubs are dogwood, crab apple, forsythia, rhododendron, some roses, and viburnum.

PUSSY WILLOW (*Salix discolor*)

This one has dainty, pearly catkins. Cut twigs for indoor decoration in January and February, place them in water, and watch them unfold. Children find these delightful. The French pussy willow (*S. caprea*) produces silver-pink catkins that are deliciously honey-scented.

PYROSTEGIA (*Pyrostegia venusta*)

This showy, tender climbing shrub from Brazil produces rich, crimson-orange, tubular flowers in large drooping panicles. The name is derived from *pyr,* "fire," and *stega,* "roof, "and refers to the upper lip of the flower. This high climber is an absolutely marvelous choice for covering the rafters of a large greenhouse or for growing outdoors on arbors in the South.

QUINCE (*Cydonia oblonga*)

The quince, a shrub or small tree, is one of the loveliest members of the Rose family. Though it is grown mainly for the fruit, the rosy flowers that bloom early in spring are very attractive. They are long-lasting when cut and give a delightful Japanese-style effect placed in a low bowl.

The pear-shaped fruit has a golden yellow color and a fragrant smell. Quince is never eaten fresh, as it is quite hard and has an acid taste, but

Quince

it is very pleasing when cooked or used in marmalades, preserves, and jelly. Plant quince shrubs with garlic; it improves the flavor of the fruit. The tree has long been cultivated but has never been popular in this country.

REDBUD, JUDAS TREE (*Cercis*)

Redbud grows wild in most of the United States. It is unbelievably beautiful in early spring when every branch and twig is covered with bright violet-red flowers. Transplant small specimens in early spring. Shrubby in growth, it seldom attains a height of more than 12 to 15 feet, and may be grown as a large shrub or small tree. Members of the Bean family, redbuds are also nitrogen-fixing trees.

The redbud and the dogwood come into blossom at approximately the same time and complement each other, one rose pink and the other sparkling white.

ROSE

See the chapter "The Queen of Flowers."

ROSEMARY (*Rosmarinus officinalis*)

This evergreen shrub of the Mint family is noted for its fragrant leaves. It has tiny, pale blue flowers and dark green leaves. In masses, blossoming rosemary looks like blue-gray mist blown over the meadows from the sea. Its name comes from the Latin *rosmarinus,* meaning "sea dew." A thick growth of prostrate rosemary planted around the flowerbed will act as a border for snails and slugs; the sharp foliage apparently hurts their soft, slimy skin.

Cooks use the plant in seasoning and its oil is used in perfume. The oil is secured by distilling the leafy tips and leaves. It gives the characteristic note to Hungary water; eau de cologne cannot be made without it. (See the chapter on Cosmetics and Fragrances.)

Rosemary oil is in all pharmacopoeias; it should not, however, be taken internally. The flowers are a stimulant, antispasmodic, emmenagogue (promoting menstruation), and rubefacient (causing redness of the skin). The leaves are rubefacient and carminative (cleansing).

ROSE OF SHARON, SHRUB ALTHEA
(*Hibiscus syriacus*)
The hibiscus shrub is distinguished by rose, purple, white, or blue flowers about 3 inches wide. The flowers appear in late summer when few other shrubs are in bloom. It does well even under unfavorable conditions, in either the city or the country, and is a good shrub for the gardener who has little time.

SARSAPARILLA (*Smilax*)
This group of woody or herbaceous vines has hardy, tuberous roots and veined evergreen leaves. The vines grow in temperate and tropical climates and bear small clusters of red, blue, or black berries. Some species yield the drug sarsaparilla, which was once widely used as a spring tonic. It is also used as a flavoring for soft drinks and medicines.

True sarsaparilla (*Smilax officinalis*) is related to asparagus. Native Mexicans have long used the roots of the vine in a concoction they believe cures impotence.

SOURWOOD, LILY OF THE VALLEY TREE
(*Oxydendrum arboreum*)
Closely rivaling the dogwood in interest and the beauty of its flowers and foliage, this plant is to summer what dogwood is to spring. The small, fragrant, bell-shaped flowers resemble lily of the valley and are borne during July and August in showy clusters 8 to 10 inches long. The attractive leaves assume deep red and scarlet tints during autumn and form a contrast with the interesting seedpods. The plant is slow growing, which is considered an asset under certain conditions.

SPINDLE TREE (*Euonymus europaeus*)
This deciduous shrub or small tree is wild in parts of Europe. The leaves color well in fall and its red fruits with orange seeds are very attractive in autumn. The fruit has a paralyzing action on aphids. The wood of the tree was once popular for butcher's skewers.

SUMAC, LEMONADE SUMAC (*Rhus integrifola*)

The parts used are the flower heads, picked early to midsummer, and the fruits as they begin to turn a bright red. Dry for future use. To prepare a tasty summer drink, steep a heaping teaspoonful of the ground flowers and/or fruits, fresh or dried, in a cup of hot water. Cover 5 or 6 minutes with a saucer. Stir and strain. The native peoples of upper North America sweetened this drink with maple syrup, the tart fruits being soaked in water until needed for use. The fruits of the staghorn sumac are distinguished from those of the smooth variety by being far more hairy; use less of the staghorn because they are more acid.

Bury bags of sumac leaves around the base of apple trees infested with woolly aphids. Tannin has been discovered as an active principle in sumac leaves.

SUMMER-FLOWERING SHRUBS

If you want spectacular bloom in July and August, here are some suggestions: hydrangeas (many varieties), *Amelia* × *grandiflora*, *Buddleia davidii*, *Ceanothus americanus*, *Clethra alnifolia*, *Hibiscus syriacus*, *Holodiscus discolor*, *Hypericum densiflorum*, *Indigofera amblyantha*, *Itea virginica*, *Lespedeza bicolor*, *Perovskia atriplicifolia*, *Sorbaria kirilowii*, *Stewartia ovata*, and *Vitex incisa*. These do well in the hot and dry areas of the southwestern states.

TULIP TREE (*Liriodendron tulipifera*)

This near relative of the magnolia has showy yellow blossoms that resemble garden tulips in size and form. A good shade tree with fast growth, it is ideal for young folks who have just purchased their first home or for older people when they move to treeless suburbia. In the lumber trade, tulip tree is called yellow poplar.

UVA URSI, BEARBERRY, WILD CRANBERRY, BEAR'S GRAPE, SAGCKHOMI (*Arctostaphylos uva-ursi*)

This low-growing evergreen has pretty pink flowers from April to June. Plant in early fall or spring in land that is loamy and free of lime. Herbalists use a tea made from the leaves in treating diabetes, Bright's disease, and all kidney troubles.

WEIGELA (*Weigela*)

This lovely shrub flowers from May through July. The rosy blossoms resemble foxglove in shape, are borne in immense quantities, and are attractive to hummingbirds. Give them a moist soil and full sun, away from competition of tree roots. As they bloom on twigs of the preceding year, prune after flowering.

WILD CHERRY (*Prunus avium*)

The avid tent caterpillar likes to eat wild cherry trees, and landowners sometimes try to get rid of the trees for that reason. However, entomologists have discovered that if an insect is deprived of its native feeding plant and learns to eat another plant, it will never return to its original feeding plant. What that means is that if all the wild cherry trees were destroyed, the tent caterpillar would go to other trees, principally apples and pecan. Worse than that—it would never go back to the wild cherry.

Keep this in mind and don't destroy the wild cherry trees; they are valuable as trap plants for concentrating the tent caterpillars where they can do little harm. Even when completely defoliated, the wild cherry instinctively protects itself against permanent damage. In about 3 weeks it may be again in full leaf. The enemies of the tent caterpillar, the calosoma beetles and braconid wasps, as well as other insect friends, help keep tent caterpillars under natural control.

WINTERFAT (*Ceratoides lanata*)

A bit of an oddity, this shrub grows 3 feet tall and is covered with woolly hairs, white but becoming rust-colored with age. In fall the twigs are covered with woolly white fruits, resembling lamb's tails, and are wonderful for dried arrangements.

WITCH HAZEL (*Hamamelis*)

This very fragrant hardy ornamental blooms at a time when few other shrubs are blossoming outdoors. Its bright yellow flowers are not injured even in zero temperatures. On a cold, frosty morning witch hazels "shoot" their seeds as the seedpods crack open with a snap.

The well-known medicinal lotion is derived from an extract of the plant dissolved in alcohol.

YLANG-YLANG, FLOWER OF FLOWERS
(*Artabotrys hexapetalus*)

The name comes from the pennant-shaped, long-petaled flowers, and means "flower that flutters."

The ylang-ylang is not a graceful tree, but its peculiarly shaped blossoms make up for any shortcoming in its stature. The flowers are noted for their heady fragrance that permeates the air for a considerable distance around them. Just one of the fleshy, 3-inch-long flowers will perfume an entire room.

As the flowers, almost hidden at first, reach maturity, the petals turn yellow. Then they gradually darken with age and their fragrance becomes proportionately stronger. The tree is generous with its flowers, which, strangely, never fall; they just dry up and gradually blow away. The birds consider these delicate tidbits.

Plant this brittle, upright tree where it will be protected from strong winds. It is fast-growing and will easily reach 25 feet within 5 years. Though listed as a rarity, it isn't difficult to find in nurseries that specialize in unusual trees.

VINES

There is always a place in every flower garden for a truly outstanding vine. In addition, vines are often used to screen out an unsightly area, to create shade, or as protection for other plants. Many vines add grace to hanging baskets.

Some suggestions for annuals are cathedral bells (*Cobaea scandens*), 30 feet; cypress vine (*Ipomoea quamodit*), various colors, 35 feet; cardinal climber (*Ipomoea* × *multifida*), scarlet, white-throated tubular flowers, 30 feet; marble vine (*Diplocyclos palmatus*), attractive leaves; and black-eyed Susan vine (*Thunbergia*).

Perennial vines of interest include coral vine (*Antigonon leptopus*), 30 feet; Chilean jasmine (*Mandevilla laxa*), fragrant white flowers; butterfly pea (*Clitoria ternatea*), double light blue flowers; Spanish flag (*Ipomoea lobata*), crimson flowers, heart-shaped leaves; perennial pea (*Lathyrus latifolius*), many colors; Madeira vine (*Anredera cordifolia*), vigorous climber, fragrant white flowers; wisteria, blue or purple clusters of spring; queen of vines; and yellow jasmine (*Jasminum humile*), a delightful evergreen bearing many clusters of bright yellow bells in summer.

Here are some further notes on vines for the yard and garden (plus an aid to recognizing one that you certainly will not want to plant but may encounter accidentally).

BEAN, SCARLET RUNNER (*Phaseolus coccineus*)

This is the king of the ornamental beans, growing over 10 feet tall with large clusters of bright scarlet flowers that blossom all summer. It is very prolific; the more pods you pick, the more the plant produces. The pods are 12 to 16 inches long with large black and scarlet beans that are absolutely delicious freshly cooked. If left on the vine to mature, the beans can be made into attractive necklaces. Pierce while still green and let dry for few days on a long hat pin. String on heavy thread with small gold beads in between.

This flowering bean is ideal for growing up the side of a porch, garage, or house as a vine for shade. The scarlet runner has unusually large leaves that maintain a lush green color all summer, and the flowers attract hummingbirds, which add to the beauty of the scene.

Scarlet runner beans are both delightfully pretty and excellent to eat. They're a fine choice if you need a vine that will grow quickly to hide an unattractive fence. Keep them away from onions and garlic.

Summer savory, strawberries, potatoes, beets, celeriac, and summer radishes are good companions, but do not plant members of the Onion family nearby.

CHINESE FLY CATCHING VINE (*Aristolochia delibis*)

The long peculiar flowers are insectivorous; their odor serves the purpose of attracting the insects required to ensure pollination and fertilization. The small leaves are attractive. Use for hanging baskets. This vine is an important medical plant in China and Japan.

CINNAMON VINE, CHINESE YAM, CHINESE POTATO (*Dioscorea batatas*)

This quick-growing vine will hide an unsightly area as it ranges up to 30 feet in a single season. In July and August it puts out profuse white, cinnamon-scented flowers borne in loose clusters. The roots are large tubers, potato-like in flavor, and considered edible in the tropics. The leaves are shiny and quite attractive. Its flowers are borne on the axils, where little tubers about the size of a pea also appear. These tubers, sown like seeds, will produce a full-size vine the second year. Cinnamon vine likes the sun but is not at all capricious as to soil. For quick growth, start the small tubers indoors in pots.

CLEMATIS (*Clematis*)

Large-flowered hybrid clematis is one of the loveliest vines known and blossoms abundantly in many colors. A wide range of cultivars is available and more are continually developed.

Also of interest is the smaller-flowered clematis called virgin's bower. There are a number of colors and varieties of these rather small, deciduous vines, highly prized for their often showy flowers. Give virgin's bower a place on fences, trellises, or posts. It likes a cool, shaded spot in rich soil and full sun to partial shade.

This is a marvelous companion plant for enlivening the rather somber branches of pine trees; Japanese gardeners often train it to grow on them. It is not a parasite and does not harm the tree.

CLIMBING VINES

Vines growing on masonry walls or on trellises on wooden walls add insulation. In summer they lower indoor temperature by protecting the outside walls from the sun's direct rays. Choose deciduous leafy vines (they will drop their foliage in winter to let the sun warm the dwelling), and plant them on southern and western walls.

In cold weather, evergreen vines on the north surface will head off the wind and keep the inside warmer. Try Boston ivy or Virginia creeper for summer insulation and for cold-weather protection. Clinging vines are not recommended for wooden walls because their stems and tendrils hold moisture, causing the wood to deteriorate. Achieve the same insulating effect with trained twining vines such as wisteria or climbing roses on trellises.

DUTCHMAN'S-PIPE (*Aristolochia macrophylla*)

This foliage vine with its handsome, heart-shaped leaves creates dense, cooling shade. It grows rapidly, reading up to 30 feet in height.

GRAPE (*Vitis*)

The sunlight on the leaves and not on the grapes determines whether or not grapes can be grown. Grapes will color normally if they have adequate leaf surface in proportion to the amount of fruit being produced. To obtain perfect bunches of grapes, they are sometimes placed in bags. If you want to do this, use brown paper bags. Grapevines need a sunny, well-ventilated location. A grape arbor is so decorative that you may wish

Grapes

to place one in your flower garden. Hyssop planted near grapevines will increase yield. Though grapes are usually propagated by cuttings, they do have flowers, beloved of bees, and fertile seeds. Wild mustard is beneficial to grapevines.

KIWI (*Actinidia deliciosa*)

The so-called Chinese gooseberry is a native of China and is commercially grown in New Zealand. The plant is a rampant grower, shooting up possibly 5 inches in one day and up to 8 feet the first year. The fruit tastes like a blend of strawberry, pineapple, and guava, and keeps well in the refrigerator.

Hardy varieties of kiwi have been developed and can be grown even in the North. Train grapelike vines on arbors, trellises, or fences for November harvesting. Grow in pairs of one male and one female vine; additional female vines may be planted with one male.

MAN-OF-THE-EARTH, WILD POTATO VINE (*Ipomoea pandurata*)

This is a very hardy tuberous vine with flowers similar to those of morning glory but larger. They have a delicious fragrance, slightly reminiscent of lemon, fresh and clean.

MORNING GLORY (*Ipomoea purpurea*)

This popular favorite possesses a simple beauty, especially in the modern versions of standard varieties. Rapid in growth, it must be provided with something to twine about; if not, it will twine on whatever is nearest, no matter what it is. Morning glory is a great success in a window box. With light support it should reach the ceiling by midsummer, blooming every foot of the way. It is great for covering trellises and arbors and for hiding unsightly areas.

Morning glory seeds will germinate sooner if boiling water is poured over them before covering with soil. This does not harm the seeds and will soften the shell, causing the seeds to sprout more quickly. Other extra-hard-coated seeds may be treated the same way.

This vine takes plant parenthood quite seriously and will vigorously reseed itself.

Morning glory has uses as a companion plant. (See the chapter "Companion Planting with Flowers and Herbs.")

POISON IVY LOOK-ALIKES

Virginia creeper is frequently mistaken for poison ivy, which it does somewhat resemble. Yet its five leaflets, compared to only three for all forms of poison ivy and poison oak, make it easily recognizable. Other innocent plants that sometimes suffer from an identity crisis are the harmless Boston ivy (*Parthenocissus tricuspidata*) and marine ivy (*Cissus trifoliata*).

The real culprit to watch for is poison ivy (*Toxicodendron [Rhus] radicans*). Usually this is a vine, but may sometimes, especially in open sun, be a shrub from a few inches to several feet high.

Virginia creeper

Poison oaks (*Toxicodendron [Rhus] spp.*), found in the South and West, is sometimes called oak-leaf poison ivy. They are shrubs, with leaflets covered by downy hairs and lobes resembling small oak leaves.

Be careful around these two, for even particles of the irritant oil wafted through the air on smoke or pollen can affect the eyes or lungs of allergic people. Jewelweed, which often grows nearby, is very good to use as a remedy against poison ivy. It relieves the itching almost at once. Boil a pot of the jewelweed and strain the juice. Keep juice refrigerated or freeze cubes of it and bag them future use.

TRUMPET CREEPER (*Campsis radicans*)

This woody, high-climbing vine is very hardy. Although a native of the woods, it is often planted in gardens. The flowers are 3-inch-long, orange tubes with flaring scarlet lobes and grow in clusters. They yield copious nectar and attract many insects. The shade of red-orange is somewhat harsh, so use this vine with discretion. The fruit is a long pod with a many-winged seed. Some people are allergic to the leaves and may get dermatitis from touching them. Although this is a beautiful vine, it can be a nuisance along roadsides in the South.

Trumpet creeper

WISTERIA (*Wisteria*)

This truly magnificent vine increases in beauty with every passing year. Drought-tolerant, it does well planted in a sunny location. *Wisteria floribunda* Alba has long racemes of pure white flowers; when in bloom, it resembles a waterfall.

Acetone extract from seeds of wisteria is somewhat toxic to codling moth larvae.

THE LIFE OF PLANTS

ALL-AMERICA

To qualify for this label, a new seed variety must be started at 30 sites across the country, each site representing a different climate and soil. Only those seeds that grow well at a wide range of sites and are a distinct improvement over the nearest existing variety win this citation.

ALLELOPATHY

Some plants release chemicals into the soil that are toxic to other plants. This phenomenon is now receiving attention as a way to control weeds. Researchers hope to breed weed resistance into commercial crops in much the same way that disease resistance is instilled.

An example of allelopathy is the so-called soft chaparral, a unique association of evergreen shrubs and trees in the semiarid land of western North America. No more than 8 feet in height, these thickets of broadleaf evergreens and stunted shrubs and trees have the amazing ability to invade grasslands and encircle themselves with dry moats of bare soil 3 to 6 feet wide.

Scientists have determined that these xerophytic (arid-climate–loving) plants release fragrant chemical compounds called terpenes from their leaves into the surround air. The soil around the shrubs absorbs the terpenes, which accumulate more rapidly during the dry season, in an amount sufficient to inhibit the germination and growth of other plants in the surrounding area. Commonly known terpenes include camphor, rosin, natural rubber, and turpentine.

BRACTS

What we sometimes think of as blossoms are often petal-like bracts such as the highly colored bracts of the poinsettia. The dogwood and the bunchberry have tiny purple flowers surrounded by white, petal-like bracts. Bracts are leaflike structures that may form a circle beneath an inflorescence.

CARNIVOROUS (INSECTIVOROUS) PLANTS

These plants trap insects for food. They usually live in moist places where they get little or no nitrogen from the soil. They must obtain it from the decaying bodies of the insects they trap. For this they have special organs and glands that give off a digestive fluid to help them make use of their food.

Some of these plants have flowers that are colored or scented like decaying meat to help them attract insects. Pitcher plants have tube-shaped leaves that hold rainwater in which the insects drown; other carnivorous plants have rosettes of leaves with sticky hairs, such as those borne by the sundews. Some plants such as the badderworts grow in water.

Venus flytrap

The Venus flytraps make interesting house-plants; in the absence of insect prey they are usually fed tiny pieces of meat, generally hamburger. Their leaves consist of two hinged lobes. When an insect is attracted to a leaf, the lobes snap shut, and the insect is digested by the plant. If you grow carnivorous plants indoors, water them with distilled or soft water to avoid toxic salt buildup.

COMPOSITE OR COMPOSITAE

This family is the largest and most highly developed of flowering plants. It consists of more than 20,000 species of herbs, trees, vines, and shrubs.

Composite plants have efficient methods of reproduction. They produce many seeds and have good methods of scattering them.

Some family members, such as calendula, chamomile, wormwood, tansy, and arnica, are used to make drugs. Chrysanthemums, asters, and dahlias are grown for their beauty; others are weeds and wildflowers, such as ragweed, goldenrod, sagebrush, thistle, or burdock.

CROSS-POLLINATION

Seeds may be produced by either self- or cross-pollination (see Pollen, in this chapter). In the former case only one plant is involved; in the latter, two. Pay attention to your nursery catalog to know whether you will need two plants for fertilization to occur.

Bayberry (*Myrica pensylvanica*), beloved of early Americans for candle making, will not bear its berries unless male and female forms keep company together.

If you have a single holly, don't discard it; buy the missing member of the pair and your holly will bear berries.

The willowy leaves of the spiny sea buckthorn (*Hippophae rhamnoides*) are a lovely silvery gray, and pollinated female bushes bear an abundance of orange berries. No need to arrange individual marriages; plant one male buckthorn amid a small harem of ladies.

FRAGRANCE

Flowers exude a powerful, seductive odor when ready for mating. This causes a multitude of bees, birds, and butterflies to join in a saturnalian rite of fecundation. Unfertilized flowers emit a strong fragrance for as many as eight days or until the flower withers and falls; yet once impregnated, the flower ceases to exude its fragrance in less than half an hour. One tropical plant (*Alocasia odora*) increases in temperature at time of flowering, repeating this phenomenon for six days from 3 P.M. to 6 P.M. each afternoon.

Fragrant flowers are usually light in color or white, with the purples and mauves coming next. Thick-textured flowers such as magnolias and gardenias are often heavily scented. The perfume of a plant is not always found in its flowers. It may be in the root, seeds, bark, the gum or oils, even in the leaves or stalk. Certain families, such as the Labiatae (lavender, rosemary, mints), are especially gifted with perfume.

This invisible quality of flowers is one of their most important assets; however, many of our modern hybrids have lost their fragrance. The so-called unimproved kinds often retain this. The modest little sweet-scented candytuft, *Iberis odorata,* is an example, The wild carnation seems to spray its admirers with its spicy incense. The scent of old-time roses, lilac, and violets are enchanting. Most of the older varieties of iris are also sweet with perfume. If you would have fragrance in your flower garden, seek out the older, less showy varieties. Note also that flowers are less scented in periods of extreme heat and drought.

GIBBERELLIC ACID

In the mid-1920s, a Japanese scientist working on rice diseases discovered a remarkable colorless substance named gibberellic acid. Less than one drop of it on a camellia bud causes the flower not only to be much larger but also to bloom much earlier than usual. This has greatly extended the season of blossoming for camellias.

Another experiment has shown that a lawn can be made to start growing in early spring at below normal temperature after being treated with gibberellic acid at a concentration of 1 ounce per acre. Scientists believe the gibberellins are natural plant products.

GREENSAND

Greensand is found along the New Jersey and Virginia coasts, among other places. This granular, half-soft marine deposit contains about 6 percent potash and is an olive green, iron-potassium silicate, also called glauconite. Greensand is available from several natural-fertilizer companies and is recommended as a major natural source of potash as well as nitrogen and phosphorus. (See Sources on page 442.)

HYBRIDS

The production of new flowering plants occurred only after the sexual basis of plant reproduction was understood and the principles of heredity and genetics were laid down. It was then possible to take a plant with one desirable characteristic (for example, flower size) and cross it with a plant having another desirable characteristic (such as a specific color).

Hybrid plants are formed by taking the pollen from one plant and transferring it to the ovary of another. The resultant seeds are first-generation hybrids and can be planted to produce the superior flowers, fruits, and vegetables found in seed catalogs.

The mule, a sterile cross between a horse and a donkey, is a first-generation, or F1, hybrid. If mules could reproduce, their offspring would be F2 hybrids. The "F" stands for filial or offspring, the digit for the generation.

Just as mules have greater vigor than either parent, so have hybrid plants. In general they grow faster, bigger, and more uniform, and bear more flowers and fruit. That's especially true for the F1 generation. Off-spring of F1 plants—the F2 generation—have more vigor and uniformity than regular plants, but less than F1. F1 hybrid seed is expensive because seed producers can't let bees do the pollinating, but must brush the proper pollen on by hand. F2 seed doesn't need hand pollination, so is less costly.

INFLORESCENCE

The word *inflorescence means* "a flowering" or a "flower cluster." The largest known inflorescence is that of *Puya raimondii*. The jolly green giant is a rare Bolivian plant with an erect panicle (diameter 8 feet) that emerges to a height of 35 feet. Each of these bears up to 8,000 white blossoms. This is also said to be the slowest-flowering plant, the panicle emerging only after 150 years of the plant's life. After blooming, it dies.

LIGHT POLLUTION

As if noise, air, and water pollution weren't enough, we now have light pollution! Safety lights, installed by many people in their backyards, upset the timetables of plants and cause them to confuse night with day.

The amount of sunlight that a plant needs each day is called its pho-toperiod. Plants that flower when days are short are called short-day plants, and include chrysanthemum, Christmas cactus, gardenia, kalan-choe, aster, and poinsettia. Plants that flower when days are longer are called long-day plants and include marigold, petunia, black-eyed Susan, China aster, coneflower, feverfew, calceolaria, and weigela. And then there are the "day-neutral plants," which have hormones to induce flowering whether days are long or short. In this category are African violets, roses, snapdragons, and tomatoes.

The growth pattern is altered when plants are near the night lights be-cause the plants grow when they should be "sleeping." This is detrimental to trees, particularly in northern regions, because they continue to grow in

Luther Burbank (1849–1926)

Burbank was an American plant breeder and horticulturist who developed many new trees, fruits, flowers, vegetables, grains, and grasses. Among the plants he developed are the Burbank potato, the giant Shasta daisy, the spineless cactus, and the white blackberry. He also improved many plants and trees already known.

One of the methods he employed was the selection process, which continued through many generations until a plant that was superior with respect to a single characteristic or group of characteristics became isolated. He employed this method in the development of the "stoneless" plum.

First, Burbank selected from a large collection of plums one that had a thin "pit" or stone. He permitted this tree to carry on pollination and set fruit. Then he chose from this tree a few plums that had the thinnest stones. He planted the seeds from these, and when the resulting trees fruited, he again selected the fruits with the thinnest stones. These were planted and when the resulting trees matured, Burbank again selected the fruit with the thinnest pits. After several selections of this type, he produced a plum with an extremely thin stone; this was marketed as a "stoneless" plum. In using this method, only the desirable variations are retained. Burbank also used this method in developing the Shasta daisy. There is really no magic about the selection process; anyone with time and patience can use it.

Burbank's first discovery was the potato that bears his name. As a young gardener in Massachusetts, he planted the contents of a seedpod from the rarely blooming 'Early Rose' potato. The pod produced 23 widely varying seedlings. Burbank sold the best plants to a dealer, who named the variety 'Burbank'.

fall when daylight is shorter and they should be ceasing growth to prepare for winter. New growth is moist and tender. When a tree continues to grow well into the frost season, it becomes more sensitive and is more easily injured. Lighting also makes the leaves more sensitive to air pollution.

High-pressure sodium lamps are twice as efficient (for lighting purposes) as the mercury-vapor lamps, and emit more red and yellow light. Mercury-vapor lamps, generally used on highways and city streets, give off a bluish green light that contains a few red rays and many ultraviolet rays. Natural sunlight gives off light in the visible region from blue to green to yellow to red. The red region of the spectrum regulates the photoperiod.

Chlorophyll for photosynthesis—the food-making process that occurs only in nature and which is the chief function of the green leaves of plants— is activated by red and blue. The blue region also attracts night-flying insects.

Research by the U.S. Department of Agriculture (USDA) indicates the red part of the spectrum is the growth-triggering light. During the 24-hour period of a day, the light-dark cycles trigger the flowering, branching, dormancy, bulging (as with onions), and other plant-growth responses.

In the Beltsville, Maryland, nursery where tests were conducted, it was found that plants near the sodium lamps grew more rapidly into the fall season and also grew much later than plants of a like age that had been screened from night lighting. Trees that had been exposed to the light suffered severe winter dieback the following spring.

The sensitivity of 17 species of trees to security lights (night lights) was rated by USDA horticulturists. Trees tested having the highest sensitivity were Norway maple, paper birch, eastern catalpa, sycamore, American elm, and zelkova. The intermediate sensitivity group included red maple, gingko, honey locust, golden rain tree, Japanese pagoda tree, and littleleaf linden. Trees exhibiting low sensitivity were American holly, sweet gum, Austrian pine, Bradford pear, and willow oak.

These lights also have an adverse effect on gardens, because plants need to sleep as well. Generally speaking, plants are affected within a radius of about 25 feet of the lights.

Laws affecting night lighting of commercial property vary from state to state, but as our need for protection grows, lights are becoming more

Do Plants Sleep?

Plants sleep during a period called *dormancy*. They are affected by the cycles of winter, spring, summer, and autumn, as well as wet, dry, cold, and hot seasons. Many plants also close up and sleep at night.

widely used. It is good to be aware of the effect they have on growing plants, so that you can place your garden in a more favorable location or plant varieties with lower sensitivity to light.

OFFICINALIS

Any plant with *officinalis* as part of its name is or has been listed in the official pharmacopoeia, a book containing standard formulas and methods for the preparation of medicine, drugs, and other remedial substances. The plants listed may vary from country to country. For instance, the British pharmacopoeia lists some plants not found in the American. Also, some plants formerly listed in the American have been deleted down through the years.

ORGANIC MATTER

An important ingredient often lacking in garden soil is organic matter: fully or partially decayed remains of plants, animals, or animal by-products such as manure. Besides its gradual release of nutrients needed for plant growth, organic matter is valuable for several other reasons:

1. It improves soil aeration.
2. It improves the water-holding capacity of the soil.
3. It reduces soil crusting.
4. It stimulates the growth of beneficial microorganisms, some of which may destroy harmful microorganisms or prevent their growth.
5. It assists nematode control by supporting parasites, predators, and diseases of nematodes.

OSMOSIS

This is the passage of one fluid into another through a membrane between them. It occurs with both liquids and gases. This passage, or transfusion, results in a mixture of the two fluids. Osmosis takes place through a semipermeable membrane that allows certain substances to pass through and keeps others out.

Plants depend on osmosis. Minerals dissolved in water pass from the soil to the plant through root membranes. Osmotic pressure probably helps raise the sap to the high branches of trees.

PLANT MIMICRY

The orchid *Trichoceros parviflorus* grows petals to imitate the female of a fly species so exactly that the male attempts to mate with it and in so doing pollinates the orchid.

The carrion lily develops the odor of rotting meat in areas where only flies abound. Other flowers that rely on the wind for pollination do not waste their time making themselves fragrant or beautiful to appeal to insects or birds but remain relatively unattractive.

PLANT QUARANTINE

Plant quarantine is a law that regulates the movement of plants and other materials that may carry a plant disease or insect pest. The quarantine keeps the disease or insect from spreading from infested areas to those free from these hazards. Some laws list plants that may not be shipped in or out of a locality. They may also give directions for moving, packing, and labeling.

In a quarantine, officials examine all plants at the border of the quarantined area and keep out the dangerous types. Foreign plant quarantines control the shipping of plants from other countries.

POLLEN

The tiny yellow grains seen in most flowers are pollen. They are used to form seeds. Plants make the pollen in the saclike anthers of their flowers. The anthers are the male organs of reproduction. The female organs include the pollen-receiving stigma leading to the ovary, which is the egg-bearing part of the plant. Pollination is simply the transfer of pollen from the anther to the stigma. When this occurs, fertilization has taken place.

Self-pollination occurs in flowers that can transfer pollen from their own anthers to their own stigmas. Cross-pollination means that the flower must depend on wind, insects, birds, flies, or some other means to carry its pollen between flowers.

Pollen has been considered a great energy producer for centuries, and it is reportedly consumed by many athletes. It is also believed to be an aphrodisiac.

The pollen of most plants is highly inflammable. When thrown on a red-hot surface, it will ignite as quickly as gunpowder.

The wind scatters the pollen of many plants, including all the grasses and cereal grains such as corn, wheat, rice, and oats.

PROLIFERATION

In the plant kingdom, *proliferation* means new growth by cell division or buds. An example of this is the daylily, which not only makes seed and may be divided by its tubers but also makes new plants at joints of the flowerstalk, sometimes growing aerial roots. (See *Daylily* in the Flower Lore chapter.)

The so-called airplane plants also propagate themselves in a similar manner by sending out runners.

ROOTS

Roots are one of the three organs that most plants must have to grow. The other two organs are the stems and leaves. The roots of most plants grow in the ground and draw their food material from the soil, but some plants have their roots in water or even in air.

Roots hold the plant in place and supply it with water and nourishing salts from the soil. Roots that form first and grow directly from the stem are called *primary roots*. Branches of the primary roots are called *secondary roots,* and branches of these are *tertiary roots.*

Roots with different forms have special names. A primary root that grows much larger than any of its branches is called a *taproot*. When tap-

The Will to Live

Why do some plants live and others do not? Like people, plants do not always respond in an exact relationship to their environment and care. Some have a will to survive, others don't.

An occasional tulip bulb planted upside down will circle laboriously around and reach up triumphantly for the light. A cactus growing in the desert will top itself with a brilliant blossom. Tiny seedlings seem to have a built-in will to live. And, of course, some varieties of plants have it more than others; for example, it's very hard to kill sansevieria (mother-in-law tongue) or a rubber plant. And some flowers and herbs grow and spread on their own without help from gardeners.

Plants have also developed various means of self-protection. Mechanical weapons include thorns on roses, prickles on thistles, and spines on cacti. Sumac (some types) uses poisonous chemical weapons, and nettles have irritating acids.

roots grow very thick and store up food for the rest of the plant, they are called fleshy roots. A cluster of thick primary roots is called *fascicled roots*. Threadlike roots are *fibrous*. Roots may also grown on the stem or in other unusual places. These are called *adventitious roots*.

Roots are also called soil roots, aerial or air roots, and water roots, depending on where they grow. Roots that get their food from other plants are called parasitic roots.

SCIENTIFIC NAMES

These are the names botanists give to all plants. They are in Latin and are the same in all countries. Scientific names have two parts, the genus and the species. For example, the prairie rose is *Rosa setigera*. This means that it belongs to the genus *Rosa* and the species *setigera*. The relationship of plants is determined mostly by their reproductive parts.

SEEDS, SURVIVAL OF

Seeds possess an enormous potential for survival. It is not unusual for large seeds, 100 years old and more, to grow when planted. During World War II, in England, the bomb craters blossomed with flowers and other plants that had not been seen within the memory of humans. Evidently, the seeds were sleeping deep in the earth until brought again to the surface, where light and moisture caused them to germinate.

In 1982 scientists tested lotus seeds found in an Asian lake-bed deposit. They were found to be viable, and radio-carbon dating showed that they formed on lotus plants between 1410 and 1640.

Other claims are doubted by scientists. These include stories of viable wheat found in the tombs of Egyptian pharaohs, and Arctic lupine seeds found in the Yukon and believed to be at least 10,000 years old.

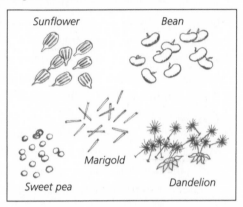

A seed is a little bundle of determination-to-grow, and it is the most important part of the plant. The roots, the leaves, the flowers all exist so there can be seeds. As shown above, there's a great variety among seeds, from the airborne dandelion seeds to the relatively large beans.

SEX

The first botanist to show that flowering plants have sex and that pollen is necessary for fertilization and seed formation was a German, Rudolf Jakob Camerarius, who published *De Sexu Plantorium Epistula* in 1694. The heated controversy engendered by this book lasted a generation before it was finally established that plants had sexual organs and could in fact be elevated to a higher sphere of creation than previously thought.

Bisexual plants. These have flowers with both male and female parts.

Dioecious plants. Male and female parts are found on separate plants.

Monoecious plants. Male and female parts are borne in different flowers but on the same plant. A good example of this is the squash plant. Gardeners frequently tell me that their squash plants are blooming but not bearing. The answer is that the male flowers appear first. As the plant grows, the female flowers (which will bear the fruits) appear and are fertilized, often by bumblebees. The female flower is borne at the end of a small bulb that gradually enlarges to form the fruit.

SYNERGISTS

Usually derived from plant products, synergists are nontoxic substances that are added for a strengthening effect. Pyrethrum, for instance, is greatly strengthened by the addition of asarinin from the bark of southern prickly ash, sesamin from sesame oil, and peperine from black pepper.

TERMINAL SHOOT

This term indicates the shoot that forms the end of the main stem or of a main branch of a tree or plant. All other shoots are called side shoots or laterals.

TETRAPLOIDS

Seeds or plants exposed to radiation are often grossly changed. A drug, colchicine, extracted from the fall-blooming crocus, profoundly changes plants too, sometimes doubling the number of chromosomes in every cell so that a normal diploid becomes a tetraploid. Such plants have enormous vigor, larger blossoms, or bigger fruit.

The brown-eyed Susan, a meadow plant, was treated with colchicine and changed into the vigorous and beautiful gloriosa daisy, a tetraploid. The same has been done with ageratum, phlox, snapdragons, and zinnias.

The small French marigold has double the chromosomes of the large American type. The result of crossing is an intermediate type, a triploid that is sterile; because it fails to produce seed, it continues blooming with tremendous vigor all season.

WHICH PLANTS GO WHERE?

ACCENT PLANTS

These plants are planted singly or in small groups to provide emphasis in the yard or garden. Usually they are of distinct color or form; an accent tree, for example, might be the Italian cypress, the Atlantic cedar, or the handsome sugar or rock maple.

To avoid monotony in a perennial border, interplant with a tall, stiff species. These exclamation marks of the garden include good tall plants such as Siberian iris, lythrum, and some daylilies. Or use plants with light gray foliage to break up the monochrome of green. Silver mound artemisia, santolina, and *Dianthus plumarius* give the desired effect.

BOULDER

Tuck succulents and tiny rock plants into crevices and crannies of an ancient boulder. Rub with moss and lichen and let sun and rain be its benison. I have a planted boulder and love it.

COLD-RESISTANT ANNUALS

The term describes the first annuals to plant in spring and those that will survive nippy autumn days. These annuals are also used for winter planting along the Pacific coast and in the Gulf States.

Pansies head the list. They are hardy and available in early spring throughout much of the country; some are even ready for late-fall planting in the milder climates.

If you need fragrant annuals for your garden, try stocks. These delicious annuals are fine for cutting and last well indoors. Snapdragons resist the cold and begin to provide their colorful spikes early in the gardening season. For strong yellows or oranges, plant calendulas. The newest hybrids are compact, floriferous, and very hardy.

Larkspur or delphinium, annual poppies, centaurea, dusty miller, annual phlox or aster, primulas, cineraria, dianthus, and carnations all will perform enthusiastically for you in early spring and will continue through several fall frosts.

DESERT GARDENING

Wise desert gardeners keep lawn areas small and use rocks, gravel, and patio paving generously, along with raised beds, pools, and water plants. Choose trees that can "take it," such as cottonwoods, poplars, Siberian elms, black locusts, chinaberry trees, as well as evergreens such as athel tamarisk, Arizona cypress, and Aleppo pine, for shade and wind control. In low deserts include more evergreens, such as beefwood, black acacia, and eucalyptus. Survival and fast growth are important criteria for desert plants.

Night-flowering cactus, Mexican fan palm, yucca, and low-growing lavender lantana make good companion plants.

For easy upkeep, grow annuals and perennials like creeping rosemary, feathery wormwood, senna, African daisy, salvia, and fairy primrose. For seasonal color, plant pyracanthas, oleanders, yuccas, and showy crape myrtles. Remember: If plants can't take heat, drying winds, and alkaline soil and water, their beauty counts for little.

DROUGHT-RESISTANT FLOWERS

Cornflower, calliopsis, sunflower, morning glory, ice plant, four-o'clock, rose moss, and zinnia will grow in regions with deficient rainfall.

FENCES

Robert Frost said, "Good fences make good neighbors." They also keep out unwanted animals. A living fence such as rugosa roses may also act as a deterrent and give lovely bloom as well. And think of the vitamin C in all those rose hips.

FIRE-RETARDANT PLANTS

While no plant will completely keep a fire from advancing, some plants resist burning better than others and thus may slow a fire's progress. However, if winds carry sparks, even protective fire-retardant plants can be breached.

Useful trees and shrubs are callistemon, *Ceratonia siliqua, Heteromeles arbutifolia, Myoporum, Nerium oleander* (dwarf varieties), *Prunus lyonii, Rhamnus alaternus, Rhus* (evergreen kinds), *Rosmarinus officinalis* Prostratus, *Schinus molle, Schinus terebinthifolius,* and *Teucrium chamaedrys.*

Perennials and vines include achillea, agave, aloe, artemisia (low-growing varities), atriplex, campsis, *Convolvulus cneorum,* gazania, ice plants, *Limonium perezii, Portulacaria afra, Santolina rosmarinifolia (S. virens), Satureja montana, Senecio cineraria, Solanum jasminoides,* and yucca (trunkless varieties).

Floating Gardens of Xochimilco

The Floating Gardens, about three miles beyond the village of Xochimilco, are the brightest and prettiest sight in Mexico City. (In fact, *Xochimilco* means "place where the flowers grow.")

The flowers have been growing there since the height of the Aztec Empire, when many nobles lived near Lake Xochimilco. In those times workers built big rafts, covered them with earth, and planted vegetables and flowers. The rafts floated and the roots worked down through the earth into the water. These "floating gardens" gradually increased in size and became anchored by the interlacing roots of the plants. Now, the island rafts no longer float but are solid islands surrounded by canals. Flower vendors move about in the canals selling their colorful wares from canoes.

This type of garden was one of the earliest uses of what is now called *hydroponics.*

HANG IT ALL!

Hanging baskets never fail to attract attention, be it cascading petunias gracing an old-time porch, a gay ivy geranium attached to a lamp post, or a dainty fuchsia suspended from a tree branch. Hanging baskets have a magic all their own.

A hanging basket is really a pot plant gone glamorous and needs to be fed to keep going. Slow-release houseplant fertilizers are a handy way to do this.

Here are some suggestions of suitable candidates for various positions. Plants to grow in sun:

Flowering. Lantana, ivy geranium, phlox (annual), lobelia, dwarf French marigold, nasturtium, oxalis, petunia, bougainvillea, sweet alyssum, verbena, monkey flower *(Mimulus)*, shrimp plant, dimorphotheca, pinks *(Dianthus),* cascade chrysanthemums, pansy (early spring), Tiny Tim tomato.

Foliage. English ivy, Sprenger asparagus, donkey-tail (*Sedum morganianum*), variegated vinca, peppermint geranium, *Gynura aurantiaca* 'Purple Passion', flowering inch plant (*Tradescantia blossfeldiana*), setcreasea, siebold sedum.

Plants to grow in shade:

Flowering. Fuchsia, achimenes, browallia, coleus (trailing), black-eyed Susan vine (*Thunbergia*), tuberous begonia, columnea, episcia, flowering maple, star-of-Bethlehem (*Campanula isophylla*), torenia.

Foliage. English ivy, German ivy, Swedish ivy, grape ivy, kangaroo ivy, pick-a-back plant, variegated archangel, spider plant (*Chlorophytum*), inch plant, zebrina (*Tradescantia*), philodendron, rosary vine, Christmas cactus, strawberry "begonia" (*Saxifraga*), epiphyllums, patience plant, pothos, Boston fern, rabbit's foot fern (and other ferns).

LAWN

Lawns are the greenest show on earth. Where the flowerbeds are the picture, lawns are often the frame. A lawn is a great asset to cooler living, for an acre of grass in front of your home gives off 2,400 gallons of water every hot summer day. This has a cooling effect equal to a 140,000-pound air-conditioner (which amounts to a 70-ton machine). Along with trees, shrubs, and flowers, a lawn helps purify the air, putting oxygen back into it so we can breathe much easier.

If you live in a dry location and have difficulty getting grasses to flourish, consider establishing a lawn of fragrant chamomile (*Chamaemelum nobile*). Drought has little effect on this green plant. Sow it just like grass. When cut it will give out a fragrant odor. To get it started in spring, sow a mixture of chamomile and lawn grasses together; as the season advances, the strong-growing chamomile will take over the lawn as the grasses begin to lose ground.

LOW-LIGHT AREAS

A dim location is the perfect spot for a dieffenbachia, a tropical plant from South and Central America. It can survive on only 2 hours of sunlight a day but must have humidity. The plant grows well grouped with palms, aglaonemas, ferns, and dracaenas.

PATIO PYRAMID PLANTER

A pyramid planter can be a most interesting focal point of your garden, balcony, or patio. A lot of different fruit and flowers can be shown off to advantage. Or use it for a miniature herb garden. Even midget vegetables will grow in one of these. For indoors, the planter is mounted on casters for easy movement; watering is provided for by way of a top reservoir.

PLANTER

Want something different in a container? Often free for the gathering, sculptured by wind and wave, driftwood makes a beautiful "pot" for cactus, sedums, and other small plants. Or fill it with lacy green ferns set with their own peaty earthballs in plastic bags, punctured at the base for drainage.

For indoors, look for a fine piece of handcrafted pottery, a dainty basket, or a filigree of old iron; Western boots, well worn, also make unusual planters. An old coal scuttle is dandy for summer flowers. It does not tip over, can be punched with drainage holes, and there is lots of room for deep root growth.

Something extra special for an apartment balcony is a birdcage on its stand. Paint it black and fill it with a tumble of white marguerites or red or yellow cascade petunias. Or paint it turquoise blue and use pink geraniums, white petunias, and green ivy.

Seashells make attractive planters. If you prefer not to deface a shell by cutting or drilling to provide drainage, water the plants sparingly. Plants such as velvet plant (*Bynura*), aloe vera, succulents, sedums, and miniature English ivy will grow well in these containers. (See also Patio Pyramid Planter, above)

POND-EDGE PLANTING

Suitable plants for the edge of a pond are *Iris pseudacorus*, *Iris versicolor*, arrowhead, sweet flag, cardinal flower, flowering rush, marsh marigold, and astilbe. Tiger lilies do not like wet soil, but *Lilium canadense* and *Lilium superbum* do.

Purple loosestrife is beautiful but should generally be avoided because its invasiveness has made it a serious problem for wetlands in many areas.

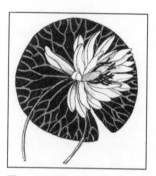

The water lily is seen in a variety of colors. Small air sacs in the leaves keep this plant afloat.

ROCK GARDEN

There are hillside rock gardens and valley rock gardens, as well as those on level sites; there are rock gardens in sun and rock gardens in shade, as well as those that include water.

All rock gardens should include dwarf material in trees and shrubs. Nurseries that specialize in choice rock-garden plants usually have a good selection. The extremely dwarf evergreen trees, for example, grow slowly and never develop out of proportion in a miniature landscape.

Make the rock garden attractive in winter by using a number of evergreen rock plants (many of which are technically shrubby or subshrubby). Examples of perennial candytuft and *Iberis sempervirens*. The rock roses (*Helianthemum*) are brilliant in early summer with myriad satin-textured blooms resembling little wild roses in pink, crimson, orange, gold, white, and scarlet. The silver saxifrages are very useful for creating the all-the-year-round picture of the true Alpine rock gardens. For cooler, shadier rock gardens try the mossy saxifrages. The genus *Dianthus* includes many splendid plants well adapted for sunny locations and well-drained sites. There are also bellflowers or campanulas, thymes, stonecrops or sedums, primroses or primulas, houseleeks or sempervivums, gentians, bulbs, and groundcovers.

ROOF TERRACE, PLANTS FOR

Consider sun and shade just as for ground-level gardens, and study the prevailing wind direction before you make plant selections and decide on locations. Use good soil mix, and you can successfully grow begonias, lobelias, nicotianas, and impatiens. Japanese yews make nice accents.

SANDY SOIL

Sandy soil, sometimes found near the seashore, need not be a problem if the right flowers are planted. Among annuals adaptable for sandy soils are calendulas, California poppies, sweet alyssum, marigolds, nasturtiums, portulaca, cleomes, and petunias. Shrubs may be Russian olive, *Rosa rugosa,* hydrangeas, rose acacia, tamarisk, and Siberian pea tree.

SCATTER PLANTS

Instead of using herbs alone in a stiffly formal herb garden, try scattering them throughout your flower garden. For their protective qualities include marjoram, oregano, lavender, santolina, blessed thistle, chamomile, lovage, chervil, lemon balm, and bergamot.

SHADE

Two delightful plants for shade that team well with each other are native bleeding heart (*Dicentra eximia*), with lacy foliage and pink, nodding, heart-shaped flowers; and lily of the valley (*Convallaria majalis*), a dainty, fragrant perennial with tiny white flowers. A third plant for shady spots is hosta (funkia), a plant much valued for its large, decorative leaves.

Achimenes, which comes in various colors, is an excellent pot plant for shady locations on the north. It blossoms almost continuously from late spring until fall frost.

STEPPING-STONES

For a flagstone walk, plant creeping thyme (*Thymus serpyllum*) between the stones in a sunny position. The plant seems to thrive from being walked on and gives off a pleasing fragrance with each step you take.

STREET NOISES

Trees and hedges muffle street noises. The best trees and shrubs for this purpose are dense evergreens, which also give year-round privacy. Hemlock, yew, and arborvitae are good if they will grow in your area. Keep the lower branches from dying off by pruning carefully; most noise comes from near the ground. If you have a choice of sites when you build, choose one above street level to keep noise at a minimum.

STUMPED BY A STUMP?

You can make it into a thing of beauty if not a joy forever. Sometimes an old tree must be cut down because it is rotten, yet the stump may not be dug up and removed. Scoop out the middle of the stump and fill it with good soil. In fall plant tulip bulbs. After they have bloomed in spring, succeed with petunias, snapdragons, or some other colorful flower.

SWIMMING POOLS

You can have a carnival of color without the pain of constantly cleaning up if you choose wisely for your poolside planting. And don't plant anything bristly, sharp, or thorny. Plants should also be as litter-free as possible.

Shrubs. Here are some beautiful but practical suggestions: camellia, *Crassula argentea, Fatsia japonica, Griselinia, Juniperus, Raphiolepis, Viburnum davidii*.

Trees. *Cordyline, Dracaena, Ficus auriculata, Ficus lyrata, Firmiana simplex*, musa, palms, *Schefflera, Strelitzia*, tree ferns.

Vines. *Beaumontia grandiflora, Cissus* × *Fatshedera lizei, Solandra maxima, Tetrastigma*.

Perennials. *Agapanthus, Agave attenuata, Aloe saponaria, Alpinia zerumbet*, artichoke, *Aspidistra elatior*, canna, *Clivia miniata*, cyperus, gazania, hemerocallis, *Kniphofia uvaria, Liriope, Ophiopogon*, philodendron (treelike types), *Phormium*, sedum, succulents, yucca, *Zoysia tenuifolia*.

Remember that the litter produced by plants (and this should be reduced as much as possible) should be large enough to be removed by hand rather than being allowed to pass into the pool's filter.

WINDOW BOXES

Some popular plants suitable for many types of window boxes include geranium, lantana, dwarf marigold, nasturtium, petunia, salvia, sweet alyssum, ageratum, verbena, and trailing vines (for sun). For shade, use achimenes, tuberous and wax begonias, impatiens, fuchsia, and torenia. For colored foliage, use coleus and caladium; for a trailer, use German ivy.

WINTER FLOWERS

Winter gardens call for tough plants, so don't think in terms of man-made hybrids. Alpines are the obvious choice. In their natural habitat there are two seasons, winter and August. This means they must bloom, ripen seed, and store food between the melting of snowbanks in late July and the new snows of September. In our less orderly climate, Alpine bulbs still obey their mountain timetable. If you choose a naturally warm site, you may have "spring in January" with pastel-shaded snow crocuses in bloom.

Brilliant orange *Crocus korolkowii* is one of the earliest blooming of all temperate-climate flowers, often blooming before winter's snow has melted, during a prolonged January thaw. It may even disappear beneath a late winter snow and suddenly reappear after the first sunny day.

Want Instant Garden Color?

Try annual bedding plants such as petunias, geraniums, salvia, ageratum, zinnias, and marigolds. For best effects, mass flowering annuals in large clumps of color rather than in a staccato line of single plants.

COMPANION PLANTING WITH FLOWERS AND HERBS

ALLIUM (*Allium*)

Onions belong to the Lily family, Liliaceae. Actually, *allium* is a Latin word for "garlic." Vegetable alliums are chives, garlic, leek, onion, and shallot, all of which are excellent protective companions for roses.

The ornamental alliums are more decorative to plant with roses and will also provide excellent protection from mildew and black spots as well as aphids and many other pests. They thrive on the same care and culture as onions and are very easy to grow. Flowering onions come in many colors besides blue and purple—greenish white, yellow, rose, and dark red. They like plenty of compost but do well even on dry soil. Some of the larger species such as *Allium giganteum* grow at a height of 5 feet and have a blossom head up to 8 inches in diameter. These should be staked in windy climates. Alliums are winter hardy and may be left in place year after year.

An interesting ornamental allium is *A. senescens glaucum,* a low-growing plant with silvery blue leaves that curl and twist in a style suggestive of Japanese art. The 2-inch umbels of soft rose pink are profuse in August and September. This plant not only makes a fine protective groundcover for roses but is also appropriate for edging rock walls or for the front of the border. Add to its useful qualities its hardiness and drought tolerance. Plant seeds 10 to 12 inches apart.

Alliums also repel moles. Here are some other beneficial services of the Onion family:

Onion. Repels cabbage butterflies and helps all members of the cabbage family.

Chives. Good companion for fruit trees and tomatoes.

Garlic. Good against fruit tree borers.

Members of the Onion family were so valued in ancient times that it is said the builders of the pyramids were paid in leeks, onions, and garlic. Onions are not only healthful; certain members of the family such as garlic are said to be an excellent aid in preserving a youthful complexion.

BORERS

Garlic planted around fruit trees will repel borers, but this is best done when trees are young and newly planted. Nasturtiums and radishes, planted around the squash hills, are also good for foiling borers in squash plants.

BUCKEYE (*Aesculus pavia*)

The flowers of the dwarf or red buckeye attract and kill Japanese beetles.

BULBS

Flowering bulbs such as crocus and colchicum (which are poisonous if eaten) are very attractive but often are simply allowed to come up without an accompanying groundcover. They are much prettier if low-growing companion plants such as white alyssum, *Phlox subulata, P. divaricata,* armeria, saponaria, common thyme, and vinca minor are used as a framing groundcover.

BUTTERCUP FAMILY

Root secretions from these plants poison soil for clover by retiring growth of nitrogen bacteria. Clover will disappear in a meadow where buttercups are increasing. Delphinium, peony, monkshood, columbine, and double buttercup belong to the same family. Unless the soil is very rich, other plants will not grow well in their vicinity. This is a strong, vital family, but they grow only for themselves.

CALENDULA (*Calendula officinalis*)

See *Marigold* in this chapter.

CHAMOMILE (*Chamaemelum nobile*)

This plant of the Aster family has strong-scented foliage and flower heads that contain a bitter medicine principle. It is used as an antispasmodic, stomachic, and perspiration ingredient for breaking fevers.

Chamomile in small quantities increases the essential oil in the peppermint plant, but as the proportion of the chamomile plants increases, the amount of peppermint oil diminishes—too much of a good thing!

Chamomile tea is effective against a number of plant diseases, especially in young plants. It can be used to control damping-off in greenhouses and coldframes. Make tea by soaking dried blossoms for a day or two in cold water.

The powdered heads of rayless chamomile (*Matricaria matricarioides*) are fairly toxic to diamondback moths.

The flower heads of scentless false chamomile (*Tripleurospermum inodorum,* formerly *Matricaria inodorum*) are said to be as effective as commercial pyrethrum in controlling face flies.

CHINABERRY (*Melia azedarach*)

This shade tree repels grasshoppers and locusts. Make a repellent tea with either live or dried leaves. The powdered fruit is somewhat toxic to European corn borer larvae.

CHRYSANTHEMUM (*Chrysanthemum*)

The name is derived from the Greek *chrysos,* "gold," and *anthemon,* "flower." There are 100 species of both annuals and perennials, some of which are known as pyrethrum. Chrysanthemums are protective to strawberries.

Chrysanthemums themselves may be protected by an all-purpose spray made from three hot peppers, four large onions, and one whole bulb (large) of garlic, ground together. Cover with water and allow to stand overnight. The following day, strain the mixture through a fine sieve. Add enough water to make a gallon of spray. This spray may also be used on roses and azaleas. Bury the mash around roots of plants.

COLUMBINE (*Aquilegia vulgaris*)

These beautiful plants are very attractive to red spiders in some areas, so keep them in the flowerbed and out of the vegetable garden. Those columbines grown in humus-rich soil are less susceptible to damage by leaf miners.

Columbines are strong growers themselves but do not assist other plants and often are detrimental. They are heavy feeders, so if you want to grow them as ornamentals, give them plenty of compost and feed other plants grown in their vicinity equally well. They will grow well in combination with rhubarb if both are given plenty of well-rotted cow manure.

Native Americans considered flowers of wild columbine (*Aquilegia canadensis*) a highly effective tranquilizer.

CALLIOPSIS (*Coreopsis tinctoria*)

Coreopsis in the flowerbed is useful as an insect control for nearby plants. It also is an attractive annual (of the Compositae family) whose yellow, red, or maroon flowers grow on tall, slender stems. It looks like a daisy and may have one or two layers of petals.

CORNFLOWER (*Centaurea cyanus*)

Cornflower, in the proportion of 1 to 100 seeds, is noted for its salubrious effect when grown with rye. Also called bachelor's-button or blue-bottle, it also is beneficial to other small grains if grown in about the same proportion. The flowers supply bees with abundant nectar even in the driest weather.

It was a custom among Russian peasants to decorate the first sheaf of the rye harvest with a cornflower wreath, after which it was placed in front of an icon.

Cornflower

DANDELION (*Taraxacum officinale*)

Dandelions, even if they grow in thick patches on your lawn, are not competing with grasses because of their long, very deep taproots. These transport minerals, especially calcium, upward from deeper soil layers (even from beneath hardpan, which they can penetrate), and deposit them nearer the surface. They are, therefore, returning to the soil minerals that have been lost through seeping downward.

Dandelions are closely associated with clover and alfalfa, which also prefer good, deep soils.

Juliette de Bairacli Levy states in her *Herbal Handbook for Everyone:* "This is one of the most esteemed plants of the herbalist. . . . It is blood-cleansing, blood-tonic, lymph-cleansing. Also has external uses for treatment of warts and hard pimples. A diet of greens improves the enamel of the teeth."

Dandelion helps other flowers grow. It stimulates fruits to ripen faster. The roots, dried and ground, are sometimes used as a coffee substitute. In early spring the unopened buds are delicious cooked with leeks, lightly seasoned with butter, salt, and pepper. The hearts may even be eaten raw if the leaves are tied together and the heart blanched. They are very rich in vitamin A.

A pale yellow dye can be made from dandelion blossoms; a deeper yellow-brown color is obtained when the dye is made from the plant's roots.

DUSTY MILLER (*Centaurea cineraria*)

Plant this around prized flowers to repel rabbits and other animals. William J. Park, former president of Park Seeds, who gave me this suggestion, considered the variety 'Diamond' to be the most effective and the prettiest.

ELDER (*Sambucus*)

Elder is noted for repelling certain insects. An old method of trapping cutworms consisted of placing compact handfuls of elder sprouts in every fifth row or hill of cultivation and tamping them down. Cutworms gathered in this trap material, where they could be regularly collected.

Elder is a powerful patriarch of the plant world; wherever it grows, it discourages other herbs. Animals usually dislike the rank elder taste.

Elder leaves are effective against moles if placed in their runs; they find the odor offensive. Branches of elder have been used against maggots. Bruise them first to increase the odor, then rake the leaves across the seedbed after sowing.

FEVERFEW
(*Chrysanthemum parthenium* syn. *Tanacetum parthenium*)

Sometimes this plant is called pyrethrum. It is not the same plant, but it has insect-repellant properties of its own, perhaps because of the spicy scent of its foliage.

FOUR-O'CLOCK (*Mirabilis*)

Japanese beetles, a pest on peach trees and many other plants, like the foliage of the four-o'clocks and are apparently unaware that in eating it they are committing suicide. Dig up the four-o'clocks in fall, and if carefully overwintered, they can be used again the following year. The foliage of this plant is also poisonous to humans, so don't let children take a bite of it.

FOXGLOVE (*Digitalis*)

The tubular purple or white flowers grow about 2 inches long in a 12- to 24-inch, one-sided cluster. The summer-blooming flowers are often spotted within.

Foxglove has a growth-stimulating effect on nearby plants and is a good companion for pines. It does well in a forest border and in open woodland for naturalizing. Foxglove tea in the vase is said to prolong the life of other cut flowers.

The flower is the most valuable source of the powerful heart stimulant digitalis. Natives of North America knew and used foxglove for heart prob-

lems before it was known to Europeans. In England the extract was recommended by so-called witches (or herb women) for this purpose before it was recognized by physicians. The drug is cumulative in its action and should never be used as a medication except under the direction of a physician.

Some people plant soybeans and corn as a lure or trap crop for deer; others plant castor beans and foxglove to repel them. An effective fence has to be at least 7 feet high.

Severe poisoning comes from eating the fresh or dried leaves, which do not lose their toxicity by cooking. Children have been poisoned by sucking the flowers and swallowing their seeds.

GERANIUM, SCENTED (*Pelargonium*)

Geraniums will repel cabbageworms and are good to plant among roses, grapes, and corn against Japanese beetles. Use the white variety near corn.

Among the diverse scented geraniums are the peppermint geranium (*Pelargonium tomentosum*), which has velvety, grapelike leaves and small white blossoms; the lemon-scented (*P. crispum*), with long-stemmed, conspicuous flowers of a deep rose color and crispy, fruit-fragrant leaves; and nutmeg and apple geraniums, which have small, almost rounded, soft gray leaves distinguished by the scent of spice or apple. Try making apple jelly flavored with rose geranium leaves.

GLADIOLUS (*Gladiolus*)

Peas or beans and gladiolus have inhibiting effects on each other.

HERB OF GRACE, RUE (*Ruta graveolens*)

This is a very bitter-tasting herb, strongly aromatic, and once important for medicinal purposes but no longer taken internally. Contact can cause a rash if you are allergic to it.

Plant rue with roses to foil the Japanese beetle. It is also helpful grown with fig trees. Cats detest rue, so rub it on your upholstery furniture to keep them from clawing. Use against fleas in the dog's bed, and a few sprigs hung in a room will drive out flies. Rue and basil are incompatible.

INSECTICIDAL FLOWERS

Asters, chrysanthemums, cosmos, coreopsis, nasturtiums, and French and African marigolds are good to plant throughout the flower garden.

JAPANESE BEETLE (*Popillia japonica*)

This bronze-blue, iridescent beetle feeds on all kinds of ornamentals. To control, a rotenone preparation is recommended. In lawns, treat the turf with a bacterial spore dust, which will infect the grubs with milky spore disease.

Beetles are sometimes caught in traps filled with geranium oil. They may also be lured to feed on trap crops of African marigold, evening primrose, four-o'clock, or woodbine. They are poisoned by leaves of cantor bean and blossoms of white geranium.

JIMSONWEED (*Datura stramonium*)

This weed spreads usually by having its seed carried by birds. It is very poisonous, causing a kind of intoxicated state, but has a certain medicinal value. It is helpful when grown with pumpkins.

LARKSPUR (*Consolida ambigua*)

Common larkspur's alkaloids—delcosine and delsoline—have been found effective against aphids and thrips.

LILY OF THE VALLEY (*Convallaria majalis*)

This delightfully fragrant garden flower, in Germany called the Mayflower, grows wild in the southern Allegheny regions. It likes a rich, humusy soil and partial shade. Do not put narcissus with it in a vase as it will cause the narcissus to wither, perhaps because the leaves and flowers of lily of the valley are poisonous.

MARIGOLD (*Tagetes*)

Nematodes, wormlike inhabiters of the soil, are microscopic in size, and the root-knot types injure many garden plants. These tiny worms like sandy, warm soils and have been known to devastate entire gardens. Chemical soil disinfectants kill them but kill the plants as well, so the discovery of a protective flower is very valuable.

Marigolds are very beneficial in discouraging nematodes that attack potatoes, strawberries, roses, and various bulbs, especially if these strong-scented beauties are grown for several seasons in ground where nematodes are suspected. Experiments have shown that marigolds suppress meadow nematodes for up to three years and control other nematodes for one or more years without injuring the plants.

An easy way to use marigolds for nematode control is to rotate plantings of marigolds with crops that are susceptible to nematode injury. To lessen competition it is also wise to interplant marigolds 2 or more weeks after the plants that they are grown to protect.

Marigolds control nematodes with sulfur-containing substances called thiophenes that are produced in the roots and kill the pests when released in soil. Both French and African marigolds have similar root excretions. The chemicals are produced slowly, so the marigolds must be grown all season to give lasting control. Interplanting them may not greatly help garden plants during the first season, but the benefits become apparent in the following years as the nematode population is reduced.

Tomatoes interplanted with marigolds will grow and produce better. Plantings with beans help protect against the Mexican bean beetle. They also help deter weeds and may be planted as a crop against invasions of ground elder, bindweed, and ground ivy. The older types with strong odor in both foliage and blossom are considered the most useful.

Pot marigold (*Calendula officinalis*), or calendula, planted in the vicinity of choice evergreens, will repel dogs. This is an old-fashioned herb whose dried flowers were used by our grandmothers to flavor soups.

MORNING GLORY (*Ipomoea*)

The Indians liked to grow wild morning glory with corn—probably one of the earliest examples of companion planting—since they believed that it gave the corn added vigor. It is believed that morning glory seeds also will stimulate the germination of many types of melon seeds.

Kentucky Wonder beans and morning glory planted on the same trellis grow and blend together. Both will last all summer if kept well watered.

NARCISSUS (*Narcissus*)

A planting of African marigolds (*Tagetes erecta*) a year before planting narcissus bulbs will defeat certain nematodes that often attack the bulbs. (See *Marigold*.)

NASTURTIUM (*Tropaeolum*)

Nasturtiums planted with squash will keep away squash bugs, but be sure to give the flowers a head start since the squash grow more quickly. If aphids appear in the nasturtiums—a sign that there is a lime deficiency in your soil—dust the plants with lime and they will disappear.

Nasturtiums sown in a greenhouse will help repel whiteflies. When planted near broccoli in the garden, they will keep down aphids and benefit potatoes, radishes, cucurbits, and any member of the Cabbage family. Under apple trees they will protect against woolly aphids.

Nasturtiums, planted nearby or made into a spray, repel a range of harmful insects from vegetables and fruits. They also improve growth in the neighboring crops.

Use sprays made from nasturtium leaves on the same crops that benefit from the nasturtium plants. Add a small amount of soap powder; the sprays will adhere better.

PEPPER (*Capsicum*)

A California study found that natural juices squeezed from succulent plants such as green peppers were effective in protecting other plants from viruses. They work against diseases transmitted by insects or wind. The sprays were found effective against tobacco mosaic virus, potato virus, and several other viruses carried by aphids. Strangely, the compounds do not kill the viruses but change the plant so it is not susceptible.

Old-time gardeners planted hot peppers among their flowers to discourage insect pests. Some of the lovely and unusual ornamental peppers would be excellent choices.

PESTS AND THEIR CONTROLS

These herbs and flowers act as troubleshooters for various pests.

Pest	Plants
Ant (and the aphid that ants carry)	Pennyroyal, spearmint, southernwood, tansy
Borer	Garlic, tansy, onion
Cutworm	Tansy

Pest	Plants
Eelworm	(see Nematode)
Flea beetle	Wormwood, mint, catnip
Fruit tree moth	Southernwood
Gopher	Castor bean
Japanese beetle	Garlic, larkspur (poisonous to humans), tansy, rue, geranium (use white geranium)
Leafhopper	Petunia, geranium
Mexican bean beetle	Marigold, rosemary, summer savory, petunia
Mouse	Mint
Mole	Spurge, castor bean, mole plant, squill
Nematode	Marigold (African and French), salvia (scarlet sage), dahlia, calendula (pot marigold), crotalaria
Plum curculio	Garlic
Rabbit	Allium family
Rose chafer	Onion family, geranium, petunia
Slug, snail	Prostrate rosemary, wormwood
Squash bug	Tansy, nasturtium
Striped pumpkin beetle	Nasturtium
Tomato hornworm	Borage, marigold, opal basil
Whitefly	Nasturtium, marigold, nicandra (Peruvian ground cherry)
Wireworm	White mustard, buckwheat, woad

PETUNIA (*Petunia*)

This is a member of the Nightshade family, Solanaceae. The word "petunia" is derived from *petun,* a South American name for "tobacco," since the tobacco plant belongs to the same family. Petunia protects beans against beetles.

RATTLEBOX (*Crotalaria*)

From its seedpods, which rattle when shaken, crotalaria derives its common name. Prized not only for its racemes of yellow sweet-pea–like flowers but also because it is toxic to nematodes.

Crotalaria retusa is 3 feet high with maroon reverse petals. *C. spectabilis,* the species tested for nematodes control, is 4 to 5 feet tall with smaller flowers.

ROSE (*Rosa*)

Roses do not like boxwood because its outspreading, woody roots interfere with the roots of rosebushes. But garlic, onions, and other members
of the Onion family, including ornamental alliums, are beneficial.
(See *Allium* in this chapter).

SUNFLOWER (*Helianthus annuus*)

Sometimes called *maiz de Texas* (Texas corn) or *tourne-soleil,* the common
sunflower is an American plant that has been widely cultivated and much
improved. It is a soil-improver when grown in moderation through certain
farm crops. However, sunflower seed will not germinate well where grass
is growing in close proximity.

Sunflowers are now believed to produce leaf substances that inhibit
other species. This may be due to a defense mechanism of the plant.
Needing little nitrogen itself, the sunflower produces a substance that
inhibits nitrogen-fixing bacteria in the soil, thus delaying the day when
grasses and other succeeding plants may take over.

Sunflowers and potatoes have an inhibiting effect on each other that
results in both being stunted. Potatoes also are more likely to be infected
by phytophthora blight. Sunflowers and pole beans should not be planted
together since both compete for space and light with resulting poor
growth.

Sunflowers also have their good points as companion plants. It has
been found that corn and sunflowers are protective to each other and

insects are reduced on each. Cucumbers benefit when sunflowers are
grown near them to provide a windbreak. In my hot climate I grow
sunflowers on the west side of the
cucumber patch to provide shade in
the afternoon. Both cucumbers and
sunflowers like a rich soil, so I dig in
plenty of compost to prevent one
from starving out the other.

Sunflower

TANSY, PARSLEY FERN
(*Tanacetum vulgare*)

Tansy

Tansy has a very strong, bitter aroma. The Latin *tanacetum* derives from a Greek word indicating immorality because the dry blossoms do not wilt. The distilled oil repels flies and mosquitoes. The plant has been used against intestinal worms (*Oxyuris*) and in wine against stomach and intestinal spasms. The Russians used it as a substitute for hops in beer and rubbed it on the surface of raw meat to protect it from flies. Tansy planted near the entrance to the house deters ants. Tansy is a friend of both vegetable and flower gardeners, since it will repel ants, borers, cucumber beetles, Japanese beetles, and squash bugs.

TRAP CROPPING

Some plants earn their living by repelling potential trouble, others by luring insect pests away from more valuable plants. Japanese beetles are attracted to white and pastel zinnias, white roses, and odorless marigolds; and nasturtiums attract aphids.

WALLFLOWER *(Erysimum cheiri,* formerly *Cheiranthus cheiri)*
A spray of rhubarb leaves will protect wallflowers against clubroot. Boil the leaves in water and sprinkle it where the wallflower seeds are to be sown. Wallflowers are believed to be beneficial to apple trees.

WEEDS

Some weeds have been found helpful in flowerbeds. Lamb's-quarters gives added vigor to zinnias, marigolds, peonies, and pansies. Growing among rosebushes, a carpet of low-growing weeds from the despised Purslane family improves the spongy soil around their roots. Lupine, a legume, helps corn and other cultivated crops. Morning glory planted near corn enhances root vigor.

GARDENING TIPS AND TECHNIQUES

For the Soil's Sake

COMPOSTING FOR SOIL ENRICHMENT

The importance of abundant soil humus is to create a favorable environment for root growth. A soil in good tilth has a granular structure that readily permits rain and air to penetrate and also retains moisture for long periods to permit maximum growth. It's good garden practice to use compost in combination with mulching.

You can start a compost pile at any time of the year. Build your pile in a pit, trench, freestanding stake-and-chicken-wire form, or any large container such as a plastic trash barrel. Build the pile with layers of organic matter in such a way that they will decompose easily.

Use grass, weeds, leaves, stalks, branches, and wood chips. Use any organic matter, but kitchen scraps should be vegetable materials such as carrot tops, lettuce, and coffee grounds; don't include meat or dairy products. A shredder to grind up the material will appreciably shorten the length of time for the compost to ripen. Put down a 6-inch layer of organic matter, and sprinkle it with finely ground agricultural lime. Add a 1-inch layer of soil to introduce the microorganisms needed for speedy decomposition.

Layer these materials up to about 4 feet high and keep the entire pile moist. Make the top of the pile slightly concave so that moisture will seep into the pile. Turn the pile to reintroduce air into the compost and speed decomposition of organic material.

Is all this worth it? Yes. A 4-inch layer of compost worked into the soil to a depth of 6 inches will almost guarantee gardening success with both flowers and vegetables. If you don't have enough for that, you can concentrate your compost where it's most needed (around the roots of transplants, for instance.)

EARTHWORM CASTINGS

The castings (droppings) of the earthworm are rich in nitrates, phosphates, potash, and calcium—all elements necessary to plants. They also contain trace elements of sulfur, boron, zinc, copper, manganese, chlorine, iron, molybdenum, aluminum, and selenium—all needed to keep a plant healthy.

Gardeners everywhere are beginning to realize the value of earthworm castings. They report that their use results in vegetables of better taste and larger size with yields often doubling. Tomatoes especially benefit from castings.

Castings will not burn even the most delicate plant. There is no danger of adding too much to your soil. Worm castings can be found packaged, ready to use, in some garden centers, and worms are easy and fun to raise at home.

Here are some suggestions for using castings:

Indoor plants. For established pots, add 1 cup of castings to a 6- or 8-inch pot and water thoroughly. Increase proportionately for larger pots. Castings will be absorbed into the soil during normal waterings.

Sick or dying houseplants. Remove 1 to 2 inches of the old planting mix (being careful not disturb the plant any more than necessary). Replace with castings.

Potting or repotting. Mix one-third castings with two-thirds peat moss, or any good soil. Add enough of this mixture to set the plant at its original depth. Pot the plant, fill around it with the castings, and press gently to firm the soil. Water thoroughly.

Root stimulator. Use castings to make up the planting rootball for newly planted fruit trees, rosebushes, and berry vines. This application for castings (where other fertilizers are too strong) provides a head start for the plant and will produce remarkable results even for years to come. Seeds also germinate much faster in castings.

NITROGEN-FIXING PLANTS

Colonies of nitrogen-fixing bacteria form on the roots of members of the Bean family (Leguminosae). These organisms take nitrogen gas from the air and convert it into nitrates that can be used by plants.

It is far better to encourage these minute creatures to manufacture nitrates than to apply nitrogen-containing chemical manures. They charge nothing for their work, and their own death makes the earth more productive. To increase nitrates in the soil, grow peas and beans in different parts of the garden each year.

Since the entire family of Leguminosae has the same properties, grow sweet peas and lupines in the flower garden for the same reason. When clearing ground in autumn, never burn the roots of leguminous plants. Break them up instead and add them to the compost heap. The tiny nodules on the roots contain the nitrogen.

pH

The pH scale measures the acidity or alkalinity of a substance. It's an important measurement for gardeners, because soil pH within the proper range is needed for good results in the flower or vegetable garden.

Soil tests for pH may be made at home using one of the test kits sold by garden centers and catalogs; or a soil sample may be sent to a professional laboratory. Ask your county extension agent for information on local soil-testing facilities.

To get a reliable soil sample, use a clean spade to cut a 1-inch-thick, 7-inch-deep slice of undisturbed soil. Take similar samples from several points in the garden and mix all in a clean bucket. Then dip out a pint jar of the mix to take to the laboratory, being sure to label it with your name and address. The lab will charge a small fee to cover costs of chemicals and a technician. The report will include not only the pH level but also recommendations on how to correct an overly acid or alkaline condition. Other tests for nutrient levels can also be done on request; ask for information.

Here are the pH preferences of some commonly grown flowers: chrysanthemum 5.7 to 7.0, daylily 6.0 to 7.0, iris 5.0 to 6.5, ageratum 6.0 to 7.5, begonia 5.5 to 7.0, marigold 5.0 to 7.5. They will attain their best growth in this pH range and are less likely to do well above or below.

Off to a Good Start

ASTROLOGICAL ASPECTS

For centuries farmers have plowed and planted according to the signs of the zodiac. The same method can also be used for flowers. Use them in conjunction with a good gardening almanac that gives the correction for time changes in each part of the country.

Planting Flowers by the Moon

Plant	Moon Phase (by quarter)	Sign
Annuals	1st or 2nd	Libra
Asters	1st or 2nd	Virgo
Bulbs	3rd	Cancer, Scorpio, Pisces
Bulbs for seed	2nd or 3rd	Cancer
Chrysanthemums	1st or 2nd	Virgo
Clover	1st or 2nd	Cancer, Scorpio, Pisces
Coreopsis	2nd or 3rd	Libra
Cosmos	2nd or 3rd	Libra
Crocus	1st or 2nd	Virgo
Daffodils	1st or 2nd	Libra, Virgo
Dahlias	1st or 2nd	Libra, Virgo
Deciduous trees	2nd or 3rd	Cancer, Scorpio, Pisces
Flowers for beauty	1st	Libra
for abundance	1st	Cancer, Pisces, Virgo
for sturdiness	1st	Scorpio
for hardiness	1st	Taurus
Gladiolas	1st or 2nd	Libra, Virgo
Golden glow	2nd or 3rd	Libra
Honeysuckle	1st or 2nd	Scorpio, Virgo
Iris	1st or 2nd	Cancer, Virgo
Lilies	1st or 2nd	Cancer, Scorpio, Pisces
Moon vine	1st or 2nd	Virgo
Morning glory	1st or 2nd	Cancer, Scorpio, Pisces, Virgo
Pansies	1st or 2nd	Cancer, Scorpio, Pisces
Peas, Sweet	2nd	Cancer, Scorpio, Pisces, Libra
Peonies	1st or 2nd	Virgo
Peppers, ornamental	2nd	Scorpio, Sagittarius
Perennials	3rd	Cancer, Pisces, Libra
Petunias	1st or 2nd	Libra, Virgo

(continued)

Plant	Moon Phase (by quarter)	Sign
Poppies	1st or 2nd	Virgo
Portulaca	1st or 2nd	Virgo
Roses	1st or 2nd	Cancer
Sunflowers	2nd, 3rd, 4th	Libra
Trumpet vine	1st or 2nd	Cancer, Scorpio, Pisces
Tulips	1st or 2nd	Libra, Virgo

During the increasing light (from new moon to full moon), plant annuals that produce their yield aboveground. (An annual is a plant that completes its entire life cycle within one growing season and has to be seeded anew each year.)

During the decreasing light (from full moon to new moon), plant biennials, perennials, bulbs, and root plants. If you wish to save flower seed, let flower heads mature for as long as possible before gathering but not so dry that the seedpod shatters. Gather seed in a dry sign such as Aries, Leo, Sagittarius, Gemini, or Aquarius.

CLONE

Recently the word *clone* has been appearing often in the news. Actually, plants have been cloned for thousands of years. The word itself comes from the Greek and means a twig or slip. Cloning plants is very easy because many plant cells have the power of regeneration.

With onions, for instance, I put cloning to work. When my onions sprouted, softening gradually in the process, I used to throw them out. No more. I plant these onions in fall, sometimes in the garden, sometimes in my rose bed. From one onion a cluster of five or six will grow, delicate in flavor, and giving me early table onions the following spring.

The first year I pulled them all, but in succeeding years I left several clumps as an experiment. In time they grew large. I harvested them at maturity with the rest of my crop. They kept very well but in time sprouted again and were replanted. A good onion—like an old soldier—never really dies! It doesn't even fade away.

COLDFRAME

A coldframe on wheels allows the gardener to place it in sun or shade as the season requires. Make the sides of the frame out of ¼-inch Plexiglas. Attach with screws to 2 × 4s supporting a plywood bottom. The top is a Plexiglas sheet with a 2 × 2 attached underneath to hold it up and allow air to pass through. When turned over it is sealed shut, keeping out animals and insects. Use old tricycle or bicycle wheels for the frame. If there's a late freeze coming, you can wheel this coldframe into the garage at night to keep it warm.

CUTTINGS

These are vegetative portions of plants used for reproduction. A cutting may consist of the whole or part of a stem (leafy or nonleafy), leaf, bulb, or root. A root cutting consists of the root only; other cuttings have no roots at the time they are made and inserted.

To ensure success, you must make cuttings when the tissues are in the right condition, prepare them properly, insert them in the right rooting medium, and keep them in a favorable environment until they are able to regenerate themselves as new plants. Detailed directions are given in many gardening guides; see your local library or garden center for help in getting started.

Take a cutting of the plant part to be used of the correct size so the regeneration of new parts is encouraged; this may mean the removal of leaves, or parts of leaves.

Many different rooting media are used for cuttings. One of the most popular is sharp sand; or use sand and peat moss, or vermiculite and sandy soil. Cuttings of some plants such as the African violet may be rooted in water.

For plants in all rooting media, prevent the tissues from drying before the new plant is established. This avoids loss by disease and encourages rapid reestablishment. The introduction of special root-inducing hormones and the use of bottom heat may speed this process. In the greenhouse, heat is usually provided by hot-water pipes, electric heating cables, or fermenting manure.

DIVISIONS

In midsummer, many perennial garden flowers begin a rest period. Divide at this time to form new plants. Suitable candidates include peony, German iris, Oriental poppy, madonna lily, painted daisy, phlox, and columbine. Usually plants that flower in spring and early summer may safely be divided in late summer and fall. Those flowering in summer and fall should be divided in early spring before new growth appears.

Crown division is one of the easiest methods. To make a crown division, lift the plant carefully and remove some soil from the roots. Cut the crown into several pieces with a knife. Use the individual sections of vigorous plants to make new plants.

How often to divide? Peonies may remain in the same spot for many years. Divide Shasta daisies and phlox every three years. Replant daylilies and iris every five years. Divide chrysanthemums and hardy asters every two to three years in spring. Some plants, such as Oriental poppies, do not adjust well to moving; transplant these only when they lose vigor from overcrowding.

FLOWER POTS

Let your dishwasher sterilize your flower pots. First scrub your pots clean to remove soil and accumulated salt. Put clay pots on the bottom rack, and plastic ones on top with pan lids on top to weight them. Nest several sizes to maximize space. The pots will come out clean, sterilized, and ready to use.

FORCING

Forcing ensures a supply of fruits, flowers, or vegetables earlier than they would be available if cultivated in the usual way. Forcing necessitates the use of warmth, supplied for large operations by hot-water or steam pipes, electric heating, or by a hotbed made of fresh manure and leaves. For home use, though, the warmth of a heated room is all that's needed.

Pussy willow (*Salix*) is a favorite for spring forcing. Cut the ends of the branches in January or February, place in water, and watch them unfold their large, closely packed catkins. Black pussy willow (*S. gracilistyla* Melanostachys) is unusual and truly different; its catkins are so dark that they appear almost black against the red twigs.

MINI-GREENHOUSE

To root African violets, rex begonias, roses, and small evergreen cuttings, use a greenhouse made from a clean, label-free, 3-pound peanut butter jar turned upside down on its lid. Fill the lid with moist gravel. Then place a pot containing the moist, sterile rooting medium, and cuttings on the gravel. Tip out any excess water. Twist the jar down over the lid to seal.

PASTEURIZING SOIL

To pasteurize potting soil for seed starting and houseplants, spread soil—moist, not wet—evenly on an odd cookie sheet or tray. Heating for 30 minutes in a 140°F oven is optimum for killing undesirable pests and will not destroy all beneficial plant organisms; 30 minutes at 180°F kills most weed seeds and all plant pathogenic bacteria and fungi; tempera-ture above 185°F may damage the soil's chemical structure. Use a reliable oven thermometer.

Pots and occasionally tools also need decontamination. A solution of 1 part bleach to 10 parts water will sterilize them.

Soil mixes should not be high in manure or compost because the heat will release too much fertilizer at one time and be toxic to plants.

PELLETED SEEDS

Tiny seeds such as those of petunia are available from some garden catalogs in a pelleted form. Pellets—in the case of petunia seed, about the size of shot—contain a little plant food as well as seed disinfectant. Easy to sow in flats, they often germinate and grow much better than common seeds. Seedlings suffer less from damping-off. Pellets also allow you to space the seeds, thus eliminating the disagreeable job of thinning.

PLANTING ERRORS

Most plant failures are caused by improper planting.

Far too many of us place a $10 plant in a $1 hole. Prepare the best possible foundation for your plant. If the plant is bare root, make the hole large enough for the roots to spread out naturally.

Plant purchases are often made on impulse with no thought to where the plant will be placed, or what its ultimate size will be, or how it will adjust to other plants in the immediate vicinity. The urge to use plants that will give a quick effort often results in overcrowding.

Flower seeds often are sown too thickly. Pull out or transplant the surplus plants as soon as they begin to crowd each other.

SEED CATALOGS

Everything in a seed catalog is "superb," "magnificent," and usually resistant to something. But just as in an insurance policy, you've got to watch the wording.

For instance, there's that phrase "reseeds itself." That one you can believe, especially if it refers to something like bluebells-of-Scotland.

"Likes full sun" is also to be believed.

"Grows in shade" needs a little interpretation. What it actually means is "filtered sunlight," a spot dappled with a few shadows now and then, as under a tree. Few plants grow well in full shade.

"Naturalizes well" means that if you give the plant half a chance, it will take over. So beware.

"Grow in clumps of three or more" means just that. Grown alone, it will barely make an imprint on your consciousness, yet grown in a small group, the plant can be very lovely.

"May be divided" means you must—if you don't want the plants to choke themselves to death.

"Once established" means that getting this plant to do what you want may take a long, long time.

SEEDS

Almost all the plants mentioned in this book appear in the catalogs listed in the Sources section. You can buy seeds and nursery stock with confidence from these dealers.

SEEDS, SOWING

To evenly distribute tiny seeds, such as portulaca, place them in a clean squeeze bottle, the kind used for mustard. Turn the bottle upside down, press the tip gently against the soil where you want to seed, and squeeze lightly. A few seeds will come out, and when the bottle is lifted, soil covers them.

To make it easier to see small seeds shaken into a seedbed or furrow, place them in a clean salt shaker with enough talcum powder to coat them. This method also saves seed as you get more even distribution and less need for thinning.

Here's a method to plant petunia seed directly into peat pellets without failure. Take a wet pencil tip, pick up one tiny seed at a time, and apply it to the moistened peat of the pellet. You can use two seeds per pellet and remove one if both sprout.

SEEDS, SPROUTING

When starting seeds in pots, use a plastic coffee can lid to cover the pot until the seeds sprout. It works better than fragile plastic wrap to hold in moisture and also prevents quick cooling or heating when the weather is changeable.

Some kinds of flower seeds sprout faster if they are tortured. One tough nut is the flower named canna. It is called Indian shot because the seeds of canna are round, heavy, and oily like buckshot. Nature made canna seed coats resistant so that not all of them would sprout the first year. You can food Mother Nature by cutting through the coats to admit water. Use a triangular file or nail clippers. Soak the nicked seeds in warm water overnight, blot dry, and plant in warm soil; the seeds should sprout in 2 to 3 weeks.

Scald the seeds of the hibiscus called mallow. Bring water to a boil, turn off the heat, drop in the hibiscus seeds, and leave them in the water overnight. The hot water won't kill the seeds, but it will cut through the natural oils.

SEEDS, STORING

Until recently, home gardeners would do little to prolong the life of left-over garden seeds, especially those naturally short-lived such as onion, parsnip, delphinium, and larkspur.

The following is an inexpensive method of storing leftover garden seed from open-pollinated plants. (Saving hybrid seed is not recommended, as they seldom come true.)

1. Unfold and lay out a stack of four facial tissues.

2. Place 2 heaping tablespoons of powdered milk on one corner. The milk must be from a freshly opened pouch or box to guarantee dryness.

3. Fold and roll the facial tissue to make a small pouch. Secure with tape or rubber band. The tissue will prevent the milk from sifting out and will prevent seed packets from touching the moist desiccant.

4. Place the pouch in a wide-mouth jar and immediately drop in packets of leftover seeds.

5. Seal the jar tightly using a rubber ring to exclude moist air.

6. Store the jar in the refrigerator, not the freezer.

7. Use seeds as soon as possible. Discard and replace the desiccant once or twice yearly. Dried milk is hygroscopic and will quickly soak up moisture from the air when you open the jar. Therefore, work quickly when you remove seed packets, and recap the jar without delay.

This method is of special value to gardeners who make repeated plantings of short rows to keep a constant supply of flowers or vegetables.

TRANSPLANTING TIPS

Flower and vegetable plants can crowd each other when seeds are planted too closely together. Crowding delays maturity, stunts growth, and distorts the roots of carrots and other root crops. At the same time, there may also be skips; transplant enough seedlings to fill the skips, then discard the surplus.

Here's how to transplant so there is as little shock as possible.

1. Start when the plants are small; if they have four to six leaves, they're big enough.

2. Transplant at sundown on a cloudy day. Wind can injure as much as sunlight.

3. Wet soil thoroughly around the roots of seedlings that are to be moved. Do this a few hours beforehand so that the plants will be plump with water.

4. Dig transplanting holes before you uproot any seedlings. Fill the holes with water and let it soak in.

5. Shove a trowel in deeply to pry up plants. Move seedlings with as much soil around the roots as possible.

6. Move one plant at a time. Transplant quickly; don't delay.

7. Immediately soak the soil around each transplant. Don't wait until you have completed the row. For a week thereafter, sprinkle the transplants at least daily.

8. Never apply garden or houseplant fertilizer—liquid or dry—around newly transplanted seedlings. Their roots are too damaged to take it up.

9. If you must move large seedlings with many leaves, trim back half the foliage to reduce the leaf area through which water is lost. Use a shovel to move a big rootball and try to keep it from breaking up when you set it in place.

Growing Concerns

CROP ROTATION

Crop rotation in the flower garden cuts down on certain insect pests. Spatial relationships are also important among flowering plants because crowding will reduce vigor.

CULTIVATION

In preparation for planting, the soil of the flowerbed should be in a good state of tilth, with plenty of organic matter added, preferably in the form of compost, for those flowers that require it. This is particularly important in the South and Southwest, where the hot sun tends to burn the humus out of the soil. Replenish the bed with compost each season.

A small area may be spaded; for larger flowerbeds a rotary tiller is invaluable. The type with the tines to the rear is easy for elderly persons to handle, permits deep or shallow cultivation, and adds greatly to the pleasure of gardening.

DISEASE PROBLEMS

Plants are attacked by fungi, rusts, and other diseases for a variety of reasons. Extreme weather conditions, either drought or excessive rains, weaken plants. Peony leaves turn brown around the edges in wet weather. Or insect pests may injure them so that fungus diseases and rots find an opening. Other insects infest plants with certain virus diseases.

Raw organic materials used as fertilizer can cause a plant to succumb to whatever comes along because it cannot digest the crude material provided. The plant's chemistry is altered in such a way that it actually may attract insect pests. Well-decomposed compost is as good for your flowers as it is for your vegetables.

INTENSIVE CARE

An intensive-care spot for plants can be a sheltered area under shrubs where houseplants can safely spend the summer. This is a good treatment for plants that have been on display in dark or smoke-filled rooms, in drafts, or exposed to too much sunshine. Plants can often be brought back to health and become attractive again if given a period of rest under more ideal conditions.

An intensive-care area can also be a place near the porch or patio to start seedlings or root cuttings, or a coldframe for starting young seedling plants.

Keep extra plants growing in a secluded area for replacement purposes as the life spans of earlier plants are reached and they must be pulled. Young zinnias make a fine summer replacement for more delicate spring plants that must eventually be removed. Keep some young shrubs or evergreens growing in case you lose one by death or disfiguration. Prepare for such misfortunes by starting extra specimens in your intensive-care "unit" in case of need.

MULCH

A good mulch can often double the time a flowerbed can go between waterings, and it also has other advantages.

Organic mulches act as insulators because of their low heat-conducting properties. On the other hand, a mulch such as pea gravel is a good heat conductor. Dark gravel mulches tend to warm light-colored soils, and organic mulch tends to keep soil cool. Use this principle to slow down or speed up plant growth in summer.

Organic mulches decrease weed seed germination. Mulch limits the splashing of mud onto flowers and foliage, in turn reducing possible disease development. And well-chosen mulches are more than useful; they can also be attractive.

PINCH A PLANT, PINCH BACK, PINCH OUT

These terms have the same meaning and are used by gardeners to describe the removal of the growing tip of a shoot to ensure the development of side shoots. The term pinching out is also used to describe the complete removal of small side shoots, as is done when tomatoes are trained to a single stem and when the stems of chrysanthemums, fuchsias, and other plants are trained to form "trunks" of standard (tree-form) specimens.

Tip pinching makes a plant grow bushier; just pinch out the growing tip of the tallest stem close to a leaf joint.

The following annuals are improved by proper pinching: ageratum, carnation, cosmos, marigold, phlox, petunia, salvia, snapdragon, verbena, and dwarf-type dahlias (grown as annuals).

SEASON STRETCHERS

Extend the growing season a couple of weeks or more by using hot caps or plastic row covers for flowers or vegetables that do not grow above 18 inches high. Also, study your seed catalogs and choose some of the many quick-maturing varieties.

SPRAY DAMAGE

When spraying plants with fungicides to control disease, insecticides to control pests, and fertilizers for foliar (leaf) feeding, there is always a danger of damaging the leaves with the spray fluid. This may occur if the spray is not sufficiently diluted, if the plants are in a soft condition, or if the weather and atmosphere are unfavorable. Also, plants growing in a smoky atmosphere may be damaged owing to the liberation of copper by the acids in the atmosphere.

Whenever possible, choose a still day for spraying, so that the spray material may be directed where it is needed. This prevents the spray from drifting in the wind and endangering other plants. Do not use oil sprays or other types of spray on fruit trees at blossoming time. The spray might injure bees, bumblebees, or other insects pollinating the fruit.

WATER

We know that seeds and the roots of growing things need water, but the water also carries nourishment. Sometimes elements for plant nourishment are in the rain itself; sometimes plant food previously put in the garden is dissolved by the rain. The most fertile soil and the best climate conditions cannot produce a single green leaf if there is no moisture.

It is important to use water wisely, especially in areas with little rain. To conserve moisture, soak furrows before planting seeds, and then cover with dry soil. Though invisible to the eye, the moisture is there where the seeds need it to germinate.

When setting out new nursery stock, especially in dry weather, dig the hole and water deeply several times before planting the shrub or tree. Then the water will be available to the roots. Splashing a little water on top after setting won't do the job. When the hole is half-filled with soil, pack firmly and fill with water. As water soaks down, fill in with soil and leave a saucer-like depression, putting a layer of loose mulch in the depression.

Mulch is important over all the garden: It prevents a crust from forming and lets water soak down easily and more slowly, preventing runoff. Check soil under mulch to see if watering is needed.

For vine crops such as tomatoes or even decorative vines, it's good to sink a perforated tin can on one side of each plant. Whenever the plants need water, fill the cans. Once you start watering, continue until there is a good rain, or you may lose your plants.

Hints for the Yard and Garden

BANANA PEELS
Plant, don't pitch! Tear the peels or cut them with scissors into small pieces and bury them around your roses. Peels provide 3.25 percent phosphorus and 41.76 percent potash. But don't overfeed; three peels per bush at a time is about right. Stockpile extras by freezing them in half-gallon ice cream containers.

CLOTHESPINS
Use snap or spring clothespins to train fuchsia and small vines in pots and baskets. Catch the stem through the indentations and clip the jaws to other stems, pot rims, or stakes. Also use to hang along stems as weight if you are working toward a trailing plant. Remove after a few weeks and the stem will stay in place.

In the garden, use snap clothespins—the type with a spring and two grooves—to train grapevines on the trellis. The innermost groove should fit over the wire and the outer one around the stem. As the vine grows, adjust the pins. Remove at pruning time when they are no longer needed. This idea works well with espaliers or with anything you need to train to wires.

ESPALIER
For small plants to train as miniature espaliers, try the annual balsam. Pinch off the side shoots so that plant will grow tall and slender; you will be rewarded with a mass of blossoms. The espalier form is still used for training fruit trees on frameworks or against a wall, although it is not as popular as it once was.

FLAGSTONE FLATTERY

A flagstone walk gives interest and drama to the flower garden. To lay the stones in concrete, first wax the upper surface of each flagstone with liquid or paste wax. Concrete will not adhere to a waxed surface, so it will be easy to clean off any spills or smears when the job is finished.

FOAM, PACKING

Save that packing foam next time you receive a package. Because it is lightweight, noncompacting, and water-repelling, package foam is good material for hanging baskets and planters.

FROST PROTECTION

Evergreen branches, which are spongy, are excellent frost protection for perennials and herbs.

GRASS

To keep neat edges on grass along flowerbeds, use a straight edge spade to cut into the edges, keeping Bermuda, carpet, and St. Augustine grasses in check.

KNEE PADS

These can be as simple as iron-on pads, inside or out, for your blue jeans, or a pair of rubber pads with adjustable straps. They make small garden chores like weeding much easier.

LABELS

Many flowers look so much alike when they are not in bloom that it is easy to make a mistake when they are dug for replanting or sale. Iris, for instance, are almost impossible to tell apart. Write on the leaves with a magic marker pencil, or for a more permanent label, cut old venetian blinds in 1-foot lengths and write the name of the plant with a wax pencil.

When potted, the plant may be inexpensively labeled with a plastic knife. Print out the name of the plant with a name-tape printer, press it onto the knife handle, and push the blade into the soil. This is neat, inexpensive, and long-lasting, especially with potted plants.

Plant labels can also be made from 1- by 3-inch strips cut from used bleach jugs. Scratch or write on these with a large needle inserted in a 2-inch section of wooden broom handle. Next, rub dark shoe polish over the label, and wipe off the excess. This makes the writing both visible and permanent.

Somewhat unusual but neat and practical are the gutter spikes available in 7- or 9-inch lengths from hardware stores. To prevent them from rusting, dip them in rust-resisting enamel and let dry. Then attach the labels with copper wire or plastic twist-ties and drive the spikes into the ground beside the plants.

MILK

It is sometimes helpful to spray milk over plants suffering from mildew, mold, or tobacco mosaic.

SHOCK ABSORBERS

Many garages will give away discarded shock absorbers for the asking. Place them at the end of garden rows. When watering, slip the hose around them to prevent it from whipping over plants.

SHOE BAG

Get a see-through plastic shoe bag and store the small separate items such as gloves, plant tags, and twist-ties in the pockets. This is a time- and temper-saver, and ideal for the greenhouse.

SMOKING

Smoking is not only dangerous to your health, but it's dangerous to your plants' health as well. The tobacco mosaic virus also affects members of the Tomato family and can be communicated by the hands of a smoker working in the garden.

SUGAR

Sugar kills nematodes by drying them up. A 5-pound bag of sugar per 100 pounds of soil will kill nematodes within 24 hours. Helpful fungi are the enemies of nematodes, and plenty of humus in the soil will promote their growth.

SUNSCREENS

The sun's harsh rays dehydrate and dry out your skin, reinforce the lines that come from squinting, and increase the risk of some skin cancers. Sunglasses protect your eyes from glare and discourage frown lines around the eyes. When working in your garden, use sunscreen and wear a hat to keep the sun off your face.

TOOL KEEPERS

Drill a ¼-inch hole in the handle of a garden tool and hang the tool over a headless nail driven into the wall. To get a large number of tools in a small area, space the nails close together. Also paint tool handles orange—the color scientists consider safest for hunting caps and jackets. The color is readily discernible to the eye when you lay a trowel or clippers down somewhere in the garden or grass.

GARDEN CREATURES

Insects

ANTS

Ants are often a nuisance in the garden or in houses. They are repelled by pennyroyal, spearmint, southernwood, and tansy. Indoors, repel them with cucumber slices.

If you have ants in your kitchen, julep mint (a spearmint) or tansy planted near the kitchen wall or entrance will help keep them away.

APHIDS (*Plant Louse*)

A most familiar small pest in the garden and on houseplants, aphids come in green to match plant stems or in red or black. They are sucking insects and affect nearly all plants at one time or another. They attack houseplants in late winter when the plants' resistance is low.

The lady beetle and her weird-looking larvae are always on the lookout for aphids and help keep them under control. Another insect preys on aphids with such voracity that it is named the aphid lion.

Garlic, chives, and other alliums, coriander, anise, nasturtium, petunia, pennyroyal, spearmint, southernwood, and tansy will repel aphids.

BEE FLOWERS

A bee is said to make three journeys in order to bring one drop of nectar to the hive; 25,000 foraging trips are thought to be necessary to gather the raw material for 1 pound of honey.

Important honey plants are clover, alfalfa, mustard, cabbage, buckwheat, willow herb, cotton, mesquite goldenrod, acacia, blueberry, willow, maple, linden, locust, pear, plum, apple, and cherry.

Almost all single flowers produce a certain amount of nectar, but the following flowers produce nectar profusely and should find a place in every beekeeper's garden: wallflower, arabis, forget-me-not, borage, all members of the Bellflower or Campanula family, the mauve catmint (*Nepeta* × *faassenii*), heather, heath, honeysuckle, thyme, hollyhock, crocus, scilla, chionodoxa, snowdrop, heliotrope, cleome, lavender, lemon balm, cornelian cherry, daphne, barberry, winter aconite, *Clematis paniculata,* mock orange, sunflower, bearberry, robinia, asclepias, hepatica, *Rhamnus frangula, Limnanthes,* mignonette, phacelia, scabious, stonecrop, and the Michaelmas daisies.

In addition, all the small fruits, including currants, loganberries, raspberries, strawberries, and blackberries, are valuable, and particularly the gooseberry, owing to its early flowering. And after being visited by bees, these fruits will set much better crops.

Bees are not especially attracted to fragrant flowers, and their marked preference for those of blue color, which are so often scentless, bears this out.

Willow flowers, passionflower, sunflower, and the inconspicuous blossoms of the English ivy are said to be intoxicating to bees.

CUTWORMS

Climbing cutworms crawl up the stems of plants at night to feed, so just because you don't see them doesn't mean they aren't there! Tansy is repellant to cutworms. Or collar stems of newly set plants, letting the cardboard collar extend both above and below soil line for about 2 inches. Occasionally a plant will be cut off even inside the collar; if this happens, dig down and find the cutworm before placing another plant.

A mulch of oak leaves or tanbark placed in strips in the beds and spread on garden paths will repel cutworms, slugs, and the grubs of the June bug.

EUGENOL

A chief constituent of oil of cloves, eugenol is an effective attractant for baiting insect traps. Star anise and citronella grass are also useful for this purpose.

FLEA BEETLES

This pest is repelled by wormwood, mint, and catnip, or you might try interplanting susceptible crops near shade-giving ones.

GRASSHOPPERS

Unlike bees, grasshoppers damage crops and are of no value in pollinating plants. The biological means of controlling grasshoppers, which are becoming an ever-increasing problem, is nosema grasshopper spore. *Nosema locustae* is a protozoan that specifically attacks grasshoppers and some species of crickets. Dissolve the spore in water and add to a bran mixture, then disperse over the garden or yard area. As the grasshoppers feed, they become infected and slowly die. Research has shown a 50 percent drop in population in a month. Furthermore, nosema is passed on from one generation to the next through the egg mass, and is also transmitted when infected grasshoppers are eaten by healthy ones.

Grasshopper

ICHNEUMONID WASPS

These are parasites of moth and butterfly larvae. The adult wasps feed on pollen and nectar, and often from puncture wounds made in host larvae.

INSECT PREDATORS

Ladybugs and praying mantises are known for their good work in the flower garden, but did you know that the larvae of the firefly (lightning bug or lampyrid beetle) benefit growers by feeding on slugs and snails? Assassin bugs feed voraciously on caterpillars, Japanese beetles, and leafhoppers. Damselflies eat aphids, leafhoppers, treehoppers, and small caterpillars. Flower and robber flies are excellent pollinators and the larvae eat aphids, leafhoppers, and mealy bugs.

INSECTS, POLLINATION BY

Many insects are helpful as pollinators. Odor usually repels or attracts them, but color also plays a part. Many insects do not see red (which attracts hummingbirds, for instance) but can see ultraviolet, which we cannot see.

WASPS

The parasitic wasp trichogramma is an efficient destroyer of the eggs of many moths and butterflies, which are leaf eaters in the larval stages. It is also effective against the eggs of cabbageworms and loopers, corn earworm, and geranium budworm.

Wasps, often ill-tempered and prone to sting, are very effective against a variety of insect pests. They're helpful as pollinators, too.

Encarsia wasps are tiny and parasitic to whiteflies. Aphytis wasps are live controls for scale insects. Fig wasps, which live in caprifigs, carry pollen from the male flowers to the female flowers of the Smyrna figs. The calimyrna, a variety of the Smyrna fig, is widely grown in California.

Black-eyed peas and other field peas grown widely in the southern states are largely pollinated by wasps. Gather the pods early in the morning while the wasps are still lethargic from the cool of night.

Insect Controls

COMPANION PLANTS

Many plants are protected from insect damage by the presence of other plants grown nearby. (See "Companion Planting with Flowers and Herbs" for a list of insect pests and the herbs and flowers that repel them.)

ROTENONE, DERRIS

Long ago the Chinese discovered the insecticidal value of the derris, or tuba root. When the crushed roots, stems, and leaves of derris are thrown into lagoons and streams, fish float to the surface, insensible. Despite the extreme sensitivity of the fish, however, derris in normal concentration is believed relatively nontoxic to domestic animals and humans.

Rotenone is an insecticide derived from derris and certain tropical plants. This contact and stomach poison is often mixed with pyrethrum. It can be used on both plants and animals for insect control. It is notably effective against leaf-eating caterpillars, mosquito larvae, and aphids. The addition of oil of teaseed as a synergist increases its toxicity to squash bugs.

Derris is widely used on farms in animal dips to control lice and ticks. Applied to young bean plants, it makes the leaf growth less palatable to Mexican bean beetles.

Devil's shoestring (*Tephrosia virginiana*) is the only native plant that contains rotenone. This is a common weed in the eastern and southern states, and its roots may contain as much as 5 percent rotenone. Rotenone is safely used on all crops and ornamentals, but the period of protection is short.

RYANIA

This plant-derived insecticide was discovered in 1943. It is a powder made by grinding up the roots of the South American plant *Ryania speciosa*. Ryania has little effect on warm-blooded organisms but is useful in controlling corn borers, cranberry fruitworm, codling moth, Oriental fruit moth, cotton bollworm, and other insects.

While ryania may not reduce the number of harmful insects present, it protects the crop by making the pests sick enough to lose their appetite. Some species are not killed outright by it but are induced into a state of "flaccid paralysis." Ryania is recommended when there is an unusually large insect infestation and the gardener feels that nature needs a helping hand.

SMOKE

Control aphids, ants, and mites in the greenhouse with smoke from oak leaves, which are not poisonous, do not kill soil bacteria, and leave no harmful residue. Dried stems and leaves from canna plants and pepper-grass may also be used.

SPRAYS, HOMEMADE

Here are some insect controls you can easily make yourself from ingredients in the house and garden.

Onions. Red spiders and various aphids, especially those attacking roses, are routed out by onion spray. Grind up onions in a food chopper or an electric blender, add an equal amount of water, strain the mixture, and use as a spray. Bury the mash in the flowerbed or garden.

Hot peppers. Chop up hot pepper pods (wear glove and eye protection). Mix with an equal amount of water and a little soap powder to make the materials stick. This makes an effective spray against ants, cabbageworms, spiders, caterpillars, and tomato worms. Dry hot pepper, ground up and dusted on tomato plants, offers protection against many insects. Dry cayenne pepper sprinkled over plants wet with dew is good against caterpillars.

Combinations. Use a combination of several materials as an all-purpose spray. For instance, grind together three hot peppers, three large onions, and a whole bulb of garlic (peeled and chopped). Cover the mash with water and allow to stand overnight. Strain the following day and add enough water to make 1 gallon of spray. Use on roses, azaleas, chrysanthemums, beans, and other crops three times daily if infestation is heavy. Repeat after a rain. Bury mash under rosebushes.

Rhubarb. Boil rhubarb leaves in water and sprinkle on the soil before sowing wallflowers and other seeds as a preventive against clubroot. It is also useful against greenfly and black spot on roses.

Tomato leaves. These have insecticidal value as they contain an alkaloid similar to digitalin and more active than nicotine. An alcoholic extract of this substance is very effective against aphids on roses, pears, beans, and other plants. A spray of macerated tomato leaves soaked in water also frees rosebushes of aphids and eggplant of caterpillars.

Elderberry leaves. An infusion made by soaking elderberry leaves in warm water may be sprinkled over roses and other flowers for blight and also to control caterpillar damage.

Soap. A simple soap spray is often effective against aphids, thrips, mites, and other garden pests. Mix 1 to 2 teaspoons of Ivory Liquid, Shaklee's Basic H, Tide, or a small chunk of Fels Naphtha with 1 gallon of warm water and apply with a plastic squeeze bottle. Be careful with beneficial insects such as ladybugs and mantises, since the soap may harm them also.

TREE TANGLEFOOT

Tree Tanglefoot is a nondrying, sticky compound that forms a barrier against climbing and crawling insects such as caterpillars and cutworms. It is effective used on trees.

Butterflies

PLANT A BUTTERFLY BUSH

The buddleia, or summer lilac, is called the butterfly bush because it attracts monarchs and swallowtails, including the tiger and zebra, with their brilliant coloring.

Many plants that attract birds and bees also attract butterflies, but there are certain flowers, shrubs, and vines particularly loved by these "flying flowers." Such plants are attractive for their nectar.

Butterflies instinctively choose very different plants for egg-laying. These are mostly herbs, weeds, and certain trees. They include umbelliferous plants such as dill and parsley; weeds such as clover, goldenrod, milkweed, and dandelion; and trees such as willow, poplar, birch, and hackberry.

Besides liking flowers whose nectar content is both ample and suitable, butterflies also have color preferences. They have a passion for yellow and purple blossoms, also notable in their likes and dislikes of weeds. Thistles are violet-mauve, clover is mauve to rose-purple, dandelions and goldenrod are yellow.

Butterflies don't care much for roses, particularly white ones, but will rush to the nearest purple lilac bush, preferring the purple to the white. They also like marigolds.

Gardeners who would attract butterflies should choose plants that will encourage butterflies to feed, not breed. Among these garden attractants are wallflower, alyssum, sweet William, sweet rocket, candytuft, mignonette, zinnia, and the beautifully scented phlox flowers. Or grow portulaca; butterflies love it.

As you will note from this selection, butterflies prefer the simpler blooms, some intensely perfumed, to the hybrids that have been bred away from their natural development. For this reason, a butterfly-attracting garden is easy to grow.

A PROBLEM BUTTERFLY

In the strictest sense, butterflies aren't harmful. They cannot bite, chew, or sting. In the butterfly stage, they pollinate many flowers. Even in the caterpillar stage they are seldom numerous enough, in most areas, to do much damage.

One in particular, however, can be a real pest: the white or imported cabbage butterfly. This one is usually found wherever cabbages are grown (also other members of this family such as broccoli, collards, and Brussels sprouts). It is generally so abundant that its eggs, caterpillars, and chrysalides are readily discovered.

This familiar butterfly is white with black dots on the wings and blackish front angles on the fore wings. They flit freely about over fields, meadows, and gardens, sipping the nectar of various early flowers through their long, coiled tongues. From time to time they light on the leaf of a cabbage or other plant of the Mustard family to deposit their small, pale yellow eggs, which remain attached by a sort of glue.

About a week later the egg hatches into a tiny caterpillar that is very destructive. You will note its presence by the lacing of the leaves. A safe control of these worms is *Bacillus thuringiensis* (BT), sold at garden centers under various trade names—Dipel, Biotrol, or Thuricide. Do not hesitate to use it. It kills only these caterpillars; otherwise it is harmless.

ENEMIES OF BUTTERFLIES

The worst enemies of butterflies are the flies and wasps that lay their eggs on the caterpillar or inside the body. When the eggs hatch, the larvae eat the caterpillar. Other insects, such as dragonflies and mantids, eat great numbers of butterflies and caterpillars. Spiders catch them in their webs or lie in wait inside flowers. Birds, frogs, toads, and lizards feed on them.

Butterflies have no strong body parts to use as weapons against attack and are easily killed by their enemies. As a group they survive because of their high rate of production. A female butterfly may lay several hundred eggs during her lifetime. Though only a few live to adulthood, they carry on the species.

Butterflies are helped by protective coloration to escape from their enemies. The anglewings, for example, have bright colors on the upper surfaces of their wings but dull brown or gray underparts. When their wings are folded, only the dull underparts show.

Some butterflies, such as the monarch, feed on milkweeds, which makes them unpleasant-tasting food for birds. And they advertise their bad taste by warning coloration. Other butterflies, such as the viceroy, are protected from attack by having coloration similar to that of the monarch.

Birds

Of all the roles birds play in the garden, the least known is probably pollination. Certain flowers depend on certain insect friends to carry their pollen from blossom to blossom so they may set fertile seed; other flowers depend on the hummingbird. Only its tongue, which runs out beyond its long, slender bill and can turn around curves and reach the drops of nectar in the tips of the wild columbine's five inverted horns of plenty. This bird also seeks honey from monarda or bee balm, coral honeysuckle, jewelweed, cardinal flower, and many others.

Birds also play a valuable part in disseminating the seeds of many flowers. Mistletoe, for example, is spread by birds scraping their bills on the bark of trees, after they have feasted on its berries.

But flowers are not always kind! The cuckoopint, or spotted arum, of Europe, a relative of our jack-in-the-pulpit, actually poisons messengers carrying its seed; the decaying flesh of the dead birds affords the most nourishing food for its seed to germinate in. Birds help keep the balance of nature by trimming down the insect population. They are caretakers of the ground floor, eating grubs and beetles; they destroy grubs in the bark of trees; and others, like purple martins, catch flying insects.

LANDSCAPING FOR BIRDS

For their beauty, their song, and their ability to catch insects, birds are an asset to flower and vegetable gardens. To attract birds to your garden, furnish them with food and shelter. Hedges and dense shrubs, as well as trees, provide birds with nest sites and protection.

Most birds need open water of some kind, such as a conventional birdbath or a small pool with stones in the shallow edge. They drink and bathe, then use the dry tops of the rocks for preening sites.

Birds like variety in their diet—so remember this when deciding what plants to use in your wildlife landscaping. Create a varied pattern by intermingling plant species, sizes, and shapes. Give them a choice of food sources—seeds, nuts, fruits, berries, and flower nectar. Use plantings of annuals such as coreopsis, marigolds, sunflowers, and petunias for additional bird foods.

Some birds are almost exclusively seed eaters; other are "switch-hitters," eating insects, worms, and other animal foods as well as seeds. Among the seed eaters are finches, nuthatches, titmice, sparrows, siskins, towhees, juncos, jays, Clark's nutcracker, and of course, doves, pheasant, and quill.

Depending on your location and altitude, here are some planting suggestions for birdseed: coreopsis, cosmos, sunflowers, verbenas, and thistles as small flowering plants; burnet and croton as dove favorites; trees such as spruce, fir, birch, pines, oaks, and paloverdes. A hackberry or sugarberry tree has red, edible seeds that are relished by birds in late fall and winter.

Native grasses having seeds beloved of birds are bluegrasses (*Poa* spp.), grama grasses (*Bouteloua* spp.), bluestems (*Andropogon* spp.), wheat grasses (*Agropyron* spp.), vine mesquite (*Panicum obtusum*), and Indian rice grass (*Oryzopsis hymenoides*).

SUPPLEMENTARY FEEDING

Feeders stocked with fruits and grains are welcome food sources in late winter after fruits from your plantings have been depleted.

To attract wild birds, try a traditional mix of sunflower seed, oats, millet, and yellow corn. Sunflower seed alone is the best all-around food for attracting the greatest number of desirable birds such as cardinals, chickadees, blue jays, grosbeaks, nuthatches, finches, and titmice.

Thistle seed draws large numbers of songbirds where they have not been attracted previously, such as goldfinches, purple finches, pine siskins, and redpolls. Tender, husked sunflower hearts also provide a wholesome food for songbirds such as chickadees, cardinals, finches, and grosbeaks.

If you wish to attract cardinals, try safflower seed. Mix it with sunflower seed at first and soon the cardinals will be won over to it, while the other birds remain indifferent. Then you'll have a feeder strictly for these colorful birds.

HUMMINGBIRD FLOWERS

In our western garden alone there are 15 different species of tiny iridescent hummingbirds, the ruby-throated being the only one widespread in the East. Hummingbirds have excellent color vision and are easily attracted by bright red flowers, to which they fly from great distances. However, once in your garden, they will visit flowers of any color in their search for nectar and small insects.

Without feathers, the smallest of the hummingbirds is no larger than a bumblebee. This bird's long, slender bill is especially suited for sucking nectar from deep-throated flowers such as honeysuckle and the trumpet flower.

Hummingbird flowers are long, tubular, contain copious nectar, and are often borne sideways or drooping rather than upright. Hovering before the flowers, the tiny creatures insert their long bills and tongues while whirring their wings more than 3,000 times a minute as they feast.

Good flower choices for hummingbirds also provide color all season and include red columbine (*Aquilegia elegantula*); Indian paintbrush (*Castilleja integra*), blooming in spring; scarlet-bugler (*Penstemon barbatus*); skyrocket (*Ipomopsis aggregata*), summer; and hummingbird trumpet (*Zauschneria latifolia*) in fall. Some species, such as Indian paintbrush, scarlet hedge-nettle (*Stachys coccinea*), and autumn sage (*Salvia greggii*), begin blooming in early spring and are stopped only by fall frosts. Other possibilities include desert beard-tongue (*Penstemon pseudospectabilis*), shocking pink flower; scarlet runner bean, very showy scarlet flowers; scarlet larkspur, red-orange flowers; toadflax, violet to magenta; Rocky Mountain columbine, blue and white; *Penstemon palmeri,* pink to salmon pink; and foxglove, purple flowers.

Animals

CATS

Although cats sometimes do catch birds, their hunting instinct also leads them to keep the garden clear of snakes, mice, rats, grasshoppers, and tarantulas.

Cats that rarely have the opportunity to go outside enjoy greenery to nibble on. Oats fulfill this need, providing a safe, healthy distraction from nibbling on houseplants, some of which may be poisonous. Catnip, a perennial mint, is also a favorite of felines, who are stimulated by its leaves.

Grow catnip as a treat for your cat. And if the cat claws furniture, try rubbing crushed rue on it; the cat will leave it alone. Caution: *Rue gives some people dermatitis.*

CIRCUS IN TOWN?

Check around; you may find exotic animal manure available for the taking. If fresh, it's a valuable repellent for animal pests. Let it age a bit before using around plants, or dig it in freely between rows. Bengal tiger and lion manure may be just what you need to make your garden a roaring success!

DOG-GONE!

Man's best friend can be a nuisance sometimes. He can leave yellow patches on the lawn, ruin shrubs, and dig in your best flowerbed. Moth-balls or naphtha flakes around the beds will discourage him, but shouldn't be used where small children can pick them up. Or pound small sticks a foot or so apart in the region you're trying to protect.

RABBITS, COTTONTAILS

Their name comes from a fluffy, snow-white underside of the tail. They like to hide in heavy thickets or dense grass. Though often seen in daytime, they usually come out at night to gather their food. When there are too many of them, they can become serious pests by eating growing hay, vegetables, grapevines, and young fruit trees. And they also like to eat young flower plants. If you find your plants being eaten and no trace of insects, cottontails may be infesting your garden by night.

Rabbits, cute as they are, can level both flower and vegetable gardens during a one-night visit. They're hard to discourage, but try one or some of these: dried blood, blood meal, cayenne pepper, or wood ashes.

Onions, a thin line of dried blood, or blood meal sprinkled around the edges of the garden may discourage them. Or try powdered aloes, wood ashes, ground limestone, or cayenne pepper.

SQUIRRELS

If squirrels dig up your flower bulbs, lay a section of chicken wire on the soil surface of the planted areas. Secure with rocks around the edges. This also discourages cats, which like to dig in newly cultivated soil.

Protect trees from squirrels with a guard around the trunk. Encircle the tree with a slippery, smooth metal, shaped in the form of a downward cone. Be sure it is wide enough to prevent squirrels from jumping over it.

More Garden Creatures

FROGS AND TOADS

Frogs and toads are avid consumers of garden pests. It is estimated that a toad eats up to 10,000 insects in about three months, and about 16 percent of these may be cutworms. It also gobbles grubs, crickets, rose chafer, rose beetles, squash bugs, caterpillars, ants, tent caterpillars, armyworms, chinch bugs, gypsy moth caterpillars, sow bugs, potato beetles, moths, flies, mosquitoes, slugs, and sometimes even moles.

In spring gather frogs and toads for your garden from around the edges of ponds and swamps. Pen them in for a day; otherwise, their homing instinct may urge them to leave. Give them some shelter such as clay flower pot turned upside down with a small hole broken out of the side. Bury the pot in a shady place several inches in the ground. And don't forget to give them a shallow pan of water.

GOLDFISH

If you have a pool in your garden, stock it with fish. Goldfish are excellent for consuming mosquito larvae, and so are the small top-minnows called mosquito fish, *Gambusia affinis*.

Toads, like snakes, work for the gardener by eating insects. You can introduce toads to your garden, and they'll remain if they have lots to eat and some shelter, such as a piece of a clay pot.

SNAILS AND SLUGS

These land-based mollusks have grayish, wormlike, legless bodies, ½ to
4 inches long when fully grown. They hide in damp, protected places during the day. At night they chew up leaves, leaving a glistening trail
of slime. Inverted cabbage leaves made good traps for both slugs and snails; or spread wood ashes on the ground. Toads, which eat them, are good garden aids.

SNAKES

The harmless, so-called garden variety of snakes catches and kills many injurious insects and is itself not poisonous. Protect it. In the Southwest, be very careful around berry patches, which are attractive to rattlesnakes. Tarantulas, the big hairy spiders, are also attracted to berries.

Snakes can be destructive, but most of them are friends of the gardener, feeding on abundant insect populations.

SOW BUGS OR PILL BUGS

These aren't insects but crustaceans, related to the shrimps and lobsters. They have oval, dark gray, flattened bodies and seven pairs of legs and are up to ½ inch in length. They hide under logs, boards, or crop refuse, and in other damp places. When disturbed, they roll up and look like pills. They feed—and how they feed!—on tender parts of plants and newly emerged seedlings. Look for and eliminate their hiding places. Toads, including the horned toad (really a lizard), are helpful.

SPIDER MITES

Spider mites attack many flowers, being particularly troublesome on columbine. A spray made from onions controls them (see *Sprays, Homemade,* in this chapter). Experiments at Purdue University made use of an old-time favorite, a combination of buttermilk and wheat flour, to destroy spider mites by immobilizing them on leaf surface, where they then "exploded."

SPIDERS

Just because the note on the spray can says "Kills Spiders," it doesn't mean that you should. Spiders make good predators and only two common species, the black widow and the brown recluse, are really to be feared.

Spiders are rather wonderful—the nimble crab spider, for instance, named for its ability to scurry both sideways and backward. The little hunter can turn white, pink, or yellow to blend with vegetation. In a mini-jungle of stalks and stems, you may see a green lynx spider snatching up a victim as it trails a dragline behind it. This is a safety thread, anchored at intervals, which most spiders put down as they move about.

Spiders are particularly helpful during a heavy infestation of grasshoppers, killing large numbers by trapping them both in their aerial webs and on the ground.

TORTOISES

These turtles live in arid regions. They are large, timid, harmless animals. But they do eat plants and may ruin low-growing tomatoes by taking a bite out here and there. If you find one in your garden, do not kill it; remove it to another area.

GROWING WILDFLOWERS FROM SEED

Wildflowers are truly wonderful materials for the gardener, especially the mixtures that can be grown from seed. They are usually labeled by climate or geographical area and contain between 6 and 12 different kinds of flowers. Many such mixtures are available from catalogs and garden centers.

Wildflower seeds may be scattered on the ground, but it is best to give nature an assist and rake them in lightly to provide some protection from wind and rain. For better results, till the soil and cover the tiny seeds with a thin layer of peat moss. Keep the seeds moist for about six weeks.

On steep slopes where moisture is difficult to retain, sow the seeds into a top covering of very coarse gravel or lava rock. The seeds will sprout in between these materials, which will help to keep the soil moist and to hold the small seedlings in place, thus giving their roots a chance to take firm hold.

Temperature

Most wildflowers germinate readily in a temperature of 60° to 75°F. Temperatures higher than this may be harmful to some species. This temperature sensitivity is nature's way of preventing seeds from germinating during hot, dry periods when it would be difficult for the seedlings to survive.

Plant hardy perennials during spring or fall for best results. Seeds should be planted in a protected area to minimize the danger of being washed or blown away. Plant late enough in fall to ensure that germination will not take place until the following spring, or early enough so that seedlings are well established before the first frost. Plant in spring after danger of frost has passed. Get the seed in the ground if you can before a rain, or water the seeds so that they have sufficient moisture to germinate.

Dormancy

Seeds that fail to germinate under favorable conditions are said to be dormant. This state of dormancy is not accidental; plants survive in nature because of certain built-in timing mechanisms that delay germination until seedlings have the best chance for survival.

Some seeds will not germinate if exposed to cold temperatures. This causes a delay in germination, most usually until spring, when rainfall and other conditions in the environment are favorable. Wildflowers having "cold-temperature dormancy" may be planted outdoors in late fall, or be treated by a procedure called moist-chilling wherein seeds overwinter in your refrigerator for 1 to 3 months. Here is how this is accomplished:

Soak seeds that respond to moist-chilling in water at room temperature for 12 to 24 hours. Then mix them with a sterile, moistened medium such as sphagnum peat moss, vermiculite, or sand. Place the medium in a plastic bag or similar container that is not airtight. Store this in the refrigerator at 40° to 50°F, not in the freezer, for 3 to 6 weeks. Keep the medium moist but not wet. When the chilling period is completed, sow seeds at once at relatively cool temperatures.

Seeds that require only 3 to 4 weeks of moist-chilling may be germinated in a greenhouse if nighttime temperatures are in the range of 40° to 50°F.

Generally speaking, if this type of treatment is necessary, the packet will list the instructions. Most wildflower seeds do not need such exposure to cold temperatures and will sprout without special treatment.

Mass Plantings

If you would like to make mass plantings of wildflowers, here is a simple way to achieve good results:

Till the soil to a depth of 6 to 8 inches. The seedbed should have a loose, crumbly texture and good drainage. You can improve the air- and water-holding capacity of the soil by mixing in peat moss or other available organic material. Broadcast the seeds evenly and cover with a thin layer (not more than ¼ inch) of peat moss. If peat moss is not used as a top covering, rake the seeds lightly into the soil. If a few seeds are not entirely covered, don't worry about them; it is best if the seeds are not deeply planted.

Use a fine spray of water to moisten thoroughly. Keep the planting evenly moistened for 4 to 6 weeks; thereafter, waterings may be gradually reduced.

You may also sow seeds into a single layer of coarse gravel or lava rock (1- to 1½-inch size). This is a good way to plant steep slopes or areas difficult to keep moist. The seeds germinating in the cracks and crevices will be in contact with moist soil and also be protected from the elements. If natural rainfall is the only source of moisture, plant seeds in spring just before anticipated periods of rain.

Dry Areas

If you plan to plant a dry area, buy a mixture containing annuals, biennials, and perennials, most of which will sprout in 10 to 21 days at a temperature of 55° to 70°F. While such a mixture is best adapted to dry climates, most of the flowers will adapt to moist climates in sandy, well-drained soil. Perennials will survive cold winters in northern climates. Such a mix might contain baby's breath, chicory, coneflower, wild blue flax, gaillardia, penstemon, poor-man's-weatherglass, California poppy, prairie aster, and yarrow.

Moist Climates

The following mixture is best suited for moist climates, but will survive in dry climates if watered regularly. Recommended are baby blue-eyes, columbine, coreopsis, dame's rocket, larkspur, ox-eye daisy, scarlet flax, and wallflower. These perennials will survive cold winter climates. In mountainous regions above 8,500 feet elevation, there is usually ample moisture for this mix.

Harvesting Seed

To save seed of your wildflowers, break off stalks or seedheads, taking care not to disturb the root system. Timing is critical; if seeds are harvested too early, their viability may be seriously impaired. A change in color (often from green to brown or black) and a tendency to disperse seed are reliable indications of maturity. After removing your plant material, dry it thoroughly and either crush or shake to remove the seeds. Clean by sifting your seeds through a series of screens to remove dirt, chaff, and other unwanted material.

Wildflowers Adaptable to Large Areas

Most of these wildflowers are particularly well suited for restoration of large areas of land. They are easy to grow and, in most instances, highly adaptable to different climates and soils. These include sweet alyssum (*Lobularia maritima*), prairie aster (*Machaeranthera tanacetifolia*), baby blue-eyes (*Nemophila menziesii*), baby's breath (*Gypsophila elegans*), black-eyed Susan (*Rudbeckia hirta*), catchfly (*Silene armeria*), chicory (*Cichorium intybus*), columbine (*Aquilegia caerulea*), prairie coneflower (*Ratibida columnifera*), purple coneflower (*Echinacea purpurea*), lance-leaved coreopsis (*Coreopsis lanceolata*), plains coreopsis (*Coreopsis tinctoria*), cornflower (*Centaurea cyanus*), ox-eye daisy (*Chrysanthemum leucanthemum*), dame's rocket (*Hesperis matronalis*), wild blue flax (*Linum lewisii*), scarlet flax (*Linum grandiflorum* var. *rubrum*), gaillardia (*Gaillardia aristata*), firewheel gaillardia (*Gaillardia pulchella*), gayfeather (*Liatris spicata*), standing cypress (*Ipomopsis rubra*), and rocket larkspur (*Delphinium ajacis; Consolida ambigua*).

INDOOR PLEASURES

Houseplants

Cosmetic care of indoor plants also helps keep them healthy. Remove dead or old flowers and yellowed leaves; these can harbor and encourage insect and disease problems.

Wash foliage periodically. To wash off your houseplants, put them in the shower, turn on a gentle, tepid spray for a minute, then leave them a few hours so excess moisture can drip off. You'll be surprised to see how this perks up plants, especially ferns. Dust and grime block out light and clog the stomata cells, which allow the transfer of gases from plants to the atmosphere. Showering or syringing also helps reduce some insect populations such as spider mite and mealybugs, and the cooling effect stimulates growth.

Propagate spider plant by removing a small "spider" and planting it separately.

Starting a burro's tail plant is easy. Just poke one of those rounded leaves into a pot of soil.

The dracaena plant has leaves much like those found on a palm tree.

It's called dumbcane because if you bite its roots, you'll be speechless for days.

However, be cautious of syringing espiscias, gloxinias and most other gesneriads, and succulents when water is cold; spot damage will occur on leaves. Time your hose-down so water will not be on your plants at night, or fungal disease will result, especially during cool months. Collect and use rainwater if your tap water is high in iron, carbonates, or other dissolved minerals.

FOLIAR FEEDING

Spraying or applying fertilizer to plant foliage is effective with houseplants as well as outdoor plants. Take care not to discolor or otherwise damage home furnishings when you spray. Apply only at recommended concentrations and when plant is not in its rest period.

ISOLATION WARD

Place a new pot plant somewhere by itself for a week or so before you put it with your other plants. This will give you a chance to ensure that it is free of insects or disease without running the risk of infecting your other plants.

KITCHEN

Use flowers to brighten your kitchen. Put them on counter space with overhead lights or under a window. Top off a room divider cabinet with flowers. Light, humidity, and good ventilation are usually in abundant supply, and the proximity of the kitchen sink encourages good watering habits. Keep your green friends well away from cooktops and ovens, where heat and fumes may damage them. Bathrooms provide another opportunity for humidity-loving plants. Choose moisture lovers such as devil's ivy, English ivy, arrowhead vine, hollyfern, and baby's tears. Avoid plants such as cacti that need dry conditions. Plants enjoy a sudsy bath; wash them with a little mild soap and cool water.

Provided you have a fairly good light where your planter is situated, grow a variety of begonias, spider plants (*Chlorophytum*), ivies, wandering Jews, ferns, and herbs—and, of course, grow such vines as grape ivy and philodendron. Foliage plants will give you more year-round satisfaction than flowering kinds. Try a selection of dieffenbachias, aspidistras, snake plants, ferns, dracaenas, pandanus, and coleus.

LAYERING AID

When air-layering plants, place wet sphagnum moss or other rooting medium in old nylon hose to wrap around the prepared spot on the plant. Wrap again with a plastic bag and tie it down. The nylon hose keeps the rooting material in place and the roots grow right through it.

LIGHTING HINT

The color of house walls affects light intensity for indoor plants. Flat white, not glossy or semiglossy, is the most efficient reflector of available light.

MISTAKES TO AVOID

Here are some suggestions for preventing problems with houseplants:

Overfeeding. Don't fertilize houseplants too often. Once a month with a dilute liquid fertilizer is sufficient. Remember, don't fertilize a plant to make it grow; fertilize it because it is growing.

Overwatering. Underwater rather than overwater. More plants are killed by overwatering than by anything else. A dry surface is a poor test of water needs because the indoor atmosphere dries the surface quickly.

Poor lighting. Proper lighting is difficult to achieve in city homes and apartments. The best way to overcome this is to grow most plants under the new full-spectrum fluorescent lights. Flowering houseplants must have light, so reserve the sun-filled windows for them. Foliage plants prefer indirect light without sun. Only a few can get along in dark hallways.

Improper humidity. Proper humidity is very important. Most homes lack humidity in winter due to radiator and heating units that dry the air. A humidity of between 50 and 60 percent is best for most plants.

Too much heat. Most homes are warmer than the recommended temperature for houseplants. A cool room is much preferred to a warm one for plant growth. Plants require air circulation even in cold weather, but avoid cold drafts or air from central heating systems blowing directly on them.

Remember the plants are not like animals or people. They cannot refuse the food given them, nor can they move to a more favorable location.

WICK WATERING

When you vacation, provide your larger container plants with wicks leading from the pot to a water supply in a coffee can fitted with a plastic lid. Punch a hole in the lid, then thread the wick through. The lid cuts down on evaporation so water supply lasts longer.

Cut Flowers

CUTTING GARDEN

In an out-of-the-way corner, plant a cutting garden so that there will be fresh flowers for the house without the necessity of robbing the display flowerbeds. No need to pay lots of attention to design or aesthetics— simply grow neat rows of those annuals that bloom abundantly in colors and forms you want for decoration.

If your color scheme calls for pinks and blues, raise larkspurs, Canterbury bells, asters, bachelor buttons, felicia daisies, and stock. Add some dusty miller for its gray foliage; it's most compatible with pink tones.

For vivid reds, yellows, and oranges, grow marigolds, plumed celosia, red-hot poker, geraniums, gloriosa daisies, and gazanias. The taller varieties are best. Coleus make a fine foliage filler with these flowers.

Poppies, both the Shirley and the Iceland poppies, are great additions to mixed bouquets. They are long lasting if you sear the stem ends when you cut them.

For airy filler, grow some annual baby's breath or dill. And plant a few rows of strawflowers, statice, and bells of Ireland to cut and use fresh or dried in your home during winter.

If you have space only for a tiny cutting bed, try tall zinnias and snapdragons. Their white, yellow, orange, red, and pink colors blend well and their forms contrast nicely.

CUTTING POINTERS

A Cornell University extension bulletin gives the following advice:

1. Cut the flower stems. A freshly cut stem absorbs water freely. Use a sharp knife or shears and cut either on a slant or straight across.

2. Follow special procedures for special cases. Some flower stems exude a milky fluid that plugs their water-conducting tubes. To prevent this, place about ½ inch of the stem in boiling water for 30 seconds or char the end of the stem in a flame.

3. Remove excess foliage and eaves that will be below water. Excess foliage increases water loss; submerged foliage decays and hastens fading.

UNUSUAL FLOWERS FOR THE CUTTING GARDEN

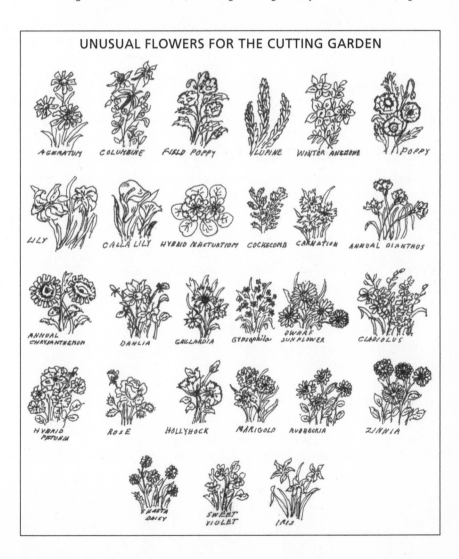

4. Place the stems in water at 110°F. Warm water moves into the stem faster and more easily than cold water.

5. Use a commercial flower food in the water. These foods combine sugars, acidifiers, and a mild fungicide to lengthen the life of cut flowers.

6. Wrap a piece of paper or plastic around the flowers after you have put them into warm water. This will prevent rapid air movement over the flowers and reduce water loss. After the flowers become crisp (about 2 hours), you may arrange them and they will continue to take up water. Sometimes wilted flowers can be restored by repeating this treatment.

7. Wash vases with soap and water after each use to remove bacteria. When bacteria multiply, they can often clog the water-conducting tubes of the flower stems.

8. Avoid excessive heat. Do not place flowers in direct sunlight, over a radiator, or in a draft.

9. Double the life of your flowers by placing them in a cold room or refrigerator at night, or when you are not at home.

10. Do not mix flowers with fruits or vegetables. Many fresh fruits and vegetables produce enough ethylene gas to shorten flower life.

SIPPING-STRAW STILTS

When your cut flowers have stems too short for a tall vase, insert the stems into plastic drinking straws, then cut them to the length you want.

TRANSPORTING CUT FLOWERS

Carry cut flowers such as iris in containers made from beverage cans. Punch six or seven holes in the top with a can punch, partially fill with water, and insert the stems. The can holds the flowers upright.

DRYING FLOWERS FOR LASTING BEAUTY

Dried arrangements brighten a home or office during the cold months of fall and winter. During fall, gather weeds, vegetables, fruits, flowers, and foliage, as well as dried grasses, from the roadside, market, and garden. There is an abundance of material to select from. By treating this plant matter with certain special agents, you can preserve its beauty and color almost indefinitely.

Many plants will dry naturally as part of their life cycle. Tall brown stalks of mullein, teasel, thistle, heracleum (cow-parsnip), and nonpoisonous sumac, cattails, and dock may be used without treatment if picked when dry. Dip cattails, however, in undiluted shellac to keep them from shattering.

Look for interesting pods on the ground beneath black locust. Sweet gum balls and cones of needled evergreens are attractive subjects. Also save some dried okra pods from the late garden.

Yarrow from the garden dries naturally, as do seed clusters of ornamental onions or leeks, but catch them before they start to deteriorate. Pick such materials a little prematurely; they can finish drying in the house. Hang them from coat hangers in your attic or in an unused room, or simply prop the stems in a deep, empty container.

Method for Hanging

Group three or four stems together and tie tightly. This is necessary because stems lose moisture as they dry, shrinking in size, and may loosen enough to slip out. Elastic bands or twist-ties are better than string to keep materials bound. Hang thick stalks such as mullein and large stems such as cockscomb singly. Suspend the bunched materials or stalks upside down; this keeps the stems straight and the flower heads upright. When hanging flowers, remove the foliage.

Hang the materials in any manner that will allow free passage of air to all surfaces. Bunches or stalks may be hung on a line as you would clothes on washday, or on a rack. You will need less space if you attach three or four bunches to a wire coat hanger. Clothespins are handy for attaching bunches to a line or hanger.

Choose any warm, dry spot for hanging, provided it has free circulation of air. Use a kitchen, garage, attic, or shed if it is convenient. Don't shut the plants up in a closet or expose them to direct sunlight while they are drying.

Weather conditions at time of drying determine the number of days needed, but generally the majority of plants will dry in 8 to 10 days.

Plants that can be successfully dried indoors by hanging include the everlastings listed in this chapter, as well as acacia, artemisia, bird-of-paradise, rubber plant leaves, magnolia leaves, bells of Ireland, cacti, ornamental grasses, palm, pepper berries, nandina berries, snowball hydrangea, and sunflowers (with petals removed). Globe thistle, liatris, and Queen Anne's lace can be dried by hanging but retain better color if buried in a drying agent. Many dried materials will last much longer if sprayed a time or two with hair spray.

Drying Flowers with an Agent

Fragile garden flowers dry best buried in a special agent such as silica gel, sold by craft suppliers under various brand names. Follow the supplier's directions.

Alternatively, you can make your own drying agent by mixing together equal parts of borax and yellow cornmeal, or use dry sand. Proceed as follows:

Pick your flowers on a dry day, selecting only those in good condition, preferably just before maturity. Process without delay.

1. Cut stems to within 1 inch of flower heads, and strip off any remaining foliage.

2. Make a false stem by dipping the end of a length of 22-gauge wire in white glue. Insert this into the base of the flower head close to the stem and tape the two together.

3. Line a shallow box of suitable size with waxed paper.

4. Prepare the drying mixture of borax and cornmeal. Add 3 tablespoons of uniodized salt per quart to the mixture if you want better color preservation.

5. Cover the bottom of the lined box with a ½-inch layer of mixture.

6. Place the flowers face up in the box, bending the wires far enough below the heads so they lie flat. Mound up the mixture beneath the flower heads to cushion.

7. Sift the mixture gently between petals, adding gradually until the flowers are just covered but not deeply buried. Do not crowd and do not dry more than a single layer of blossoms at a time.

8. Leave the box uncovered. Check in six or seven days. Flowers should not be left in the mixture when dry.

9. When drying is complete, gently brush away the mixture. Then slide your hand carefully under each head and lift it on your outstretched fingers. Place the flower on top of the mixture and allow it to remain for at least another 24 hours to firm up the petals. Without this treatment, the petals will shatter. When the petals are completely cured, spray them lightly with a plastic coating for permanence.

10. Keep your dried flowers in closed boxes on tissue paper until you are ready to use them.

11. Store the drying mixture in tightly covered tins to be reused.

Flowers to dry by burying include acacia, China aster, bells of Ireland, celosia, daffodil, dahlia, delphinium, echinops, everlastings, gladiolus, liatris, lilac, lily, marigold, peony, Queen Anne's lace, snapdragon, stock, tulip, wandflower, and zinnia. Roses may be dried if picked when only about two-thirds open; if in full bloom, they will fall apart. Also try drying the buds of small "sweetheart" roses, especially nice for potpourri jars.

Foliage

It is possible to preserve lovely autumn foliage with glycerine. Leaves will achieve a rich brown shade but still remain soft and pliable. If the foliage has been in water before processing with glycerine, the method may not be successful, so keep this in mind when making your selections.

Types of foliage that can be treated successfully include those of house-plants, as well as garden and florist materials. Here are some you may wish to try: acuba, aspidistra, beech, boxwood, cocculus (snailseed), copper beech, elaegnus, euonymus (*Euonymus alatus* and *E. japonica*), ivy, laurel, leatherleaf viburnum, leucothe, magnolia (*Magnolia grandiflora* and *M. virginiana*), mahonia, papyrus, peony, plum, red maple, sorrel tree, winter hazel.

There are no hard-and-fast rules as to what foliage can be preserved in glycerine, but it is best to use leaves or branches that absorb water freely and are entirely crisp and fresh.

Make a solution composed of one-third glycerine to two-thirds water. Fill a container with this to a depth of 4 to 5 inches. Slash the stem of the leaf or branch with a knife or mash it with a hammer for about 1 inch at the cut end so that absorption will be more complete. After placing the stem or branch in the solution, allow it to remain until saturated. This is easy to determine, as the color of the leaves will change. The foliage absorbs the solution more readily during the warmer months.

Store the container in a dark, cool place for about 3 weeks. When you notice beads of glycerine on the leaves, absorption is complete. Remove the branches and hang them upside down, allowing the solution within the stem to work down to the tip.

Adding a few drops of chlorine bleach to the mixture will usually prevent mold from forming. The solution is reusable.

Everlastings

It's easy to have colorful cut flowers year-round with the simple-to-dry kinds known as everlastings. Cut your material before the flower is fully open, remove the leaves, and hang heads-down in an airy location. These are lovely in arrangements and used in framed pictures and other crafts. Here are some suggestions:

Strawflower (*Helichrysum*) is an easily grown annual in a delightful range of colors from white and pastels to deep reds and yellows. It grows about 30 inches high, and can also be had in dwarf varieties.

Ammobium alatum, a hardy annual, has silvery flower heads.

Asclepias fruiticosa has brown to greenish yellow fruits on 3-foot stems. Use the annual, which fruits in August, for both flowers and arrangements.

Statice is available in both annual and perennial varieties. Use fresh or dried for light, airy bouquets.

Echinops, or globe thistle, is a splendid everlasting but does best dried with an agent. Globe thistles in shades of purple grow wild in many regions.

Xeranthemum annuum has papery single or double flowers in white, purple, or mixed shades. Do not pick until the flowers reach the papery stage.

Helipterums, frequently sold under the names of rhodanthe and acroclinium, are among the prettiest of the immortelles and are easy to grow. Pick the flowers as soon as they are open and, again, dry head-down. Acroclinium (sunrays) has rich shades of salmon, apricot, pink, rose, and cerise with white and creamy tones. Pick in bud to dry for winter bouquets, or use as fresh cut flowers. Plants branch and bloom freely and grow to a height of 24 inches. All of the helipterums have soft, silvery foliage, although the foliage itself is not useful as a dried flower. *H. rhodanthe*, also sometimes sold as *H. manglesii* or Swan River Everlasting, has showy, nodding, 1½-inch flower heads of handsome rose pink.

Globe amaranth (*Gomphrena globosa*) will thrive in any soil and stands drought well. It retains its form and color in drying.

Amaranthus caudatus, or love-lies-bleeding, is very pretty.

Catananche caerulea or Cupid's dart, with lovely blue flowers, must be picked and dried as soon as the flowers open or they will fade.

Trachymene coerulea has many lavender-blue flowers on long stalks.

Physalis, the old Chinese lantern, and lunaria, or honesty, are excellent in mixed, dried bouquets.

Bells of Ireland (*Moluccella laevis*) has wonderful 18-inch sprays of tiny flowers with large green calyxes. It is as pretty and green in a summer arrangement as it is silvery and dried in winter.

The yarrows (*Achillea* spp.) and the silvery sprays of lamb's-ears (*Stachys byzantina*) should be dried upside down.

Wood rose (*Merremia tuberosa,* syn. *Ipomoea tuberosa*) is a must have! The "wood roses" are the dried flower calyx (seed cases) of a yellow-flowered perennial morning glory native to the West Indies and other tropical regions. It blooms the second year when grown from seed. The five so-called petals are usually the sepals rolled back from the central budlike ball. The wood rose looks like a rose carved out of wood, stiff and polished to a beautiful satin brown. These "roses" make spectacular corsages and are lovely in dried arrangements.

DYEING WITH NATURE'S COLORS

Why use natural dyes? First, because they are so beautiful, and second, because of the wonderful feeling it gives you to say, "I did it!"

Natural dyes can be used on many types of material: yarn, cloth, macramé objects, crochet work, tie-dye fabric-silk, wool, cotton, jute—sometimes even wood. However, you will have your best results with cloth or yarn made of natural material such as wool or cotton. Experiment with others if you like, but try a sample first.

Some natural dye colors will be fast, others less so. Colors can be repeated, but don't count on the results being exactly the same; sometimes even commercial dyers have problems. Quantity dyeing is possible if large enough quantities of natural dyestuffs are available and your container is large enough.

Do all your dyeing in enamel kettles. Aluminum, tin, and iron pots change colors. Also, do not use the mordants (mentioned later) in pots that you are using for cooking because some of these are poisonous. Keep a separate pan for dyeing and keep mordants out of the reach of children. Actually, many of the flower and vegetable dyes do very well without mordants, but the colors will not be so permanent and sometimes not so bright.

Dyes from the Garden and Kitchen

There are probably hundreds of flowers, both garden and wildflowers, that can be used in dyeing, including coreopsis (*Coreopsis auriculata* and *C. calliopsidea*), dock root (*Rumex*), goldenrod (*Solidago*), goosefoot (*Chenopodium*), hedge-nettle, betony (*Stachys*), herb Robert (*Geranium robertianum*), red-flowered orchid cactus (*Epiphyllum*), pansy (*Viola tricolor*), pearly everlasting (*Anaphalis margaritacea*), and iris (*Iris*). For an attractive yellow-green, try sunflower seeds. After boiling them, spread the seeds outdoors somewhere and let the birds have a feast. (See the chart later in this chapter for further possibilities.)

Fruits and vegetables from your garden can also be used. Many vegetables that make good dye can be served at the table and the cooking water saved for use as the dye bath. Overcook spinach just slightly and it becomes a fancy purée for the table and a lovely green dye for your wool. Purple cabbage cooked for 1 hour is still edible and the broth is left for dyeing. From the cabbage you will derive several shades of green, depending on the size of the cabbage and the quantity used.

Save the water from boiling your beets or beet tops. It will dye your fabrics pink or yellow-green. Beet dyes are fugitive and the color may fade a little, but the material will still be attractive.

Oranges are a real treasure. Use the orange juice or the pulp, boil the peel for 1 hour, and dye your yarn or cloth a bright orange!

Blackberries, huckleberries, blueberries, strawberries, and raspberries all produce delightfully strong colors. For brightest purple try the raspberries; the equivalent of two 10-ounce frozen packages will dye about ½ pound of yarn. Frozen grape juice gives a lively purple and is easy to use. The bottled grape juice gives an even darker shade. Soak a box of currants overnight and simmer for a pearl gray or lavender color. For a lovely rose to lavender color, try cranberries—and use the berries with sugar for sauce or preserves.

Any of the nut hulls—particularly those of walnuts—make good dyestuffs, and the nut meats are used in cakes or cookies. Or look in the barbecue section of the supermarket for hickory chips and soak these overnight for the same rich brown that walnuts and hickory nuts give.

Cinnamon, turmeric, ginger, saffron, paprika, curry powder, and mustard give vivid dyes in yellows and reds. Experiment with oregano and chopped chives for other colors and effects.

Instant tea or coffee is quick acting and gives interesting shades of tan or brown. Boil coffee grounds in a cheesecloth bag with wool for a rich chocolate brown. Rose-hip tea produces a rosy tan that is delightful.

The canned-goods shelf holds possibilities too. Canned spinach, beets, okra, and blueberries give pretty colors when simmered half an hour, including the liquid in the can. Strain out the vegetables, drop your thoroughly wet wool into the dye pot, and you will have enchanting color. Try pie cherries for a soft rose.

If you achieve a color that is too pale or doesn't please you, overdye it in another dye bath. The color will probably come out more interesting.

Mordants for Fast Colors

Mordants are a must if you plan to use dyed material in an article that will be washed frequently. A mordant is a mineral salt that binds the color, making it sunfast and washfast. If mordanting seems a bit difficult, use ½ cup of white vinegar or lemon juice or 1 tablespoon of salt in the dye bath. The materials should then be washed gently by hand rather than machine-washed.

Dyer's mordants may be purchased at a drugstore, a chemical supply house, or, occasionally, at a health food store. To use them, mix the pre-scribed amount of the mordant crystals in a jar with about 1 cup of water. Shake them well to dissolve the mordant and add the solution to the dye bath. The easiest ones to find and use are the following:

Alum (potassium aluminum sulfate). This does not change the color of the dye but binds the color to the material. Use with cream of tartar (found in the spice section of the supermarket). The quantity for ½ pound of yarn is 2 teaspoons of alum and 1 teaspoon of cream of tartar.

Chrome (potassium bichromate). This brings out greens and yellows. Mix 1 teaspoon of chrome and 1 teaspoon of cream of tartar for ½ pound of yarn. Cover your dye pot, because chrome is weakened by exposure to light.

Copper sulfate (blue vitriol). Copper sulfate intensifies green dye. Use 1 teaspoon to ½ pound of wool.

Iron (ferrous sulfate or rust). This "saddens" or darkens colors. The effect may be had by boiling a handful of rusty nails to a cheesecloth bag.

Tin (stannous chloride). Tin is good for brightening colors. It may be added to an alum bath during the last half of the cooking process to pick up bright color. Rinse in soapsuds after dyeing to keep tin from hardening wool.

Experiment for surprises. Sometimes a plain household ammonia rinse will give a unique shade. A lovely, soft rose from cranberries changed to a bright chartreuse after this treatment! After mordanting, wash your cloth or wool in a washing machine.

The Basics of Dyeing

Here are the basic steps for dyeing half of material—for more, just multiply:

1. If using skeins of wool, tie them loosely in two places so they will not tangle—ditto for dyeing thread.

2. Soak skeins or cloth in water for at least an hour to keep wool from streaking or blotching in the dye bath. This also helps the dye to absorb more evenly.

3. Put the dye bath (your vegetable, flower, or fruit broth) into an enamel pan. Add sufficient water to make 3 quarts. Add the material and simmer 1 hour, or until the desired color is achieved. Remember that colors look darker when wet and allow for this. Keep the material submerged by poking with a wooden dowel or spoon. Wool is light and tends to float. Do not stir yarn or it will tangle. *Note:* If using a mordant, dissolve it in a jar with 1 cup or so of water, shaking to mix well. Pour this into the dye bath and stir well before adding the yarn or cloth.

4. Remove the material when the desired color is reached. Pinch a little with your finger to get some idea of the color when dry. Rinse in hot water about the temperature of the dye bath (a change of temperature may shrink or mat wool). Rinse until the water is clear and hang in a shady place for drying. If you mordant with tin, rinse in soapsuds and then rinse out the soap. Each rinse should be a little cooler than the one before. Gently squeeze out water, never wring. A commercial fabric softener used at the end of rinsing helps make wool soft and fluffy.

As you get more experienced in dyeing, you may wish to try for variegated effects. Try dyeing half a length of a skein by placing a stick across the dye pot and letting half the skein hang in a cranberry bath, the other half length in a grape juice bath. Let each simmer the required amount of time. Remove and rinse. You will have lovely soft colors that blend well with each other. Other combinations give equally interesting results. Try two shades of green or brown, yellow, or orange.

Dyeing with nature's colors is fascinating and satisfying, and the colors you create will be uniquely yours. With these beautiful yarns and fabrics you can go on to make clothing or decorative objects. If you have dyed wool yarn, for instance, use it for knitting, weaving, crochet, or needlepoint.

Dyes from Nature

For a soft shade of rose, use beets for dyeing.

Here are some suggestions for using nature's colors. Most of these work best with wool. Where indicated by **w/c** you may also dye cotton:

Common Name	Scientific Name	Plant Part Used	Dye Color
Acacia	*Acacia* spp.	Flowers	Yellow or gray, maize yellow to light golden brown
		Pods	Moss green or tan
Althea shrub or rose of Sharon	*Hibiscus syriacus*		Medium to dark blue
Anemone, blue	*Anemone* spp.		Teal blue or light green
Bottlebrush	*Callistemon* spp.	Flowers	Tan to greenish beige
Brassbuttons	*Cotula coronopifolia*		Deep brassy gold
Butterflybush	*Buddleia davidii*	Flowers	Orange-gold or gold-green or golden brown-wool or jute
		Leaves and stems	Olive green— in iron pot Various greens or black
Cactus	*Opuntia robusta*	Purple fruit, steeped	Magenta to rose
Camellia, red, in iron pot	*Camellia* spp.		Medium gray to dark gray
Chamomile	*Anthemis nobillis*		Various gold yellows, aromatic
Canterbury bells, purple	*Campanula medium*		Medium green or pale blue
Chrysanthemum, maroon	*Chrysanthemum* spp.	Flowers	Variations of gray-turquoise **w/c**

Common Name	Scientific Name	Plant Part Used	Dye Color
Daffodils, yellow	*Narcissus pseudonarcissus*		Bright yellow or deep gold
Dahlia	*Dahlia pinnata*	Seedheads	Bright orange
Daisy, gloriosa, black-eyed Susan, brownish	*Rudbeckia* spp.		Bright olive green to dark green
Morning glory, bindweed	*Convolvulus arvensis*		Dull green or khaki green to yellow
Mulberry, black,	*Morus nigra*	Berries	Intense red-violet to dark purple or purple, wool and jute
Dock	*Rumex* spp.	Blossoms	Rose beige to terra-cotta **w/c**
		Root in iron pot or with nails	Dark green to brown or dark gray **w/c**
Dodder and bits of pickleweed	*Cuscuta* spp. and *Salicornia* spp.		Yellow or ocher
Elderberry, blue or black, one pot and mordant method	*Sambucus* spp.	Berries	Mauve
Eucalyptus, blue gum	*Eucalyptus globulus*	Leaves	Deep camel tan
		Leaves in iron pot	Light to dark green or charcoal gray **w/c**
Fennel	*Foeniculum vulgare*	Flowers, leaves	Mustard yellow or golden brown, aromatic
Flax, New Zealand	*Phormium tenax*	Flowers Seedpods	Brown Bright terra-cotta
Foxglove, purple	*Digitalis purpurea*	Flowers	Chartreuse
Geranium, red	*Pelargonium hortorum*	Leaves	Dark purple to gray **w/c**
Goldenrod, in iron pot	*Solidago* spp.	Flowers (fresh)	Mustard, tan-orange, or brown-olive
Goosefoot, in unlined copper pot or with cupric sulfate	*Chenopodium* sp.		Dark green or green-gold
Grape, Concord-type	*Vitis labruscana*	Skins and ferrous sulfate	Dark blue
Hawthorn	*Crategus* sp.	Blossoms	Variations of yellow-green or gold-brown

Common Name	Scientific Name	Plant Part Used	Dye Color
Hedge-nettle, betony	*Stachys* spp.		Chartreuse green
Herb Robert, red robin	*Geranium robertianum*		Light golden brown to rich brown
Hibiscus, red, and tin	*Hibiscus* spp.		Purple
Hollyhock, red, in iron pot	*Althaea rosea*		Brown
with tin crystals	*Althaea rosea*		Wine color
Honey bush, and tin crystals	*Melianthus major*		Violet
Indigo, blue, in pot	*Indigofera tinctoria*	Leaves	Blue
Iris, dark purple	*Iris* spp.		Various violet blues
Iris, purple, fleur-de-lis, and tin crystals	*Iris germanica* and other species		Various dark to light blues
Klamath weed, and ammonia	*Hypericum perforatum*		Mustard gold or raw siena
Knotweed, doorweed, matgrass	*Polygonum aviculare*		Creamy yellow, brighter yellow or brassy yellow
Ladies' purse, yellow	*Calceolaria angustifolia*		Maize yellow to gold or deep orange, wool and jute
Laurel, California, Bay, California	*Umbellularia californica*	Flowers	Greenish beige, aromatic
Lichen, brown, rock oyster lichen	*Umbilicaria* spp.		Magenta violet, aromatic
Lilac, purple	*Syringa* spp.		Light green or light blue-green
Lobelia, blue, in copper pot	*Lobelia erinus*		Pastel green
Lupine, purple	*Lupinus* spp.		Bright yellow-green or dull green
Manzanita	*Arctostaphylos* spp.	Leaves	Deep camel or rose buff **w/c**
Marguerites, yellow, Paris daisy	*Chrysanthemum frutescens*		Gold or mustard-green
Marigold, with tin crystals	*Tagetes* spp.	Flowers	Yellow-orange, gold, or dull green

w/c: Can also dye cotton.

Common Name	Scientific Name	Plant Part Used	Dye Color
Meadow rue	*Thalictrum polycarpum*		Bright yellow, fragrant
Milkweed, showy, and cupric sulfate or unlined copper pot	*Asclepias speciosa*	Leaves/flowers	Moss green or brass green
Morning glory	*Ipomoea*		Dull green or khaki green to yellow
Mulberry, black	*Morus nigra*	Berries	Intense red-violet to dark purple or purple, wool and jute
Mule ears	*Wyethia angustifolia*		Gold to brass
Mullein, and ammonia	*Verbascum thapus*	Leaves/stalks	Bright yellow or chartreuse
Nicotiana, maroon, and cupric sulfate	*Nicotiana* spp.		Gray-green
Nightshade	*Solanum* spp.		Bright yellow or dull gold or various khaki greens
Oleander, dark pink	*Nerium oleander*		Light gray-green or medium gray
Olives, raw	*Olea europaea*		Variations of maroon
Onion, red	*Allium* spp.	Skins	Gold to henna red to maroon **w/c**
Onion, yellow	*Allium* spp.	Skins, in iron pot	Yellow-green
Osage orange	*Maclura pomifera*		Intense greenish yellow or deep burnt orange
Owl's clover	*Orthocarpus* spp.	Extract	Lemon yellow or mustard or ocher
Pansy, dark blue, steeped	*Viola tricolor*		Blue-greens
Penstemon, red, one pot and mordant method	*Penstemon* spp.		Medium brown
Petunias, purple, and English walnut	*Petunia* × *hybrida* and *Juglans regia*	Leaves	Light khaki green

Common Name	Scientific Name	Plant Part Used	Dye Color
Petunias, red, and marigolds	*Petunia* spp. and *Tagetes* spp.		Various dark greens to brown
Pigweed	*Amaranthus* spp.		Moss green or brass or pale yellow
Pine needles, in iron pot	*Pinus* spp.		Olive green, aromatic
Plum, dark red	*Prunus* spp.	Leaves, and tin crystals	Violet to purple or lavender
Poinsettia	*Euphorbia pulcherrima*	Leaves	Greenish brown
Primrose, dark red, in iron pot	*Primula* spp.		Greenish yellow or bright avocado
Rabbit brush	*Chrysothamnus* spp.		Lemon yellow or gold copper
Ragwort, tansy-ragwort, stinking Willie	*Senecio jacobaea*		Bright yellow or brassy gold
Redwood, California	*Sequoia* spp.	Bark	Tan or light golden brown to terra-cotta
Rhododendron	*Rhododendron* spp.	Leaves, in iron pot	Gray-green
Rosemary	*Rosmarinus officinalis*	leaves, flowers	Various yellow-greens
Rudbeckia	*Rudbeckia* spp.		Bright chartreuse to dark green
Sagebrush	*Artemisia tridentata*		Various tan-golds, brilliant yellow, or yellow
Salal	*Gaultheria shallon*	Berries Berries, and cupric sulfate	Dark blue Various dark greens
Santolina, lavender cotton	*Santolina chamaecyparissus*		Siena gold or yellow
Scabiosa, purplish, pincushion flower	*Scabiosa atropurpurea*		Bright green or dull dark blue
Self heal, heal-all	*Prunella vulgaris*		Bright olive green
Silk oak	*Grevillea robusta*		Intense canary yellow or olive green

w/c: Can also dye cotton.

Common Name	Scientific Name	Plant Part Used	Dye Color
Snapdragon, dark reddish	*Antirrhinum majus*	On plant fibers	Pale green or tannish gold
Spicebush, and cupric sulfate	*Calycanthus occidentalis*		Light brown
Stock, purple	*Matthiola incana*		Blue or turquoise
Tarweed	*Hemizonia luzulaefolia*		Golden yellow or light yellow; aromatic **w/c**
Tea, black	*Camellia sinensis*	Leaves	Rose tan or gray or black
Tea, sassafras	*Sassafras albidum*	bark	Light terra-cotta to orange tan
Twinberries, and tin crystals	*Lonicera involucrata*		Gray
Walnut, black	*Juglans nigra*	Leaves	Cinnamon to dark brown or tan to brown; wool, cotton, jute
Woodruff, sweet	*Galium odoratum*		Soft tan or gray-green
Woolly aster, seaside	*Eriophyllum staechadifolium*		Bronze gold to golden brown
Yarrow	*Achillea millefolium* and spp.	Flowers	Yellow to maize or dark green
Yarrow, in copper pot	*Achillea millefolium* and spp.	Leaves	Chartreuse to tan-greens

w/c: Can also dye cotton.

Beautiful Easter Eggs, Naturally

Coloring eggs at Easter time is a very old tradition practiced in many countries. The methods have varied, but none is more lovely or simple than the old German custom of employing natural materials. Perhaps best of all, in these days when we are again becoming ecology-minded, these eggs may be eaten without worrying about their wholesomeness or their effect on the system. And many of these materials are right in your own flower or vegetable garden, on your lawn, or in the woods or fields near your home.

Save the outer skins of onions. Carefully peel these off as they dry and darken. Store them in a mesh bag, using a bag of fairly close weave so small particles will not be lost. If you have a good supply of both yellow and red onions, save the skins separately for greater variation of color. Do not cook them together, for the results will not be attractive.

Rainwater (or melted snow) makes the best onion-skin broth. Catch this in a glass or enamel vessel and store it in advance. Well-washed glass or plastic vinegar jugs are handy for this purpose.

Make a broth by simmering the onion skins gently for an hour or until the color of the water is quite deep. Let cool to room temperature but do not remove the skins.

While the broth is cooling, gather the materials you'll need from your garden. Find an old sheet or several old pillowcases; those that are ready to be discarded are best, as thin material absorbs well and permits the color of the skins to pass through to the egg. Tear the cloth into long strips, 1 inch wide and 1 yard long. If the material is still strong enough, tear the salvages into strips about ¼ inch wide. If not, have ready a spool of thin, soft white twine for use in tying the wider cloth securely after the eggs are wrapped, or use sewing thread.

Take your garden basket outdoors in the warm spring sunshine and see what you can find. This will depend on the Easter-time weather and on your climate zone.

Grape hyacinth makes a lovely delicate imprint; often the blue color is transferred to the egg as well.

Dandelion heads, carefully cut so they will lie as flat as possible, impart their own yellow color, and pink japonica and rose petals leave their own lovely hues. Hunt in the lawn or in a nearby field for young yarrow plants. Their fine, fernlike leaves leave an exquisite tracery. Clover leaves and ferns are lovely.

Do not overlook the decorative possibilities of dried grasses left over from winter and still holding their shape. Consider weeds with interesting outlines; some of these make markings on the eggs as pretty as cultivated flowers and plants.

Do not gather too much at a time. Some of your plants or flowers may wilt or curl before you can use them. You can always go back for more. Try, at least in your early attempts, for flowers and leaves that will lie as flat and close to the egg as possible. These leave a more definite imprint.

Remove the eggs you will use from the refrigerator and run them under warm water for a few minutes to bring them to room temperature. Cold eggs may crack in the boiling process and spoil your efforts. Choose white eggs as large as possible. Some shells may not take up color as well as others, but this does not happen often.

Now, lay a large soft bath towel on the table over which you will work. Have a pair of scissors handy for cutting cloth or plants as you need them. If you are right-handed, hold an egg in your left hand, slipping the cloth strip under the egg slightly so you can grip it with your fingers. Lay a bit of fern, leaf, or flower on the cloth and fold it upward so that it pressed securely against the egg.

As you lay each bit of flower or fern against the egg, pull the cloth over it and hold firmly. Then put on another flower or leaf, continuing in this manner until the egg is completely covered. Give a half-turn twist to the cloth when you have gone around the egg once (preferably lengthwise) so that you may also place a bit of plant on the ends. Do not be dismayed if you find this procedure awkward at first. With a little practice, even children become adept. You may not be able to cover your first few eggs completely with flowers, nor is it even desirable to do so, for the contrasting white is what makes the eggs so pretty.

After the egg is wrapped with the 1-inch-wide strip of cloth, gently tighten the wrappings by going over them once more with the narrow cloth or twine. This is to prevent the covering from coming off in the coloring bath.

Use a slotted spoon to insert the eggs gently in the warm broth. Make sure they are well covered. For a 4-quart saucepan, only cook three to five eggs at one time. After the eggs have been placed in the broth, cook them as you would any hard-cooked eggs. Heat the broth slowly, letting it simmer for 8 minutes so that the shells take up as much color as possible.

When the time is up, take the eggs out one at a time with your slotted spoon. Flick any clinging onion skins back into the saucepan. Place the eggs in water that is at room temperature or slightly warmer, and allow them to cool briefly until they can be conveniently handled.

After a few minutes, change the water to cool the eggs more rapidly, or add more cool water. Or have two pans handy and just slip them into the other one as a new batch is made.

Raw or cooked eggs that are unrefrigerated for more than two hours (an hour and a half on a warm, humid day) should not be eaten, so you need to work carefully, but as quickly as possible.

The wrappings should begin to slip off easily, but if necessary, untie them, pull off the cloth, and discard. (If you are pressed for materials, you can rinse them out in clear water, hang them up to dry, and reuse.)

After the wrappings have been removed, place the eggs on a couple of layers of absorbent paper to dry thoroughly. While they are still warm, add a bit of glamour by rubbing them with cooking oil, one that is not sticky. Use a soft cloth for rubbing. You will be delighted with the added shine, which enhances the beauty of the coloring and the brown or reddish brown background.

These eggs, with their lovely, shadowy imprints, are perfect when used as centerpieces in low bowls or trays, or in a pretty, brightly colored basket.

If you wish, write names of these "flower eggs" with a wax pencil instead of, or in combination with, the flowers before cooking them in the broth. You may also cut out small pictures of thin (easily bendable) cardboard or heavy paper and wrap them on the eggs along with the grasses or leaves. Tiny bunnies and chicks are special favorites.

You need not have any qualms whatsoever about letting youngsters enjoy these eggs. Even if an egg should crack in the cooking process and a little color get on the egg, it is perfectly harmless.

The eggs can be made several days in advance of the time they are needed. After they have been cooked, return them to the egg carton and store them in the refrigerator. They will keep just as well as any other hard-boiled eggs. The oil film may dull a little when the eggs are cold, but it will quickly become glossy again when they are taken out and returned to room temperature.

Many other natural materials can be used for coloring eggs. Some people use beet juice or even coffee grounds. In Russia the pasqueflower, *Pulsatilla vulgaris,* which imparts a green color, has been used to color Paschal or Easter eggs. In England, furze or gorse, a shrub with yellow flowers, has been used. Perhaps you will be inspired to experiment with other flowers and plants from your own garden.

COSMETICS AND FRAGRANCES

BEAUTY PREPARATIONS FROM THE GARDEN

HUNGARY WATER

Queen Elizabeth of Hungary, an eternally youthful beauty, attributed her marvelous looks to an herb tonic that became known, in her honor, as Hungary Water. Here is how it was made:

Hungary Water

12 ounces rosemary

1 ounce lemon peel

1 ounce orange peel

1 ounce mint

1 ounce balm

1 pint rose water

1 pint spirits of water (pure alcohol, grain alcohol, or vodka)

Mix together and let stand for several weeks. Then strain and use the liquid to rub into the skin after bathing.

This recipe, too, is reputed to be that of the queen. This tonic lotion was once made and used by ladies everywhere. Not only was it used for perfumery but also a few drops taken internally were recommended for nervous ailments and mental depression. Here is the recipe:

Queen of Hungary's Water

2 tablespoons dried rosemary flowers

1 nutmeg, grated

2 teaspoons cinnamon

1 tablespoon sweet cicely leaves (if available)

1 quart pure alcohol (grain alcohol or vodka)

Pulverize all the dry ingredients (a mortar and pestle is helpful) and mix well together. Add the alcohol. Let the mixture steep for 10 days. Then strain off and bottle. Apply on cloths wrung out in cold water, and place over forehead to allay headaches and sooth fevers. With fevers, also apply to the pulse of the wrists.

ROSES FOR BEAUTY

Gather roses (wild or old-fashioned roses are best) when dew has dried but before the sun becomes warm. The green or white base of the petals, known as the heel, should be clipped off, as this has a bitter taste. Press petals between two sheets of paper toweling to absorb moisture.

Glycerine and rose water, beloved of 19th-century ladies and responsible for many a beautiful skin well into old age, is again becoming popular; several of the leading cosmetic firms are now offering it. Other preparations such as Rose Milk are widely advertised. Here is a recipe:

Rose Hand Lotion

Soak ¼ ounce of tragacanth in water for four to five days. Mix 2 ounces of glycerine with 1 ounce of alcohol and add to the strained solution of tragacanth with ¼ to ½ ounce of rose water and 1 pint of water. If lotion needs thinning, add more water.

Rose water. Rose water may be purchased at herb or specialty food shops. But if you want to make your own, you can easily do so.

Rose Water

1 teaspoon rose extract

12 tablespoons distilled water

Measure carefully and use only distilled water. Mix liquids thoroughly and bottle, storing in a cool, dark place.

Rose vinegar. This can be added to warm water for a hair rinse after shampooing, or a cupful can be used in bathwater. It can also be wrung out of a cloth and placed on the forehead as a headache remedy.

Rose Vinegar

1 pint white vinegar

1 cup fragrant rose petals

Pinch of rosemary or lavender

Boil the vinegar and pour it over the rose petals. Add the rosemary or lavender. Cover tightly, and let stand for 10 days. Strain and pour into sterilized bottle.

Cook only in stainless-steel, enamel, or glass pans and stir with a wooden spoon.

Rose facial mask. This treatment is especially helpful for oily skin. The ingredients are ⅔ cup finely ground oatmeal, 6 teaspoons honey, 2 teaspoons rose water. Blend oatmeal and honey until well mixed. If desired, add more honey to make a smooth paste, blending it with rose water. Spread over clean face and neck. Leave on for 30 minutes. Best results will be achieved if you lie down and relax. Using a soft washcloth and warm water, remove and follow with cold water or astringent.

Rose soap. Save odds and ends of leftover hand soap and grate or cut up finely. Add hot water that contains about 6 drops of rose oil (see Fragrant Delights, in this chapter), using enough to cover. Place over a low flame until the soap dissolves. Pour into a clean cream carton to a depth of 1 inch and set aside to harden.

SKIN CARE

Elder flowers added to steam baths will clear and soften the skin. Freshly crushed leaves or freshly pressed juice of lady's-mantle (*Alchemilla vulgaris*) is helpful against inflammation of the skin and acne, as well as freckles. Externally used lime flowers (*Tilia*) stimulate hair growth and are a fine cosmetic against freckles, wrinkles, and impurities of the skin. Aloe vera is now widely used in many skin preparations, and jojoba preparations are also becoming popular.

You can grow the very versatile luffa gourd for washcloths and bath sponges. This beauty treatment is centuries old in the Orient.

Use the oil of sesame seeds or the juice of lemon and cucumber to soften and whiten the skin. Wheat germ oil and liquid lecithin (from soybeans) are believed helpful against lines and wrinkles.

Fragrant Delights

EAU DE COLOGNE

Incense is predominantly of plant origin, a mixture of sweet-smelling gums and balsams. It burns with a delicate fragrance. The early Egyptians burned it at religious ceremonies; the Greeks and Romans and later the early Christians adopted this practice. The burning of incense is still part of the ritual of the Eastern Orthodox Church, the Roman Catholic Church, and some Episcopalian churches. Buddhists also burn incense at religious ceremonies. And it is often burned in homes to give fragrance to a room.

Incense ingredients of ancient civilization and the early Christians were *Frankincense galbanum,* myrrh, mastic, rosemary, oplopanax, and storax.

Incense ingredients of oriental nations were cinnamon, cloves, camphor, dragon's blood, galbanum, sandalwood, and star anise.

LEAF ODORS

Leaves hold their aromatic scent far longer than flowers. Often they are sweeter in a dried state than when fresh. To release the odor of herbs and leaves, grind them in a mortar using a pestle. Mortars and pestles can be found in specialty cookware shops, and sometimes in herb and health food stores.

MOTH REPELLENTS

If you detest the odor of commercial moth repellents, try this one from Euell Gibbons's *Stalking the Healthful Herbs.* You'll need 1 pound pine needles (the needles of western pinion pine are best, he advises), 1 ounce cedar shavings, and ½ ounce of shaving from the root of sassafras.

Line a drawer with paper, sprinkle in the mixture, and cover with a thickness of cloth, something like a piece of an old sheet or a thin bath towel, fastening it firmly in place with thumbtacks. Store your woolens on top and they not only will be protected from moths but will also have a clean, fresh fragrance when you take them out to wear. Feverfew, sage, tansy, and members of the Artemisia family contain camphor and are also moth repellents.

A wide variety of plants have been used to repel moths, including oil of cade (*Juniperus oxycedrus*), lavender, costmary, wormwood, and clove; leaves of fennel, patchouli, sweet flag, fern, bracken, and rosemary; flowers of the male breadfruit tree; black pepper; Irish moss, citron, alcoholic solution of coumarin and hemp; extract from broom seed, cichona, lupine, tung oil, and elecampane. The wood of cedar has long been recognized as a moth repellent.

In addition to serving as moth repellents, botanicals have been found useful against other insects that destroy cloth. These include camphor and powdered clove against carpet beetle larvae. Clothing has been treated with a soapy emulsion of anise oil or bayberry oil to ward off insects.

PILLOWS

Flowers and herbs for scented pillows were once very popular. The delicate fragrance of herbs or blossoms is released when pressure is put on the pillow. The herbs may be mixed with the pillow stuffing, used entirely as a stuffing, or put into sachet bags and placed inside the pillow. Scented pillows can be made from calamus, lavender flowers, lemon verbena, meadowsweet, orrisroot, rosemary, rose geranium, sweet fern, or woodruff. Pillows stuffed with white pine needles are delightful. (See also *Allheal* in the Flower Lore chapter.)

POMANDER

Take a small, thin-skinned orange and press whole cloves into it until the surface is entirely studded. Roll the orange in powdered orrisroot and powdered cinnamon, patting on as much as possible. Wrap in tissue paper and place in a dry, well-ventilated area for several weeks. Remove paper, shake off surplus powder, and the pomander is ready for use. Hang by a ribbon in a closet where it will share its fragrance and aroma for many weeks. Lemons, limes, and clementines are also used for this purpose.

POTPOURRI

There are many recipes for potpourri. This is one of the most delightful.

Gather as many as possible of the following kinds of scented flowers as they become available: petals of the pale red and dark red roses, moss roses and damask roses, and acacia; heads of pinks, violets, lily of the valley, lilacs (blue and white), orange blossom and lemon blossom, mignonette, heliotrope, narcissis, and jonquils; a small proportion of the flowers of balm, rosemary, thyme, and myrtle. Spread materials out to dry.

As the flowers become fully dry in turn, put them into a tall glass jar, with alternate layers of coarse salt (uniodized) mixed with powdered orrisroot (use 2 parts of salt to 1 part orris). Pack the flowers and salt-orris until the jar is filled.

Close the jar for 1 month, then stir all up and moisten with sufficient rose water (see the recipe in this chapter) to penetrate to the lowest layer. Cap with a muslin cloth, tightly tied, and use a cotton bag when wanted to scent drawers, linen closets, and clothes hangers.

ROSE, ATTAR OF

A tiny, 1-ounce copper vial of greenish yellow fluid—the essence of roses used in the world's most expensive perfumes—is more valuable than gold! "Treated like a magic potion," according to Cyril Williams, a British perfume expert, "it's kept locked in bank vaults and fireproof, temperature-controlled safes. Small containers are insured for thousands of dollars."

The precious liquid concentrate (it takes over 100,000 roses to produce just 1 ounce) comes from the unique Valley of the Roses, where soil and climate have combined to make the finer rose scent in the world.

The rose grown in the Kazanlik Valley, Bulgaria, for the purpose of making attar of roses is *Rosa damascena* Trigintipetala. It grows 3 to 4 feet tall and has one annual flowering. It has semidouble, rose red, 3½-inch flowers with stiff yellow stamens. This rose also makes wonderful potpourri.

The Turks brought these roses from Damascus when they conquered Bulgaria nearly 600 years ago. When they departed in 1878, they left them behind. "From a distance the fields look like a pink sea," said Valentine Ruskov, a Bulgarian trade attaché in London. "The Turks used to take baths in rose-scented water and before long the distillation of attar became a cottage industry." Ruskov explained that the roses are harvested in May and June before daylight, "so the sun doesn't dry the petals, which are boiled to remove the essence." He went on to say, "Other countries have tried to duplicate the essence, but have not been successful." For home use, try the following:

Rose oil. Steep rose petals in a bland oil, such as mineral oil (or purchase essential oil of rose from a craft supplier or health food store). The petals may also be put into a crock and covered with water. Keep in a warm place and a little oil will rise to the surface. Collect it on a piece of dry cotton and squeeze into a bottle. This oil is attar of rose.

ROSE JAR

The great rose for making potpourri is also the rose used for attar of roses, *Rosa damascena* Trigintipetala. It is available in this country. Plant garlic or onions with your roses; they not only are protective but actually also increase rose fragrance when grown nearby.

Roses are the most fragrant in the sunniest, most protected spot in the garden. It is there that they develop their essential oils in the highest degree. Collect the flowers before the sun is high, on a dry day after two or three days of dry weather. Never use inferior, rain-soaked blossoms or those that have been open for a few days. Also, fragrant oils will not be present in the petals of flowers that have been in the house for a week.

Gather damask rose petals when the roses are blooming abundantly. Pack them in a glass jar that has a tight cover. The addition of tiny pink buds of 'De Meaux' will make the final product prettier and more highly scented. Between every 2-inch layer of petals sprinkle 2 teaspoons of salt. (Use common uniodized salt.) Add more layers of petals and salt each day until the jar is full. Keep in a dark, dry cool place for 1 week. Then spread the petals on a paper towel and loosen them carefully.

Mix the following ingredients thoroughly and mix well through the petals in a large bowl: ½ ounce violet-scented talcum powder, 1 ounce orrisroot, ½ teaspoon mace, ½ teaspoon cinnamon, ½ teaspoon cloves, 4 drops oil of rose geranium. Add the following very slowly: 20 drops eucalyptus oil, 10 drops bergamot oil, 2 teaspoons alcohol. Repack the mixture in the jar, cover tightly, and set aside for 2 weeks to ripen. It will then be ready for distribution into rose jars, which make wonderful birthday or Christmas gifts.

Some of these ingredients are readily available at most supermarkets, but you may have difficulty finding others. Indiana Botanic Gardens and Nichols Garden Nursery are possible sources, as are health food and herbal stores.

TRADITIONAL REMEDIES FROM THE WORLD OF PLANTS

ARTHRITIS

Eating the right fruits, vegetables, and seeds sometimes helps alleviate the pain of arthritis. These include alfalfa, tea brewed from alfalfa seeds, asparagus, celery, cherries, collards, fennel, gooseberries, kale, lemon juice, lettuce, limes, melons, molasses, mustard greens, oranges, spinach, sunflower seeds, tangerines, and watercress.

EYES

Various Native American tribes developed herb and flower remedies for sore or strained eyes. Plants are often named according to the way they were used. The prairie zinnia (*Zinnia grandiflora*) was known as "put into eyes" because the Zuni crushed the flowers in cold water and used the strained liquid as an eyewash. "Wash eye teas" is a native name for the wahoo (*Euonymus atropurpurea*).

Several tribes made infusions for sore eyes from either St.-John's-wort or St.-Andrew's-cross. Other mild infusions were made of any one of the following: alder, bark, bearberry leaves, yarrow herb, chickweed herb, blackberry root, black oak bark, chokecherry bark, or bark of buttonbush. Ginseng root was also used for sore eyes, and pounded root being soaked in cold water and then strained.

In the South, buds of sassafras were gathered and placed in cold water; the mixture then was allowed to stand several hours in the sun. The glutinous substance formed was used in treated sore eyes.

Herbalist Jethro Kloss recommends a poultice of slippery elm applied (cold) to the eyes to relieve inflammation. Other herbs he recommends for sore eyes are rosemary, borage, chamomile, chickweed, elder, fennel, gold-

---

Strewing Herbs

In northern countries, before rugs were commonly used, the floors of castles and churches were strewn for warmth with various organic materials such as rushes. Herbs, called strewing herbs, were popular, and lavender, thyme, *Acorus calamus,* the mints, basils, balm, hyssop, and santolina were widely used. Marjoram, believed to be an antiseptic, was scattered over church floors at funerals.

---

enseal, hyssop, rock rose, sarsaparilla, sassafras, witch hazel, wintergreen, yellow dock, plantain, tansy, white willow, and angelica.

FATIGUE

The Aztecs breathed flower fragrances to ease fatigue. They also recognized melancholia and loss of memory as diseases. The fragrance of flower concentrates and other ingredients calculated to retain the delicate aroma were used in body massage. A very similar preparation of flowers was also taken internally.

GINSENG, AMERICAN (*Panax quinquefolius*)

Ginseng is the most expensive botanical in the entire vegetable kingdom and surpasses even the truffle as a precious aphrodisiac. Ginseng has

been known and prized in Asia for centuries. People believe that ginseng promotes long life and virility, and have used it as a cure for many ills. Americans who use the root praise it for its tranquilizing qualities. Recently, clinical and biochemical studies have found that ginseng has a beneficial, estrogen-like effect on women.

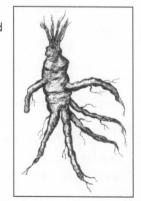

Ginseng also stimulates the nervous system, making one feel more active, more aggressive, more interested.

Ginseng

GINSENG, SIBERIAN (*Eleutherococcus senticosus*)

The plant is so called because *senticosus* means "prickly" in Latin. One day, it is said, a young Russian doctor named Gorovoy noticed deer greedily eating the leaves of a thorny plant commonly found growing wild in the Russian Far East. It belonged to the Araliaceae family and shortly joined the other araliaceous medicinal plants on trial in the laboratory. It outshone them all, rivaling or even surpassing ginseng itself. It was found to be a first-class tonic plant medicine, greatly increasing stamina in long-term administration.

Although similar in pharmacological effect, *Eleutherococcus* was found to have advantages over ginseng. Under the stress of hard physical work, both humans and animals were found to receive greater stimulation for longer periods with *Eleutherococus* than with ginseng. In hyperactive individuals it seems also to possess some calmative effect.

Actually, the name Siberian ginseng is a misnomer, for it is an entirely different plant, though having much the same effect. It is sometimes called eleuthero to make the distinction. Its qualities were not discovered until 1962 and there is no mention of it as a Russian folk medicine. It is an example of that rare species, a new national plant medicine discovered by modern research. Since the plant grows abundantly, it is far less expensive to buy than the rarer ginseng.

HAY FEVER

Knowing what and when to avoid is helpful. The eucalyptus tree spreads its misery from January through September. March, April, and May are particularly rough periods for those allergic to sycamore, English walnut, and live oak. Springtime also brings the cottonwood fuzz. Bermuda grass can keep you down from March through November. April and May are the toughest grass months, with pollen from fescues, rye, and Kentucky blue keeping your eyes watering. Johnson grass hits later, from May through August.

Weeds of all types pollute the air, starting in May and running through October. Ragweed's worst two months are August and September—and don't blame the innocent goldenrod.

Benign and Evil Herbs

Down through the centuries, some plants were thought to have a benign effect while others brought evil in their wake.

The "good" plants were southernwood, rosemary (effective against witchcraft), lavender (against the evil eye), bracken, ground ivy, maidenhair fern, dill, hyssop, agrimony, and angelica. Yellow and green flowers growing in hedgerows were believed to be especially disliked by witches.

Other herbs had the opposite effect. The herbs to call up evil spirits were vervain, betony, yarrow, and mugwort. St.-John's-wort was used to exorcise them.

How herbs were used made a lot of difference, as did the combinations. It was believed that if coriander, parsley, hemlock, liquid of black poppy, fennel, sandalwood, and henbane were laid in a heap and burned together, they would call forth a whole army of demons.

Besides flowers, trees, weeds, and grasses, there are some common year-round environmental allergy producers, including cat hair, cattle hair, chalk, dog hair, glue made from animals, horse hair, house dust, newspaper, sheep wool tobacco, and feathers. Foods that are most often allergy producers are wheat, celery, chicken, cow's milk, chocolate, eggs, oranges, peanuts, strawberries, tomatoes, and cantaloupe.

HEADACHE

The Romanies have a number of interesting flower and herb remedies:

1. A few pieces of willow bark boiled in a pan of water is helpful. It contains salicylic acid, the ingredient found in many over-the-counter headache medicines.

2. For nervous headaches, a tea made from the flowery tops of rosemary with boiling water is soothing.

3. A tea made from a few dried lime flowers cures a headache in about half an hour. Take hot, then lie down for 30 minutes and relax.

4. For a severe headache, put a pinch of dried marjoram into a teacup. Half fill it with boiling water, cover and allow to draw, and drink while hot.

HERBAL REMEDIES, TYPES OF

The information given below is not intended to replace a physician, but for those interested in herbal remedies, some simple definitions may be helpful:

Alternatives improve nutritive processes gradually, normally, and naturally. Take in the form of tea. Peppermint tea is a "for instance"; it has been used for relieving colic in babies or minor bloat in children and adults.

Antispasmodics reduce involuntary contractions often arising from nervous causes. *Chamaemelum nobile,* or true chamomile, is notable for giving relief. Chamomile flower tea has been famous since the days of the pharaohs and was as popular in past times as aspirins are today.

Carminatives and aromatics are usually herbs with a spicy scent and pungent taste useful for expelling gas from the stomach and thereby reducing flatulence. Mint, a mild carminative, has often been chewed after dinner. Other herbs traditionally used separately or together are anise and caraway, cloves, dill, ginger, and other aromatic spices. Their delicious fragrances also provide a lift to our spirits.

Demulcents soothe the intestinal tract and are usually of an oily or mucilaginous nature. Olive oil is a natural demulcent. Emollients, similar in meaning, are used to soothe the skin rather than internal membranes, often allaying the pain of irritated parts. Irish moss and slippery elm are considered both demulcent and emollient.

Diaphoretics produce sweating. Sweat baths have been used by many peoples over the centuries (the Finnish sauna, in which water is thrown on heated stones, is a familiar example). Tea of yarrow (*Achillea millefolium*) induces sweating and formerly was used as a cold remedy. Dried elder flowers and mint are both believed to be helpful when added to the brew.

Diuretics stimulate the secreting cells or nerves of the kidneys, increasing the flow of urine. Herbs valued as diuretics are dandelions and juniper berries; lemon juice and milk are also mild diuretics.

Expectorants help in the expulsion of mucous secretions from the air passages and are used to clear phlegm that accumulates in the lungs and windpipes. Yerba santa, red clover tea, and valerian are examples; many more herbs are also considered helpful.

Sage, one of the easiest herbs to grow, is commonly listed both for cooking and for medical uses. Its name comes from the Latin salveo, meaning "to save or heal." Recent research suggests that it should not be taken internally for prolonged periods, however.

Febrifuges are agents that reduce fevers. Dogwood and boneset, sometimes called feverwort, were used for this purpose.

Laxatives are taken to relieve a temporary condition of constipation, but any plants that are cathartic in action should be used with care. Dandelion greens are a tonic and mildly laxative.

Nervines act on the nervous system to overcome irritability. The most notable plant for this purpose is valerian.

Sedatives help quiet the nervous system without producing narcotic effects. The bark of the wild black cherry is added to cough medicines because of its sedative action.

Stimulants quicken vital action and digestion, raise body temperature, and increase general awareness. Culinary herb stimulants are anise, pepper, cinnamon, cloves, dill, ginger, horseradish, nutmeg, peppermint, and sage. Medicinal herbs include horehound, hyssop, lavender, lobelia, marjoram, and spearmint.

Stomachics improve stomach activity and stimulate appetite. Spearmint is an excellent example. Chokecherry, often used for making brandy, is another.

Tonics tone up our bodies and give us a feeling of well-being. They are often referred to as bitters. Chamomile has been used in this way, and so have feverwort, chicory, goldenseal, and *Verbena officinalis*.

HIGH BLOOD PRESSURE

Fruits and vegetables thought to be helpful in reducing high blood pressure are broccoli, carrots, cauliflower, celery, cherries, cranberries, cucumbers, endive, fennel, garlic, grapefruit, guavas, kumquats, melons, oranges, peaches, pears, peppers, pineapples, pomegranates, raspberries, spinach, strawberries, tangerines, and turnip tops.

HOARSENESS

This is a Romany remedy: Take a good-size turnip, wash it well but do not peel. Then cut a piece from the bottom so that it will stand upright, and cut it downward in four equal slices. Fit the turnip together again to stand up in a deep dish or soup plate, having first added a layer of honey between the slices.

When the turnip has been left standing for an hour or two, a thick syrup from the turnip juice and the honey will have formed in the bottom of the dish. This can be taken a spoonful at a time.

TRANQUILIZERS, NATURE'S OWN

Pleasant at any time, herbs are particularly appropriate at times of emotional stress. Herbs believed to be exceedingly soothing to the nervous system are chamomile, valerian, rosemary, and lavender.

For an herbal bath, put the herb or herbs on a muslin square and tie them up like a hobo pack. Toss the packet into the bath as the water runs in and allow to steep for at least 10 minutes. Also good are pine needles, fresh from the tree, wrapped and used like the herbal bath bags. And, while you are relaxing, drink a cup of warm chamomile tea, one of the oldest nerve-calming teas beloved of herb enthusiasts—far better for you than tea or coffee. Verbena tea is also highly thought of.

In his book *Back to Eden,* Jethro Kloss has some additional suggestions: celery, dill, fitsroot (skullcap with goldenseal and hops), lobelia, motherwort, origanum, pennyroyal, red clover, rosemary, rue, sage, spearmint, St.-John's-wort, thyme, verbain, wild cherry, wood betony, blue violet, sanicle, buchu, red sage, catnip, peppermint, marshamallow root, and mugwort (used in an antispasmodic tincture for quick results).

WARTS

A number of plants are credited with eradicating warts. A Romany remedy is the thick white juice exuded by milkweed plants. Other traditional cures include the white, milklike sap of dandelions and the milky sap of figs.

MORE PROJECTS
FOR PLANT LOVERS

DOLLS
In colonial times corncob dolls were enchanting little toys that children loved.

Shell off the corn and dry the cob with the shucks left on. Then pull the shucks off and back for the "hair." Slit the shucks and make them into braids; or they may be curled by holding a strip of slit shuck between thumb and a knife and pulling the shuck through, just as ribbon is curled. Try decorating a small Christmas tree with tiny dolls made from midget corn.

When digging potatoes, watch for those with unusual shapes; they often lend themselves to roly-poly comic characters or animals. Carrots with a little trimming can be made into dolls as well. Dolls with apple heads have been made for centuries; some of them are truly works of art.

DOWSING
To find out if you are a water witch, take a small tree branch with a fork in it—a branch shaped something like a wishbone. Hold the two ends of the forked branch with the palms of your hands facing up. Walk slowly with the branch pointed straight up. When you approach a source of moving water, even through an underground water pipe, you will feel a definite pull on the branch as it begins to move toward the earth. When you are directly over water (if you have the power of a water witch), the branch will insist on pointing straight downward even though you are trying your best to hold it up.

Part of an old tradition, water witches are still called on to use their divining rods. Some can even get results by holding the rod over a map of the area. Most water witches say that a forked stick from almost any tree will work if you have the power, but the preferred species are peach, apple, willow, and maple.

FISHING WITH HERBS

Oil of anise rubbed on bait will attract fish, and so will the juice of smallage or lovage and the steeped root of sweet cicely.

FLORAL CLOCK

Certain flowers, called sensitive flowers, have a natural "clock" that causes them to open and close at the same or nearly the same time each day. This effect can be used to create an interesting garden project.

The great Swedish naturalist Karl von Linné (born 1707), better known as Carolus Linnaeus, composed a floral check. One could determine the time of day by the opening and closing of certain flowers, which folded and unfolded their petals at regular hours. Here are a few of those that served for the construction of his dial:

Dandelion opens from 5 to 6 A.M. and closes between 8 and 9 P.M.; mouse-ear hawkweed opens at 8 A.M. and closes at 2 P.M.; yellow goat's beard opens at sunrise and shuts at noon; smooth saw thistle opens at 5 A.M. and closes at 10 A.M.; white water lily opens between 5 and 6 A.M.; mallow opens at 9 to 10 A.M. and closes at 1 P.M.

Another form of floral clock was popular during the 1800s. Some flower lovers planted sensitive flowers in U-shaped gardens. One point of the U was planted with the spotted cat's-ear flower, which opens at 6 in the morning. Then five more different flowers were planted in the row, each kind opening one hour later than the one before it. The passionflower, opening at noon, was placed at the center of the curve of the U. From this curve to the other point of the U, six other kinds of flowers were planted in a row. Each one closed one hour later than the flower before it. The last flower was the evening primrose, which opened at 6 P.M. in the evening. Though not very exact, of course, the flowers' internal clocks made them quite regular and predictable.

FLORAL PRINTS

Gather thin, light, colorful flowers (pansies, petunias, columbines, buttercups) just before they come into full bloom. Brush off the pollen and press the blossoms between sheets of absorbent paper (such as paper toweling). Place a heavy object on top. Replace the paper after 8 hours, and again after the next 8 hours. Then leave the flowers for 2 or 3 weeks.

Place your pressed flowers on construction paper in any design you wish. No glue is necessary, as the pressure of the glass will hold them in place after they are framed. Ferns also press well.

Pressed like this, flowers can be used to decorate Easter eggs. Touch them lightly on one side with white glue and gently press them on the egg.

Another method is to press flowers between sheets of waxed paper and touch them lightly with a warm iron to seal. These make attractive greeting cards between two sheets of folded notepaper.

To make an ink print of dried ferns or flowers, use a blockprint roller. Ink the roller and roll back and forth over the material to be used. Then place in the desired arrangement on construction paper and roll a clean roller over the design. Different colors on the same arrangement give an interesting pattern.

GROCERY PLANTS

Check your grocery bag for some good winter growing projects. While such plants started from seeds cannot be depended on to produce "true" fruit, or any at all for that matter, they are fun to grow. Lemons, oranges, grapefruit, and other citrus fruits are good for a start, and sweet potatoes make an attractive, interesting vine.

IVY TREE (*Hedera*)

Many plants change their style of vegetative growth from prostrate to erect at flowering time. This is common among annuals and perennials, but rare in vines. In ivy, the flowering process triggers the formation of erect branches on which the leaves are different from those on other parts of the vine—narrow and lance-shaped.

Ivy vines flower only when about 15 years old. The flower head is a cluster of tiny cream-colored flowers. If cuttings are made of the erect flowering branches before the flowers are produced, the erect form is retained by the plant that develops from the cuttings.

Take the cutting below the tip at the bud on a young woody stem. Insert the cutting into a loose perlite and peat moss mixture; roots will quickly form. After roots have formed, pinch back the stem to produce an interesting form with side branches. The variegated *Hedera canariensis* with dramatic green and white branches is a wonderful subject for this horticultural wizardry. Remove any leaves that revert to the vine habit.

JACK-O'-LANTERNS ON THE VINE

Pumpkins are fun to grow and fun to eat. In pioneer days they were sliced and hung from cabin roofs to dry for winter storage. They were made into soup, stew, pudding, bread, griddle cakes, and a thick sauce, as well as pie.

Pumpkins are pretty in flower and bright in fruit. To decorate them while still on the vine, start with the seeds. Choose your variety for the size you want; small pumpkins are best for small children.

As pumpkins ripen, they turn from green to yellow-orange. While they are still green but have almost reached full size, take a paring knife and carve the jack-o'-lantern face, or a child's name, or any other design, through the rind and into the flesh about ⅛ inch deep.

In a few days a callus will form along the cut lines, and the design will begin to rise up in a distinct pattern, turning light-colored against the orange skin after the pumpkin ripens. No harm comes to the interior of the fruit. Near frost time, harvest the decorated pumpkin as usual.

JEWELRY

"Beads" from the garden are a delightful bonus from many colorful flowers and their seeds, such as ornamental corn, sunflowers, and Job's tears plants. But rose-petal beads, with their sweet, mild fragrance, have always been a great favorite. Once these were much in demand for rosaries as well as necklaces, and very lovely examples were produced with contrasting mountings of either gold or silver. Here is how the beads were made.

Rose beads for a rosary. In an enamel pan, heat 1 cup of uniodized salt with 1 heaping cup of rose petals firmly packed. When this has been mashed together, stir in ½ cup of water. Add a drop of oil paint for any desired color, or omit if natural color is preferred. Reheat over low heat, stirring constantly until smooth. Roll out to ¼-inch thickness, cut with a thimble, and roll each bead in the palm of the hand until smooth and round. As each bead is rolled, string it on #24 or #26 florist's wire. Hang in a dark place until dry, then string on dental floss. Move the beads occasionally while drying to keep them from sticking.

Rose beads for a necklace. Put 1¾ cups of flour and 4 tablespoons of salt into a bowl and add a little water to make a smooth dough. Into this, press 3 cups of rose petals that have been finely chopped. Flour a breadboard and roll the dough to about ¼-inch thickness. Use a thimble to cut the dough. Roll each circle in the palm of the hand to form a smooth bead. Follow above directions for stringing and drying. When stringing, add a crystal, gold, or silver bead after each rose bead.

A drop or two of rose extract or rose oil will add a delightful fragrance to either rosary or necklace.

PLANTS WITH PARTY TRICKS

Among the amusing flowers of the plant world is the mouse plant (*Arisarum proboscideum*). This useful, low-growing groundcover is related to the arum lily. The flower is in the form of a single spathe, which by some freak of nature takes almost exactly the shape of the rounded hindquarters of a tiny mouse with a long curling tail. Hold it in your hand as if it were a mouse; friends will probably reward you with gratifying squeals.

Silene armeria is one of the catchflies, related to the campions. The stems are divided into sections by nodes, below which is a dark, sticky patch. And they do actually catch flies and many other tiny insects.

Gayfeathers (*Liatris* spp.) is an upside-down plant. Unlike other flowers that grow in a spike, these herbaceous perennials bloom from the top first, the opening buds gradually working their way down the stem.

An easy perennial to grow is *Gaura lindheimeri,* 4 feet high, which when in flower looks like a cloud of white butterflies. The flowers, which have pink buds, have only four petals. Each petal spreads out like the wings of a butterfly, while the long stamens resemble antennae.

Dictamnus albus, the gas plant, gives off a volatile gas that on a hot, still day can be ignited by a match held near it. It will flare for a moment without the plant suffering any damage.

Obedient plant (*Physostegia*) demonstrates its obedience when you touch the flowers. They can be moved up, down, or sideways and will stay where they are put.

SHARING

Gardeners are generous people, and one of their greatest pleasures is sharing—an experience all the more delightful for both giver and receiver if the gift plant is unusual.

An unusual plant should be hard to find in the rank and file of nursery catalogs or garden stores, something not too well known, yet interesting and effective wherever it is planted. If the plant is easily grown and easily separated, so much the better.

Adonis amurensis, a very early-spring perennial, blooming sometimes before the snow has entirely melted, is just such a one. The feathery, much-divided foliage dies to the ground by late June and is forgotten after that; the following spring it peeps through even before the snowdrops with bright yellow, buttercup-like flowers. Divide the long, fibrous roots with a spade so that several buds are left on the clump of roots for starting the new plant.

Another gift possibly is the double-flowered bloodroot, a native American plant not, as yet, too well known. The large, white, double flowers appear in early spring not long after the adonis has bloomed. The plant grows well in shade.

Other good gift plants include *Narcissus asturiensis,* a miniature daffodil that flowers in the very early spring along with the snowdrops. A bulb that flowers in early summer is the golden garlic, *Allium moly.* It has yellow, starlike flowers in clusters up to 3 inches across. White-flowered scillas and grape hyacinths or truly white violets are also good conversation pieces, and these, too, increase rapidly, making them ideal for sharing.

TERRARIUMS

Terrariums fascinate and are not at all difficult to care for. You can make one in a glass container (almost any size) that closes with a lid.

Mix up soil for the planting base. A good combination for foliage plants and a general woodsy scene is 2 parts peat moss, 1 part perlite, 2 parts reasonably descent garden soil (or potting soil), and 1 part sand. You can buy a mix from a nursery and adjust according to the kinds of plants you will be growing.

Clean your container thoroughly. Scrub, rinse, and air if it has been used before. Put down the bottom layer of 1 inch or so of moist perlite (the exact depth depends on the proportions of the container). Sprinkle a thin layer of charcoal on top of the perlite. Add the soil mixture and shape the basic land construction with hills, valleys, or gentle slopes.

Tentatively place the plants in the terrarium to make sure that the living design is effective. Then plant your miniature garden. In large containers, you can sink the pots to their rims. In smaller terrariums, either unpot the new plants and shake their old soil loose, or set them in the new soil and tamp it around to the same level as the pot soil. Or tap the plant loose from its pot and plant the entire earth ball. The choice depends on the quality of the original pot soil and the individual plant's sensitivity to transplanting.

Plant ferns and small evergreen seedlings for the taller growth of the terrarium. Ivy, moss, and lichens are also attractive. For a grasslike "carpet" effect, plant partridgeberry. Place each plant carefully in the soil with enough space between plants to allow for growth. Set the completed terrarium in a light place, but not where sun will strike it. With the glass container closed, this "balanced terrarium" preserves temperature and moisture inside. Open the lid if the glass clouds with moisture.

VICTORIAN BOUQUET PLANTINGS

The Victorians were lavish with flowers. They planted large formal bedding gardens for viewing and for cutting, and grew quantities of flowers that they arranged in huge bouquets of many varieties. Many gardeners are now copying the Victorian-bouquet effect with living flowers, by mixing many plants in one planting container for a full summer of floral abundance.

Choose flowers of different shapes, sizes, and forms. Arrange so that tall plants are surrounded by less stately plants. Fill out with varieties that will tumble over the side of the hanging basket or planting container. Flowers growing like this create their own sense of compatibility and usually grow well together.

Choose a container with ample volume for supporting the growth of 10 or 12 plants of four or five varieties at a minimum. To keep the plants growing well, give them plenty of water and fertilizer, or use soluble fertilizer, or use soluble fertilizer as you water.

A wave of nostalgia is sweeping the country, and this is making Victorian gardens very popular. Victorian flower arrangements, too, are seen more often.

Plants to include in your Victorian bouquet might be pansies, browallia, *Felicia amelloides,* verbena, lobelia, dianthus, zinnias, begonias, ageratum, marigolds, vinca, torenia, petunia, impatiens, coleus, and cineraria. For height, choose from such winning plants as snapdragons, geraniums, Shasta daisies, and the taller varieties of African marigold.

VICTORIAN GARDENS

The Victorian-style garden is again becoming popular. This style provides for a garden of seclusion and natural beauty. Such gardens include lawn shrubbery (including the exotics), terrace walks, and even a conservatory if room permits.

WEATHER FORECASTERS

Old sayings or clues from nature often contain truths. Signs of a hard winter include unusually large crops of nuts or acorns, heavy moss on the north side of trees, sap of maple and sassafras going down early in fall, leaves of grapes turning yellow early in the season, and thick husks on the ears of corn.

When the flowers of scarlet pimpernel close during the day, it was believed to be a sign of rain. For this reason the plant also became known as poor man's-weatherglass.

Mushrooms and toadstools are said to be more numerous before a rain.

If the down flies off dandelions when there is no wind, it is a sign that rain is on its way.

Scientists are studying plants in areas known to be earthquake-prone, believing that certain plants indicate by their behavior when tremors are impending.

PLANTS AND PEOPLE

ALLERGY SUFFERERS

Linda Alpert developed an allergy-free demonstration garden outside the Tucson Medical Center's Allergic Clinic. Her garden shows that many attractive plants can be grown in desert areas without contributing to the pollen count. In addition, the plants she chose take very little water once established. Recommended trees include desert willow (*Chilopsis linearis*) and the lysiloma, sometimes known as the fern of the desert. Both are lacy-looking with attractive flowers and grow to about 25 feet high.

Cassia, jojoba, and Texas ranger (*Leucophyllum*) are among the shrubs of choice. Flowering groundcovers include the desert primrose (*Oenothera* spp.) and desert verbena (*Verbena wrightii*). For flowers almost year-round, there is blackfoot daisy (*Melampodium leucanthum*). As might be expected, there is also an assortment of cacti, agaves, and yuccas, many of which are gorgeous in blossom.

BUSINESS

Increasingly business offices are making use of green plants. Professional space planners and interior designers are entering the picture. Their first concern is function and efficiency, but they are making very effective use of healthy, thriving plants to soften the lines of functional architecture and, at the same time, subtly direct traffic, diffuse sound, or screen certain areas.

Selecting the wrong plants can be an expensive mistake, so consulting an interior landscape specialist is often money well spent. It may be best to let the specialist not only provide the plants but maintain them as well.

For the small office, the tried-and-true sansevieria, rubber plant, cacti, African violets, and ivies are good choices.

George Washington Carver (1864–1943)

Jerry Baker didn't originate the idea of talking to plants; George Washington Carver did and went one better. Farmers' wives brought Carver their ailing houseplants. He cared for them tenderly and sang to them. During the day he would take them out to "play in the sun." When he returned them to their owners, he would tell them gently, "All flowers talk to me and so do hundreds of little living things in the woods. I learn what I know by watching and loving everything."

As a botanist and chemist, Carver is best known for the range of uses he found for the peanut, but he has many other achievements to his credit. For the leaves, roots, stems, flowers, and fruits of various plants he coaxed 536 separate dyes that could be used to color wood, cotton, linen, silk, and even leather; 49 of them were produced from the scuppernong grape alone!

CHILDREN

Teach your children the principles of gardening. These may well be some of the most important skills they will ever learn. To start, keep things simple. Let children grow the veggies and flowers they like. You might even buy them some started plants to ensure the success that builds confidence in growing things. Counsel and encourage, aid and comfort, but don't do their work. And praise their efforts, especially when vegetables or flowers are brought to the table for the rest of the family to enjoy.

HOSPITAL PATIENTS

Sending cut flowers to hospital patients recovering from surgery is a thoughtful gesture, but, according to a British medical journal, it may lead to possible infection. A concentration of dangerous bacteria may grow in vases within 1 hour after flowers are put in water. And after

3 days, some of the bacteria are resistant to commonly used antibiotics. Flowers, the article continues, should be avoided in hospital units dealing with intensive care, burns, neurosurgery, and newborn babies.

HUNTERS

Some Native American hunters imitated the scent of the deer with roots and herbs. The roots of blue wood aster and others were used to make smoke to attract the deer so that they could be shot bow and arrow. Other plants used to attract deer were large-leaved wild aster (root smoked); Canada and Philadelphia fleabane (disk florets smoked); and swamp persicaria (flowers smoked).

To sharpen their powers of observation, some hunters drank a tea made of heal-all root (*Prunella vulgaris*).

Peoples of the Great Lakes region used botanicals for trapping and fishing. The smell of the root of alternate-leaved dogwood, boiled in water, was used to disguise muskrat traps. Mountain mint was used as a lure on traps to catch mink. Other fur-bearing animals were attracted by a wash made from the roots of kidney liverwort. Traps were boiled in maple bark to deodorize them so the animal would not detect the scent of previous victims.

The roots of wild sarsaparilla, mixed with roots of sweet-scented calamus, were boiled in water to make a lure for fish. Nets were then soaked in this brew; the scent would cling to the nets even after they were immersed for many hours.

LIVESTOCK OWNERS

Botanicals have been used to relieve domesticated animals of insects. Freshly cut pumpkin or squash leaves, a decoction of black walnut leaves soaked overnight, or an infusion of pignut leaves, rubbed on horses or cattle, will repel flies. Sometimes yellow wild indigo is placed on harnesses to keep horses free of flies. Concentrations of potato water rubbed on cattle, and clove on chickens and dogs, will repel lice, and a water solution of wormwood is used to bathe small animals and rid them of fleas. In Brazil, a tincture of cocoa leaves is considered a remedy for poultry lice, while cocoa shells, used as bedding for dogs, are credited with repelling fleas.

PERSONS UNDER STRESS

Gardening gives relief from tension, fears, and worries. Many people with stressful occupations—doctors, pilots, police officers, mothers of small children, teachers, and others—escape to their gardens whenever possible. Stresses disappear in the familiar tasks of preparing soil, planting, cultivating, and harvesting. The garden is a place of healing, not just a "factory" for producing food or flowers.

PERSONS WITH DISABILITIES

Gardening is wonderful therapy for the physically and mentally challenged. Children of all ages, including children with developmental delays, love to "make things grow." And wheelchair gardening can add interest to a life that may often be dull and monotonous. Even a box or raised bed of suitable height in which to grow flowers or vegetables (or both) can brighten the hours as seeds sprout, leaves unfold, and flowers bloom.

SENIOR CITIZENS

Some retired persons are born gardeners, some learn gardening after they retire, still others have gardening thrust on them. Many retirees have moved to places where they must learn about new soils, new climates, and even new types of plants. For born gardeners the new environment presents an interesting challenge.

Those who have never gardened before become gardeners as the result of encouragement by neighbors, friends, and relatives. And then there are those who have had gardening thrust on them as a way to pass the time or because social pressure has demanded that they "keep the place looking nice." If you asked these seniors, they would probably say they garden because they like growing things, because it takes them outdoors and gives them exercise, or because it's something pleasant to do.

The handicapped, too, find a way.

STAMP COLLECTORS

One of the most popular design topics for postage stamps the world over is flowers. Switzerland was the first country to bring the beauty of a flower to a postage stamp. For many years Switzerland issued a yearly colorful series showing its native flowers: edelweiss, alpine rose, slipper orchids, and many others. Inspired by the Swiss success, other nations including the United States followed suit; now almost every country that issues stamps has honored flowers on some of them.

STUDENTS

A survey has shown that the most popular plants for students are Swedish ivy, coleus, and spider plants.

THERAPISTS

Many professional therapists are advocating gardening to alleviate mental depression. They believe that gardening is a time-proven way to stay "alive and well," both mentally and physically. If you know someone who is lonely or depressed, you might write on their behalf for information from the American Horticultural Therapy Association (AHTA), 909 York Street, Denver, CO 80206, or call (800) 634-1603.

TRAVELERS

To traveling gardeners: Resist the temptation to dig up unusual plants. Certain plants may not be shipped to certain states, and there is a sound reason for this.

Whenever a nonnative species, plant, animal, or insect is introduced to a new environment where natural controls are not present, big trouble may lie ahead. Given the right conditions, it can spread like wildfire. Just look at the case of the Mediterranean fruit fly, the present worry over fire ants in certain southern states, and the so-called killer bees. The beautiful water hyacinth, now clogging southern waterways, is another case in point. By bringing plants home and possibly letting them escape from your garden, you may be creating a monster.

First Garden Book

Published in England in 1563, the first garden book, titled *A Most Briefe and Pleasaunt Treatyse Teachynge Howe to Dress, Sowe and Set a Garden,* was written by Thomas Hyll of London. "If you want your Parsley to be crinkeld or curled," he writes, "bruise the seed, or when it comes up roll small weights on it, or else jump up and tread it down with your feed." Botany and medicine, which in Hyll's time were identical, were just beginning to free themselves from the influence of superstition and witchcraft—influences that are frequently evident in Hyll's book.

From it we learn much about gardens at the time of Queen Elizabeth I. They were laid out formally with arbors and trellises. Mazes and knot gardens were popular, and often wells were includes "for water is a great nourisher of herbs." Beds were raised for drainage, and walks were sanded "lest by rayne or shower the earth should cleave and clogge on they fete."

WHEELCHAIR GARDENERS

Choose easy-to-grow plants that need a minimum of repotting. Terrariums, midget gardens, dish gardens, and bonsai are all practical choices for wheelchair gardening.

PLANTS OF THE NORTH

AKPIK, APPLEBERRY, CLOUDBERRY, SALMONBERRY (*Rubus chamaemorus*)

This perennial sends out erect shoots from a creeping rootstock. Its flowers are solitary and its terminal leaves have three to five rounded lobes with toothed edges. The fruit is red when unripe, amber color when mature. In fall it is collected by native people, who store large quantities for winter use. The berries are eaten like strawberries with sugar and cream or used in pie or shortcake. The berries are very high in vitamin C and if frozen will retain much of their nutrient value.

ANEMONE, NARCISSUS-FLOWERED (*Anemone narcissiflora*)

The anemone grows in open meadows, along hillsides, and on alpine tundra. Its flowers appear in clusters at the top of the stem, with the white petals often being tinged with blue on the back. The early-spring growth on the upper end of the root is eaten by Aleutian Island natives and has a waxy, mealy texture and taste. *Note:* Some members of this family contain the alkaloid anemonine, which causes irritation and inflammation in sheep that feed on it.

ASPARAGUS, BEACH OR SEA (*Salicornia*)

The stem is smooth, fleshy, and jointed, with opposite branches. The inconspicuous flowers are usually three, sunk into the fleshy hollow of the thickened upper joints. The plant grows on sea beaches in south-eastern Alaska around Prince of Wales Island and Ketchikan. It is available in summer. When young, plants may be used in salads or for pickles.

BROOK SAXIFRAGE (*Saxifraga punctata*)

The flowering stalk, 4 to 20 inches high, is hairy and leafless. Small flowers with five white or purplish petals form headlike or flattened open clusters at the top of the stem. Found in moist, rocky, shady places along rivulets, road-

sides, rocky cliffs, and gulches throughout southeastern Alaska, the Gulf
Coast, and westward on the Alaskan Peninsula, leaves for salad are collected
in spring before the plant flowers and are a good source of vitamin C.

BUMBLEE PLANT, WOOLLY LOUSEWORT
(*Pedicularis lanata*)
This perennial flowers in spikes that are pink to rose, although occasionally
they may be white. The entire plant except the lower leaves is densely gray
and woolly. It is common on the tundras of the high mountains and in the
Bering Sea district. Flowers are collected in June by the natives around
Cape Prince of Wales and Shishmaref. Water is added and the flowers allowed to ferment. The root is also edible and may be gathered in fall and
prepared by boiling or roasting.

BUTTERCUP, PALLAS' (*Ranunculus pallasii schlecht*)
This pretty plant, a native of Alaska, is found growing in the saturated
sphagnum moss at the shallow edges of tundra lakes and ponds. The natives of the lower Kushkokwim Valley use the young, tender, succulent
shoots, which are available in spring and autumn. They must be cooked
before eating to drive off the poisonous anemenol contained in the plant.

COLTSFOOT (*Petasites frigidus*)
The tawny-colored flowers, appearing before the leaves expand, are not
showy. The leaves are palmate or somewhat triangular in shape, green,
shiny above the felty beneath, and may become extremely large. Widespread, the plant is usually found on tundra. The young leaves are collected and mixed with other greens. Mature leaves are sometimes used to
cover berries and other greens stored in kegs for winter use.

COWSLIP, MARSH MARIGOLD (*Caltha palustris*)
The plant is found in marshy places along creek beds and ditches, in
swamps and wet meadows. The large plant, *Caltha asarifolia,* is abundantly found in southeastern Alaska and the coastal areas of the Gulf of
Alaska westward to the Aleutians. A much smaller and less leafy plant,

C. arctica, is found throughout the Yukon and Tanana River basins. The bright yellow flower may be borne singly or in clusters. The leaves and thick, fleshy, smooth, slippery stems are best when young and tender before the flowers appear. The raw leaves contain the poison helleborin, which is destroyed in cooking. The roots are long and white. When boiled, the usual method of preparation, they look somewhat like sauerkraut.

DOCK, ARCTIC; SOURDOCK (*Rumex*)

The flowers are green or tinged with purple, numerous, and mostly crowded in panicled racemes. The plant grows in wet, marshy places along riverbanks in Canada and Alaska. Its young tender leaves, an excellent source of vitamins A and C, make an excellent salad green and cooked vegetable.

INDIAN RICE, CHOCOLATE LILY, KAMCHATKA LILY (*Fritillaria camschatcensis*)

This perennial plant has a simple stem, 1 to 2 feet, arising from bulbs with thick scales. The flowers are one to six, large, nodding, bell-like, dark wine color—often almost black—tinged with greenish yellow outside, and have three petals. The bulb of large scales is subtended by numerous ricelike bulblets. The plant grows in open coastal meadows in southeastern Alaska, the Gulf of Alaska coast, and north to Talkeetna, the Alaska Peninsula, Kodiak Island, the Aleutian Islands, and the Bristol Bay area.

Bulbs are dug in fall; they are then dried and used in fish and meat stews or pounded into flour. They are used extensively by natives of southeastern Kodiak and the Aleutians.

TALL COTTON GRASS
(*Eriophorum angustifolium*)

The flower heads of this perennial develop into 2 to 12 nodding heads of white, silky bristles called cotton by Alaskans. Tall cotton grass is found on tundra bogs and wet roadsides. In autumn tundra mice cache the underground stem for winter use. Native Alaskans call these underground stems mouse nuts and sometimes eat them with seal oil.

GARDEN PLANS

In this chapter you will find a number of plans for creating gardens that showcase flowers. Feel free to depart from these plans as much as you wish, to make the garden truly your own inspiration.

The Fragrant Evening Garden

At night, when the moon lights the garden, the gaudy garden colors disappear and leave only brilliant white and pale pastel flowers to attract the eye. Then come the sweet fragrances to refresh us. Gardens are sweetest when the air is mild and moist; in heat and drought fragrant ethers are appreciably lessened. A frost will also set fragrance free, as does a shower of rain. The perfume of a plant is not always in its flowers. It may be in the root, seeds, bark, gum or oils, leaves, or stalks. When we are tired at end of day, it is a real pleasure to entertain friends or just sit quietly and enjoy the fragrance of our evening garden.

Aloysia triphylla **(lemon verbena).** This delightful, lemon-scented shrub has unremarkable flowers, but its leaves are lovely to pinch and smell.

Cactacea. There are a number of genera among the cacti that have fragrant flowers, and one of the sweetest is the night-blooming cereus (*Hylocereus undatus*), giving unforgettable pleasure on the night that it blooms.

Cestrum nocturnum **(night jessamine).** Hardy in the South, elsewhere as a greenhouse plant. This 6- to 9-foot West Indian shrub has insignificant flowers in the leaf axils, which produce an evening fragrance out of all proportion to their size.

Dianthus **sp. (carnation).** The very word *carnation* or *pink* connotes fragrance. It's wonderful for an evening garden as a terrace edging plant.

Hesperis matronalis **(sweet rocket).** Sweetest at night, this common and easily grown plant should be in every evening garden.

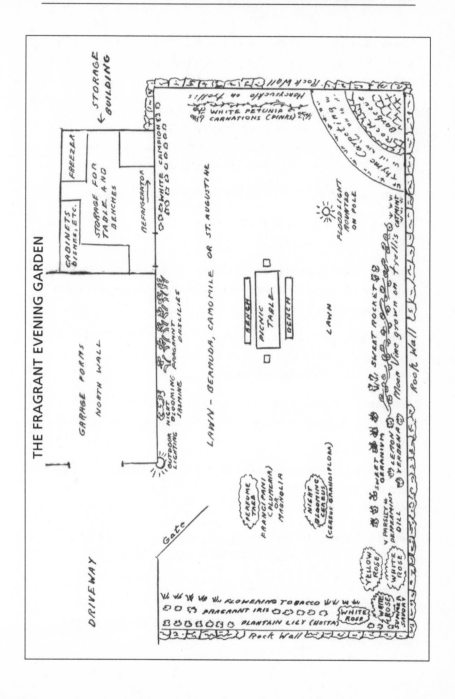

THE FRAGRANT EVENING GARDEN

Hosta plantaginea **(plantain lily).** This was a delightful feature of our great-grandmother's gardens. The large white flowers with their distinctive fragrance makes it "stand out" literally and figuratively.

***Ipomoea alba* (moonflower).** This strong-growing vine is a must for the evening garden. Its fragrant, pearly white, wide-faced, trumpet-shaped blooms are a wonderful conversation piece as they spin open evening after evening, spilling their lemon-scented fragrance on the evening air.

Lilium **spp. (lily).** Fragrant lilies will dramatize your evening garden.

***Linnaea borealis* (twinflower).** This flower is sweetly fragrant but needs special treatment in acid soil.

***Lonicera japonica* (honeysuckle).** There is no aroma lovelier than that of the Japanese honeysuckle at night. It is a rampant grower, however, and may become a pest.

***Matthiola longipetala* spp. *bicornis* (stock).** This night-scented stock is powerfully fragrant at night. It has no floral value, however, and is best sown in patches in open garden spots.

***Nicotana alata* (flowering tobacco).** This plant has a bold habit of growth, pretty white flowers, and a lovely evening fragrance. Seek the older kinds.

***Oenothera* spp. (evening primrose).** Nearly all members of this family open their flowers at night, but all are not night fragrant. *O. caespitosa* is one that is.

***Petunia* hybrids.** Not many realize just how fragrant petunias are at night. The white and pale yellow varieties are attractive in the evening garden.

***Polianthes tuberosa* (tuberose).** Grown for centuries in Mexico, and one of the most delightful of all sweet flowers, it needs a long growing season.

Ptelea trifoliata **(hop tree or water ash).** A sweet-scented shrub. It has a cloying sweetness when smelled close by but a nice fragrance if planted at some distance from the evening activity.

Reseda odorata **(sweet mignonette).** Every garden should have a small planting.

Rosa **(rose).** White and yellow roses are especially fragrant at night.

Schizopetalon walkeri. This plant has no common name. It is a low-growing annual with erect racemes of fringed white flowers that emit a delicate, almondlike fragrance in the evening.

Silene alba **(vespertina) (evening or white campion).** This easy-to-grow member of the Catchfly family has a most delightful fragrance.

Thymus serpyllum **(creeping thyme).** The creeping, carpeting kinds of thyme provide fragrance underfoot and are especially nice to plant near the barbecue.

Green aromatica. While most herbs emit odors only when crushed, they are attractive, protective plants and are easily grown with night-scented flowers. And they are hardy to have nearby to use fresh to season outdoor cookery or salads. Guests enjoy choosing their favorites. Here are some outstanding examples: Apple mint, catmint, costmary, dill, hyssop, lavender, marjoram, peppermint, rosemary, rue, sage, spearmint, summer savory, sweetbrier, sweet geranium, tansy, tarragon, thyme, and winter savory.

Sweet-scented trees include frangipani, honey locust, hop tree, juniper, linden, magnolia, sassafras, and sweet gum.

Plan for a succession of bloom. Honeysuckle blooms first in my area, followed by moonflower. Try out, experiment with various kinds of plants, change the annuals from time to time to find out what you like best and what does well for you. Where plants are placed may make a difference. Hop tree and tuberose may be overpowering close up but enjoyable at a suitable distance.

The Edible Flower Garden

Who says you have to plant edibles in rows? Picking a handful of marigold petals or lavender blossoms on your way in from a late morning stroll in the garden is a pleasure. Many edible plants are also beautiful and look great in your landscape.

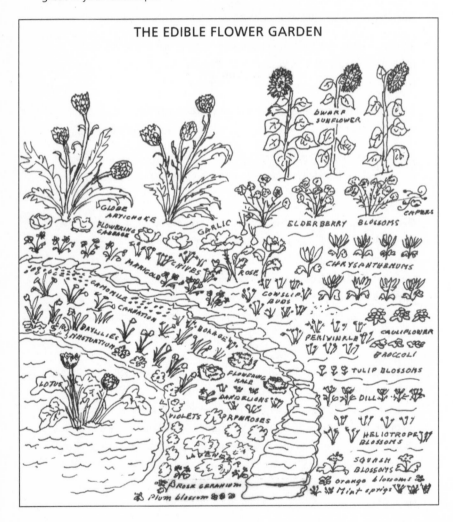

THE EDIBLE FLOWER GARDEN

Plant	Part Used
Artichoke, globe	Flower buds
Borage	Flowers, leaves
Broccoli	Flower heads
Cabbage, flowering	Salad, kraut
Capers	Flower buds
Carnation	Blossoms, petals
Cauliflower	Flower head
Chamomile	Blossoms
Chives	Flower heads, salad, float in soup
Chrysanthemum	Petals
Cowslip (American)	Buds
Dandelion	Buds, leaves
Daylily	Buds, leaves, roots
Dill	Flower heads, seeds, pickling
Elder	Flowers, wine
Honey locust	Flowers followed by pods
Garlic	Tiny bulbs suitable for flavoring follow flowers in heads
Heliotrope	Blossoms
Lavender	Blossoms
Lotus	Petals, tender young leaves, seeds
Marigold	Petals
Mint (spearmint or peppermint)	Sprigs used in mint julep
Orange flowers	Flowers, brandy
Periwinkle	Flowers
Plum blossom	Buds used for pickling
Primrose	Blossoms, petals
Rose geranium	Leaves in preserves and canning fruits
Rose	Buds, petals, hips, candied petals
Tulip	Blossoms used as containers for salads
Violet	Blossoms, leaves; candy blossoms

The Butterfly Garden

Unlike bees, butterflies can see the whole spectrum of colors but prefer the brilliantly colored, deep pink, red, scarlet, bright blue, and usually those that are very fragrant.

The butterfly garden is one that must not be tamed. Just place it at the edge of your property in full sunlight, and let it find its sown wild, natural look. Most butterfly plants are perennial or self-seeding and will be a permanent sanctuary for butterfly caterpillars.

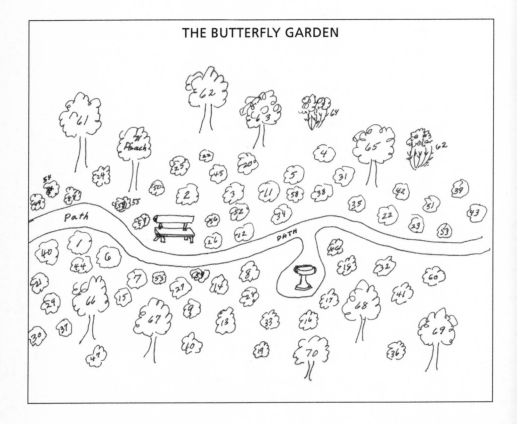

THE BUTTERFLY GARDEN

KEY

WILDFLOWERS
1 *Butterfly weed*
2 *Milkweed*
3 *Indigo bush*
4 *Dogbane*
5 *Shrubby cinquefoil*
6 *Cutleaf coneflower*

GARDEN FLOWERS
7 *Honeysuckle*
8 *Verbena*
9 *Hibiscus*
10 *Lantana*
11 *Butterfly bush*
12 *Violets*
13 *Columbine (also wild columbine)*
14 *Larkspur*
15 *Morning glory*
16 *Frog-fruit*
17 *Blue-eyed grass*
18 *Lemon mint*
19 *Salvia*
20 *Verbena (also wild verbena)*
21 *Penstemon*
22 *Spring beauty*

INSECT-REPELLENT COMPANION PLANTS
(Many are also breeding plants.)
23 *Ornamental allium*
24 *Four-o'clock*
25 *Milkweed*
26 *Parsley*
27 *Dill*
28 *Rue*
29 *Fennel*
30 *Anise*
31 *Flax*
32 *Wormwood*
33 *Borage*
34 *Petunia*
35 *Larkspur*
36 *Geranium*
37 *Nasturtium*
38 *Feverfew*
39 *Nettle*
40 *Tansy*
41 *Marigold*
42 *Garlic*

BUTTERFLY-BREEDING PLANTS
43 *Black swallowtail—parsley, dill, parsnips*
44 *Blue swallowtail—Dutchman's-pipe, wild ginger*
45 *Tiger swallowtail—tulip tree, birch, wild cherry, apple, ash, poplar*
46 *Cabbage butterfly—cabbage (undesirable)*
47 *Gossamer-winged orange tip— Mustard family*
48 *The sulfurs—wild senna, alfalfa, buffalo clover*
49 *The nymphs—fritillaries, angel-wings, and sovereigns—elm, willow, poplar, nettle*
50 *The crescent spots—snakehead or turtlehead (Chelone glabra)*
51 *The emperors—goatweed, hackberry, and various grasses*
52 *The heliconias (zebra butterflies)— passionflowers*
53 *Monarchs—milkweed*
54 *The coppers—alder leaves*

NATURAL INSECTICIDES FOR BUTTERFLY FLOWERS
55 *Leaves, stems, and spent flowers of flowering tobacco (Nicotiana)*
56 *Elderberry tea against aphids*
57 *Wormwood against slugs and snails*
58 *Citrus peels: Chop in blender with water—good against various insects*
59 *Garlic, onions, hot pepper: Chop and blend with water*

TREES AND BUSHES USED FOR BREEDING
60 *Sassafras*
61 *Spicebush*
62 *Tulip tree*
63 *Birch*
64 *Wild cherry*
65 *Apple*
66 *Poplar*
67 *Ash*
68 *Pawpaw*
69 *Alder*
70 *Peach*

The Bee Garden

One of the most charming of old-time gardens combined beauty and utility. Until rather recent times almost everyone who had a bit of land was a beekeeper. The price of honey today may be an incentive to again keep bees to enjoy one of nature's most healthful foods. My husband had kept bees as a boy in New York, and when we married and bought our first home, we established two hives at the south end of our garden, where they lived happily and prospered.

Bee plants are both beautiful and useful. Bees are not especially attracted to fragrant flowers but have a marked preference for those of blue color. Bee garden flowers should include some for honey and some for pollen, for easily available sources of pollen are second only to an abundant supply of nectar.

It is best to bathe before working with bees; sweat infuriates them.

Bee masters of old recommended drinking a cup of good beer before going among the bees.

Working with the bees should be done gently, calmly, but with decision.

"If by accident," instructs an old beekeeper, "a bee buzzes about your face, thrust your face among a passel of Boughs or Herbs, and he will desert you."

In her book *The Fragrant Garden,* Louise Beebe Wilder says, "Indeed, if you go your way among bees anointed with the bitter juices of the Herb o'Grace you will be quite safe, for no bee will come near you."

As previously mentioned, bees are not especially attracted to fragrant flowers, but apiarian lore of ancients lists a number of plants used by beekeepers to attract bees by rubbing the inside of the hive:

| Juniper | Bee balm (*Melissa*) | Fennel |
| Lime flowers | Hyssop | Anise |

Bees seldom notice the sweet-scented blossoms of the lilac, heliotrope, or rose.

Bees like flowers on which they can land and poke around. They have good color vision, preferring blue and yellow. They cannot see red but they can see ultraviolet. Using an ultraviolet filter to photograph bee flowers allows us to see them as the bee sees them. A flower that appears one color to us may have additional markings, nectar guides, in the ultraviolet. Color changes on plants help the bees save time by showing them which flowers are older and have probably already had their nectar taken. For instance, the blossoms of the horse chestnut, catalpa, or golden currant turn from yellow to pink as they age. Bees prefer to visit the fresh, yellow-centered flowers.

- Honey is not "just honey." Depending on the plant variety of the bees' feeding area, it can have many flavors.

- The most prized honey in the United States is clover honey; second is the product of orchards.

- Honey gathered later in the season is usually darker in color and stronger in flavor.

- Orange blossom honey is fragrant and delicious and highly prized.

- Alfalfa honey is mildly spicy.

- Buckwheat honey is very fragrant.

- Mignonette honey has an exquisite bouquet.

- Honey most esteemed by the ancients was made from lime flowers.

- Aromatic nectar comes from the blossoms of thyme, lavender, and rosemary.

- In France, sainfoin honey is preferred and is made from the fodder plant of that name.

- In Spain, rosemary and orange flower honey, along with peach honey, are the favorite flavors.

- Heather honey is a bright, warm amber with an unusual flavor.

- The most prized honey in Germany, called Echter Honig, is harvested from the pine forests to delight epicures. It is nearly black.

- In Portugal the bees feed on the vine flowers, which produce a delicate and fragrant honey.

- The Syrians prize honey made from lavender, wild acacia, cactus, and wild thyme.

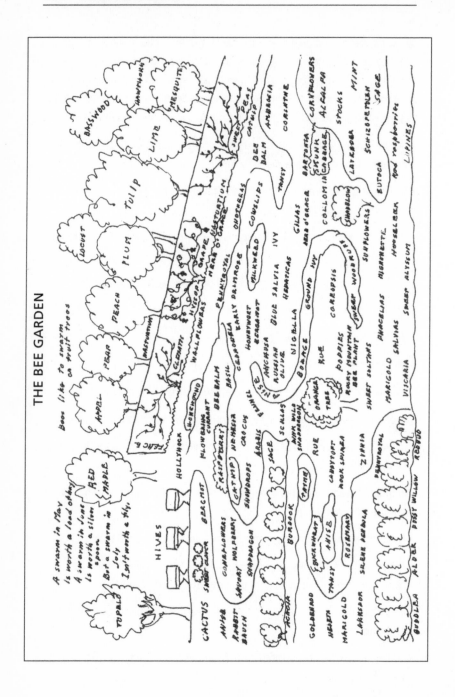

THE BEE GARDEN

EARLY PLANTS
Snowdrop
Early primrose
Red maple catkins
Crocus
Cowslip
Redbud
Scilla
Catkins of alder and
    pussywillow
Shadblow
Hepatica

**LATER-FLOWERING
TREES**
Apple
Locust
Orange trees (California)
Pear
Basswood
Palmetto (Florida)
Peach
Tulip tree (especially fine
    honey)
Tupelo
Plum
Lime tree
Hawthorn

**BEE STING REMEDIES:
RUB THE WOUND**
Bee balm
Hollyhock
Burdock
Marigold
Ivy
Rue

**COMPANION HERBS
FOR BEE FLOWERS
(MANY OF WHICH ARE
ALSO NECTAR PLANTS)**
Horehound
Lavender
Winter savory
Pennyroyal
Rosemary

Summer savory
Aromatic tansy
Borage
Basil
Mint (all mints)
Sweet woodruff
Anise
Catnip
Bergamot
Thyme
Sage

**PLANTS REPUTED
POISONOUS
OR UNDESIRABLE
TO BEES**
Moutain laurel
Andromedes
Flowering tobacco
Snow-on-the-mountain
Pieris
Wild aster
Rhododendron
Box

**PLANTS OF
OFFENSIVE ODOR**
Leek
Onion
Garlic

WILD HONEY PLANTS
New Mexico privet
Pink wild snapdragon
Salvia
Mesquite
Bee balm
Rabbitbrush
Wolfberry
Rocky Mountain bee
    plant
Skunk cabbage (the
    odor does not flavor
    the honey)
Russian olive
Rock spiraea
Phacelias

NECTAR FLOWERS
Buckwheat
Cedronella
Ambrosia
Clover, white
Nepeta
Nemesia
Clover, sweet
Mints
Coreopsis
Basswood
Thymes
Buddleia
Tupelo
Violet
Zinnia
Raspberry
Milkweed
Pinks
Collomia
Goldenrod
Broom
Candytuft
Alfalfa
Heather
Viscaria
Acacia
Cornflower
Catnip
Nigella
Gilia
Mignonette
Blue salvia
Bartonia
Lime blossoms
Phacelias
Lupine
Labiate tribe
Sweet pea
Sweet alyssum
Cerinthe
Flowering currant
Sweet sultan
Snapdragon
Ground ivy
Archangel (angelica)

THE HUMMINGBIRD GARDEN

The Hummingbird Garden

Here's a list of plants hummingbirds love.

Anise	Beardtongue, dwarf	Blueberry, lowbush
Artemisia	Bee balm	Borage
Azalea	Bellflower	Bottlebrush
Basswood	Betony, wood	Bouncing Bet
Bean, scarlet runner	Bluebell, Virginia	Bridgesia
Bearberry	Blueberry, highbush	Buckeye, dwarf

Buckeye, flame
Butterfly bush
Campion, Mexican
Canna
Cardinal flower
Catnip
Century plant
Chestnut, red horse
Columbine
Coralbells
Cornflower
Cotoneaster
Currant, red flowering
Daylily
Eucalyptus
Fennel
Feverfew
Figwort
Firecracker
Fireweed
Flax
Flowering quince
Four-o'clock
Foxglove
Fuchsia
Geranium
Gilia
Gladiolus
Gooseberry, wild
Great Solomon's-seal
Hamelia, scarlet bush

Hibiscus
Hollyhock
Honeysuckle
Honeysuckle, Cape
Hummingbird trumpet
Hyssop, wild
Indian paintbrush
Iris
Jasmine
Jewelweed
Lantana
Larkspur
Lilac
Lily
Loquat
Lupine
Maranta
Matrimony vine
Mexican mint marigold
Milkweed
Mint
Monkeyflower
Morning glory
Nasturtium
Nettle
Nettle, scarlet hedge
Nettle, scarlet
New Jersey tea
Ocotillo
Onion—ornamental
   alliums, also

Painted cup
Poinciana, royal
Poppy, Oriental
Prickly pear
Ragged robin
Rattlesnake root
Red-hot poker
Sage, autumn
Salmonberry
Scabious
Scarlet-bugler
Siberian pea tree
Skyrocket
Snapdragon
Sourwood
Spiderflower
Star-glory, scarlet
Sweet William
Tansy—controls ants
Thimbleberry
Thistle
Tigerlily
Toadflax
Tree tobacco
Trumpet creeper
Waxmallow
Weigelia
Wormwood
Yarrow

Hummingbird enemies are few—no other bird can catch them. They may be occasionally caught in a spiderweb or grabbed by a frog. Weather may sometimes be a danger.

THE WITCH'S GARDEN OF COMPANION MEDICINAL PLANTS

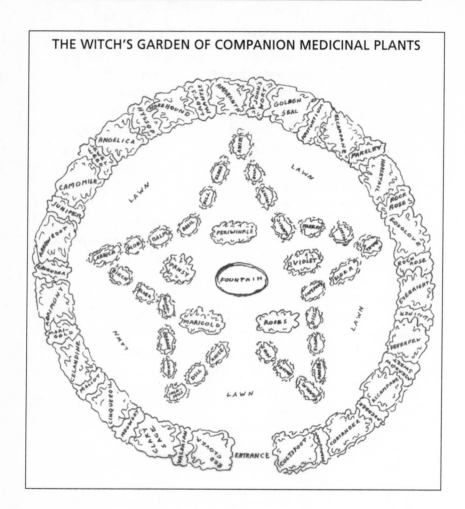

The Witch's Garden of Companion Medicinal Plants

The witch's garden of companion medicinal plants is in the form of a pentagram. The pentagram is believed to be a weapon of power. When one point projects upward, the pentagram may be used to invoke good influences and banish evil ones.

Adding Life to Our Years

Greetings to my fellow octogenarians—and good wishes to all you younger kids who hope to become one of us! As many of you know, I am 88 years old and confidently looking forward to the 21st century. How have I managed to get this far? For one thing, I picked the right ancestors; my mother lived to be 93. Even so, I could have muffed it if I had not taken care of my health. I enjoy life and eat anything I want to, including all garden vegetables enhanced by herbs, both in growing and in cooking.

Are any herbs especially helpful to the elderly? Yes, and two are at the top of my list—garlic and greens. Garlic is especially helpful to the cardio-vascular and digestive systems. It is an antiaging herb. It is antibacterial, antiseptic, bactericidal, a blood cleanser and purifier, lowers blood pressure, and is helpful in colds and coughs. It protects against many diseases. When I was a hospital volunteer, I took a small clove of garlic each day, chopping it fine and swallowing it whole like a vitamin pill, followed by chewing a sprig of parsley or a bit of apple.

Garlic yogurt is helpful in cases of dysentery, herpes, indigestion, and itchy skin.

Greens of all kinds are particularly helpful in cleansing the system, so necessary to the aging body. These include alfalfa, cress, collards, dandelion, dock, kale, lamb's-quarters, nettle, parsley, plantain, and violet. Kelp, containing alginic acid, aids the excretion of lead from our bodies.

Herbs helpful to the prostate include pumpkin seed, sunflower seed, and garlic soup.

Herbs helpful to blood pressure are blackberry leaf, onion soup, garlic soup, and sassafras. The herb most helpful at time of menopause is licorice.

To lower cholesterol, take helpings of eggplant and Jerusalem artichoke.

Aids to sleep include aniseed tea, linden tea, and onion soup.

To improve memory, try lemon balm and rosemary lemonade. For wrinkles, try baked beets, mushroom/onion sauté, and steamed asparagus. Asparagus is also a laxative.

Beets are especially noted as an antiaging food (they contain potassium, good for the heart), boiled and seasoned with mint, parsley, tarragon, basil, or dill. They are good baked with a dressing of oil and lemon juice. Pickled beets are an old-time favorite of our ancestors.

Adding Life to Our Years

Plants for the Elderly

Anti-aging	Asparagus • Beets • Garlic • Mushrooms • Onion
Antidepressant	St.-John's-wort
Appetite depressants	Fennel tea
Appetite stimulant	Parsley soup
Arthritis, rheumatism	Black cherry juice • Comfrey salad • Raw vegetables • Rosemary/comfrey tea
Blood pressure	Blackberry leaf • Garlic soup • Onion soup • Sassafras
Bones	Comfrey/nettle salad • Calcium herbs • Borage • Chickweed • Comfrey • Dandelion
Brains, to improve	Almond, rose water • Balm wine • Bamboo salad • Marigold broth • Rosemary tea
Bruises	White and yellow flower petals
Burns	Aloe vera • Marshmallow root
Cholesterol (to lower)	Eggplant • Jerusalem artichoke
Colds, coughs	Angelica sweetmeats • Cayenne • Elecampane sweetmeats • Garlic soup • Garlicked beans • Rose-hip soup
Constipation	Garlic, onion • Leek, chives • Alfalfa, cress • Collards • Dandelion • Dock, kale • Lamb's-quarters • Nettle, parsley Plantain, violet • Spinach
Dental problems	Apple eaten raw daily
Disease (to protect from)	Garlic soup • Mallows
Eyes	Borage salad • Lemongrass
Feet	Black mustard • Mullein leaf
Female foods	Angelica • Wheat
Hair	Lemongrass • Sea vegetables (kelp) • Southernwood • Various cacti
Headache	Carnation wine

Heart, to improve	6 garlic cloves per week • 9 onions per week
Hormones (female)	Angelica • Licorice • Sarsaparilla • Wild yam
Memory	Lemon balm • Rosemary lemonade
Male food	Barley
Menstrual	Amaranth • Candied angelica root • Pennyroyal • Sweet cicely
Menopause	Licorice
Muscle	Linden tissue • Sage • Wallflower
Nerves	Basil-stuffed mushrooms • Poppyseed
Prostate	Pumpkin seed • Sunflower seed • Garlic soup
Rejuvenatives	Comfrey • Ginseng • Wheatgrass juice
Respiratory system	Bay leaf • Cayenne • Garlic
Sinus	Garlic soup • Steamed zucchini
Skin	Burdock root • Pansy • Parsley • Watercress • Cabbage
Sleep	Aniseed tea • Linden tea • Onion soup
Stress	Alfalfa extract • Bee pollen • Ginseng
Sunburn	Aloe vera • Marshmallow root (external use only)
Wrinkled skin	Baked beets • Mushroom/onion saute • Steamed asparagus • Cabbage (Russian)

Use of herbs as alternative medicine is more widespread in England, Germany, and other foreign countries than in the United States. Much of what we use comes from Native Americans.

One more thought. Elderly persons need light—not direct sunlight, but filtered light. Far too many of us spend too much time indoors. We need to exercise in the cool of the morning or the evening.

Live well. Laugh often. Love much.

SUGGESTED READING

Campbell, Stu. *The Mulch Book.* New York: Workman, 1991.

Clarkson, Rosetta. *Herbs, Their Culture and Uses.* Old Tappan, NJ: Macmillan, 1990.

Gibbons, Euell. *Stalking the Wild Asparagus.* Chambersburg, PA: Hood & Company, 1988.

—. *Stalking the Healthful Herbs.* Chambersburg, PA: Hood & Company, 1989.

Harris, Charles. *Eat the Weeds.* New Canaan, CT: Keats, 1995.

Meyer, Clarence. *The Herbalist Almanac: Fifty Years of Herbal Knowledge.* Glenwood, IL: Meyerbooks, 1977.

Riotte, Louise. *Catfish Ponds and Lily Pads.* Pownal, VT: Storey Communications, 1997.

—. *Successful Small Food Gardens.* Pownal, VT: Storey Communications, 1993.

Tomkins, Peter, and Christopher Bird. *The Secret Life of Plants.* New York: HarperCollins, 1989.

Wilder, Louise Beebe. *The Fragrant Garden.* Mineola, NY: Dover, 1974.

SOURCES

Applewood Seed
  Company
5380 Vivian Street
Arvada, CO 80002
303.431.6383
www.applewoodseed.com

Bio-Dynamic Farming and
  Gardening Association, Inc.
Building 1002 B, Thoreau Center,
  The Presidio
P.O. Box 29135
San Francisco, CA 94129-0135
888.516.7797
www.biodynamics.com

Bluestone Perennials
7211 Middle Ridge Road
Madison, OH 44057
800.852.5243
www.bluestoneperennials.com

Breck's
P.O. Box 65
Guilford, IN 47022-4180
513.354.1511
www.brecks.com

Burgess Seed and Plant Company
904 Four Seasons Road
Bloomington, IL 61701
309.663.9551
www.eburgess.com

Dutch Gardens, Inc.
144 Intervale Road
Burlington, VT 05401
800.944.2250
www.dutchgardens.com

Emlong Nurseries, Inc.
2671 West Marquette Woods Road
Stevensonville, MI 49127
616.429.3431

Farmer Seed & Nursery
Division of Plantron, Inc.
818 NW Fourth Street
Fairbault, MN 55021
309.663.9551
www.farmerseed.com

Gurney Seed & Nursery Co.
110 Capital Street
Yankton, SD 57079
605.665.9391
www.gurneys.com

Henry Field Seed & Nursery Company
5100 Schenley Place
Lawrenceburg, IN 47025
812.539.2537
www.henryfields.com

H.G. Hastings Nature & Garden Center
3920 Peach Tree Road NE
Atlanta, GA 30319
404.869.7447
www.hastingsgardencenter.com

House of Wesley
1704 Morrissey Drive
Bloomington, IL 61704
309.663.9551
www.cometobuy.com/houseofwesley

Indiana Botanic Gardens
P.O. Box 5
Hammond, IN 46345
219.947.4040
800.644.8327

Jackson & Perkins
1 Rose Lane
P.O. Box 1028
Medford, OR 97501
800.292.4769
www.jacksonandperkins.com

J.E. Miller Nurseries, Inc.
5060 West Lake Road
Canandaigua, NY 14424
800.836.9630
www.millernurseries.com

Johnny's Selected Seeds
955 Benton Avenue
Winslow, ME 04901
207.861.3900
www.johnnyseeds.com

J.W. Jung Seed Company
335 South High Street
Randolph, WI 53956
920.326.5672
www.jungseed.com

Kitazawa Seed Co.
P.O. Box 13220
Oakland, CA 94661-3220
510.595.1188
www.kitazawaseed.com

Lilypons Water Gardens
P.O. Box 10
Buckeystown, MD 21717
800.999.5459
www.lilypons.com

Louis Gerardi Nursery
1700 East Highway 50
O'Fallon, IL 62269
618.632.4456

Mellingers Nursery Inc.
2310 West South Range Road
North Lima, OH 44452
330.549.9861
www.mellingers.com

Michigan Bulb Co.
P.O. Box 4180
Lawrenceburg, IN 47025-4180
513.354.1498
www.michiganbulb.com

Musser Forests, Inc.
P.O. Box 340
Indiana, PA 15701
800.643.8319
www.musserforests.com

Nichols Garden Nursery
1190 Old Salem Road NE
Albany, OR 97321-4580
800.422.3985
www.gardennursery.com

Park Seed
1 Parkton Avenue
P.O. Box 46
Greenwood, SC 29647
800.213.0076
www.parkseed.com

Plants of the Southwest
Route 6 Box 11-A
Santa Fe, NM 87501
505.471.2212
800.788.7333
www.plantsofthesouthwest.com

R.H. Shumway Seedsman
P.O. Box 1
Graniteville, SC 29829
803.663.9771
www.rhshumway.com

Richters
Goodwod, Ontario L0C 1A0
Canada
905.640.6677
www.richters.com

Spring Hill Nurseries
Tipp City, OH 45371
800.582.8527
www.springhillnursery.com

Stark Brothers Nurseries
P.O. Box 10
Louisiana, MO 63353
800.325.4180
www.starkbros.com

Stokes Seeds, Inc.
P.O. Box 548
Buffalo, NY 14240
716.695.6980
www.stokeseeds.com

Syngenta Seeds, Inc.
7500 Olson Memorial Highway
Golden Valley, MN 55427
763.593.7333
www.nk-us.com

Thompson & Morgan
P.O. Box 1308
Jackson, NJ 08527
800.274.7333
www.thompson-morgan.com

Twilley Seed Co.
(Otis S. Twilley Seed Co., Inc.)
121 Gary Road
Hodges, SC 29653
800.622.7333
www.twilleyseed.com

Van Bourgondien Bros.
P.O. Box 1000
Babylon, NY 11702-9004
800.622.9997
www.dutchbulbs.com

Van Ness Water Gardens
2460 North Euclid Avenue
Upland, CA 91784
909.982.2425
www.vnwg.com

Vermont Bean Seed Co.
Garden Lane
Fair Haven, VT 05743
803.273.3400

W. Atlee Burpee Seed Co.
300 Park Avenue
Warminster, PA 18974
215.674.4900
www.burpee.com

Wayside Gardens
1 Garden Way
Hodges, SC 29695
800.213.0379
www.waysidegardens.com

White Flower Farm
Route 63
Litchfield, CT 06759
800.503.9624
www.whiteflowerfarm.com

Wolfe Nursery 1
400 Memorial City Way
Houston, TX 77024-2513
713.984.5680

INDEX

Note: **Boldface** page references indicate illustrations. <u>Underscored</u> references indicate tables.